Sustainable Agriculture

Sustainable Agriculture

Ram Swaroop Meena

JRF, SRF, Ph.D., PDF, NET, Raman Fellow (MHRD, Govt. of India)
Ex-Visiting Scientist, C-MASC, USA
Assistant Professor, Dept. of Agronomy
Institute of Agricultural Sciences,
BHU, Varanasi (UP) - 221 005, India

SCIENTIFIC PUBLISHERS

Published by

SCIENTIFIC PUBLISHERS (INDIA)
5 A, New Pali Road, P.O. Box 91
Jodhpur 342 001 (INDIA)
E-mail: info@scientificpub.com
Website : http://www.scientificpub.com

Print: 2020

ISBN: 978-93-88043-62-5

e-ISBN: 978-93-88043-85-4

Printed in India

Preface

Sustainable development means that the needs of the present generation should be met without compromising the ability of future generations to meet their own needs. Sustainability is the key to preventing or reducing the effect of environmental issues. Agriculture sustainability is the process of making sure current processes of interaction with the Agriculture are pursued with the idea of keeping the environment as pristine as naturally possible based on ideal-seeking behaviour. Ecosystems are dynamic interactions between plants, animals, and microorganisms, and their environment working together as a functional unit. Ecosystems will fail if they do not remain in balance. This book examines our current cultivation system from production to consumption, and the urgent need to transition to long-term sustainability. The book promotes the study and application of Agro-ecology for developing alternatives to the complex problems of resource depletion, environmental degradation, a narrowing of agro-biodiversity, consolidation and industrialization of the food system, climate change, and the loss of farm land. The book uses a food systems approach, and seeks experiences in eco-friendly that are on-farm, participatory, change-oriented, and backed by broad-based methodologies of sustainability analysis and evaluation. The objectives of this book are: (1) to understand the role sustainable agricultural productivity and its importance to the sustainable soil managements, (2) to restore soil health to transforming agriculture for sustainability, and (3) to understand the matching of management rules in climatic perspective. In general, the book will be suitable for teachers, researchers, policy makers, undergraduate and graduate students of soil science, soil microbiology, agronomy, ecology, and environmental sciences. Highly professional and internationally renowned researchers has invited to contribute, authoritative and cutting-edge scientific information on a broad range of topics covering sustainable management of agriculture. All chapters are well-illustrated with appropriately placed data, tables, figures, and photographs and supported with extensive and most recent references.

The submitted chapters are reviewed by the members of the relevant field for further improvement and authentication of the information provided. I am also provide a roadmap for sustainable approaches for agriculture systems for food and nutritional security, and soil sustainability in the ecosystem.

Ram Swaroop Meena

Contents

1

Integrated Nutrient Management for Sustainable Rice-based Cropping Systems and Soil Quality

B.L. Meena[1], R. Raja[2], M.L. Dotaniya[3], Gangadhar Nanda[4] and Ram Swaroop Meena[5]*

[1]ICAR-Central Soil Salinity Research Institute, Karnal – 132 001
[2]ICAR-Central Institute for Cotton Research, Regional Station, Coimbatore – 641 003
[3]ICAR-Indian Institute of Soil Science, Bhopal – 462 038
[4]Ph.D. Scholar, Dept. of Agronomy, GBPAUT, Pantnagar – 263 195
[5]Institute of Agricultural Sciences, Varanasi – 221 005
**Formerly associated with ICAR-Central Island Agricultural Institute, Port Blair – 744 101*
**Email: blmeena2008@gmail.com*

ABSTRACT

The real challenge is to keep the pace of sustainable production under condition of natural resources should be used to generate increased output and incomes, especially for low income groups without depleting the natural resource base. Rice is a component of widely varying cropping systems. Rice-based cropping systems form an integral part of agriculture in India. Several intensive rice-based cropping systems have been identified and are being practiced by the farmers in India. Productivity of this soil has remained low and unstable owing to climate and soil-related constraints. Intensive agriculture, involving exhaustive high yielding varieties of rice and other crops, has led to heavy withdrawal of nutrients from the soil. Also imbalanced and indiscriminate use of chemical fertilizers has resulted in deterioration of soil quality or soil health. The basic concept underlying the principle of integrated nutrient management (INM) is to maintain or adjust plant nutrient supply to achieve a given level of sustainable crop production by optimizing the benefits from all possible sources of plant nutrients. It involves utilization of local sources and hence, turned out to be the rational, realistic and economically viable way of supply of nutrients to crops. So that, environmentally benefiting manner without sacrificing soil productivity of future generations. Application of organic manure in conjunction with fertilizer and inclusion of legumes gave higher yield, net income and benefit: cost ratio under rice based cropping system. INM practices also improved the soil quality of the soils to some extent. Integrated nutrient management relies on a number of factors including appropriate nutrient application and conservation and the transfer of knowledge about INM practices to farmers through extension personal. Thus, INM is the only way to restore soil quality and make Indian agriculture sustainable.

Key word: INM, Rice based cropping system, Soil environment, Soil quality, Sustainable crop production

1. INTRODUCTION

Rice is the most important crop in Asia, where more than 90 per cent of rice is grown and consumed. In tropical Asia, rice-rice constitutes an important annual crop rotation. In the subtropical Asia, rice and wheat are grown in rotation in more than 13 million hectares in the Indo-Gangetic plains of South Asia (India, Bangladesh, Nepal, Pakistan) and on similar hectares in the basin of the Yangtze river in China (Timsina and Connor, 2001). In addition to wheat, other crops grown in rotation with rice are barley, oats, maize, sorghum, legumes (mung bean, peanuts, soybean, lentil, and chick pea), oilseeds (mustard, rapeseed), potato, sugarcane, and cotton. Nutrient cycling in the soil-plant ecosystem is an essential component of sustainable productive agricultural enterprise. Rice-based cropping systems form an integral part of agriculture in India. Several intensive rice-based cropping systems (rice-wheat, rice-rice, rice-legumes, Rice-sugarbeet, rice-maize, rice-jute *etc.*) have been identified and are being practiced by the farmers in India. Although during the last three decades fertilizer practices have played a dominant role in the rice based cropping systems. While intensive agriculture, involving exhaustive high yielding varieties of rice and other crops, has led to heavy withdrawal of nutrients from the soil, imbalanced and indiscriminate use of chemical fertilizers has resulted in deterioration of soil quality (John *et al.,* 2001) in the marginal soils. Productivity of this soil has remained low and unstable owing to climate and soil-related constraints. Despite that, it has importance in food production for growing populations. Against the above background, it is imperative to devise ways to sustain and improve the overall productivity of rice-based cropping systems for better livelihood. Sustainable agriculture involves successful management of resources for increased agricultural production to satisfy changing human needs while maintaining or improving the environmental quality and natural resources (Gill *et al.,* 2008). Integrated nutrient management (INM) still play an essential role in the sustainable crop production of rice based cropping system. Thus, INM can provide the right answer for a better livelihood for the resource-poor farmers. INM integrates the use of all natural and man-made sources of plant nutrients, so that productivity and nutrient status of food increases in an efficient and environmentally benefiting manner without sacrificing soil productivity of future generations.

Integrated nutrient management can reduce plant requirements for inorganic nitrogen fertilizer, and reduced use of purchased fertilizer nutrients can result in a significant saving of scarce cash resources for small farmers. It also ensures the conservation and efficient use of native soil nutrients, recycling of organic nutrient flows, enhancing biological nitrogen fixation and soil biological activity, and addition of plant nutrients. There is increased emphasis on the impact on the environmental quality due to continuous use of chemical fertilizers. The INM system is an alternative and is characterized by reduced input of chemical fertilizers as well as combined use of chemical fertilizers with organic materials. For sustainable crop production, integrated use of inorganic and organic fertilizers has proved to be highly beneficial. Moreover, INM reduces erosion, improves water infiltration, soil aeration and plant root growth, Moreover, it minimizes the risk of downstream flooding (Gill *et al.,* 2008). Organic manure along with inorganic fertilizers might be helpful to obtain a good economic return as well as provide favourable conditions for high yield of rice based cropping system. Therefore, this chapter focuses discussion on the impact of INM of organic and inorganic fertilizer sources on sustainability of rice based cropping system and soil quality, so that INM is helpful in sustaining soil quality in rice based cropping system.

2. SOME THOUGHTS ON THE CONCEPT AND RELEVANCE OF INM

In literature at least 3 terminologies are being used often convey same meaning and Goswami (1998) defined these terms as given in Table 1.

Table 1. Basic difference in terminologies in literature

IPNS (Integrated Plant Nutrition System)	IPNS System (Integrated Plant Nutrient Supply System	INM (Integrated Nutrient Management)
It is concept "which aims at the maintenance or adjustment of soil fertility and of plant nutrient supply to an optimum level for sustaining the desired crop productivity through optimization of benefit from all possible sources of plant nutrients in an integrated manner".	Supply of nutrients to the plant from various sources- *(i)* nutrient reserves in the soil, *(ii)* organic source. It is only a method to achieve the objective of IPNS.	It is actually the technical and managerial component of achieving the objective of IPNS under farm situations. It takes into account all factors of soil, crop, water, agro-chemicals etc., besides nutrients.

3. INTEGRATED NUTRIENT MANAGEMENT AND ITS COMPONENTS

Balanced and efficient fertilizer application is essential for the increased yields and for replenishment of the larger removal of soil nutrients. Fertilizer efficiency can be improved by reducing nutrient losses between applications and uptake by the plant. INM is actually the technical and managerial component of achieving the objectives of integrated plant nutrient systems (IPNS) under farm situations.

It takes into account all factors of soil and crop management including management of all other inputs such as water, agrochemicals besides plant nutrients. The basic principle of IPNS is the maintenance of soil fertility, sustaining agricultural productivity and improving farmers' profitability through the judicious and efficient use of mineral fertilizers, organic manures and biofertilizers. Thus, the objectives of INM are to ensure efficient and judicious use of all the major sources of plant nutrients in an integrated manner so as to get maximum economic yield from a specific cropping system. INM includes all possible combinations of organic, inorganic or biological sources to maintain soil chemical, physical and biological properties. The components of INM as given in Figure 1. Among various organic sources, FYM or compost and green manuring are common source. Inclusion of legume in rice based cropping system either as a substitute or biomass to enrich the soil due to their capability to fix atmospheric N_2 and addition of organic matter, respectively. Crop residues on the other hand constitute an important source of organic nutrients. About 80.12 Mt crop residues are available for recycling with a total nutrient potential of 1.61 Mt, which can replace 0.80 Mt chemical fertilizers in India (Gill *et al.*, 2008). Use of biofertilizers is an integral part of IPNS as these contain living cells of different types of microorganisms that have ability to mobilize the nutrients from unavailable to available form through biological process. It broadly includes N_2-fixers, both symbiotic and non-symbiotic bacteria, phosphate-solubilizing bacteria and fungi. These approaches ensure the sustainability of rice based cropping system.

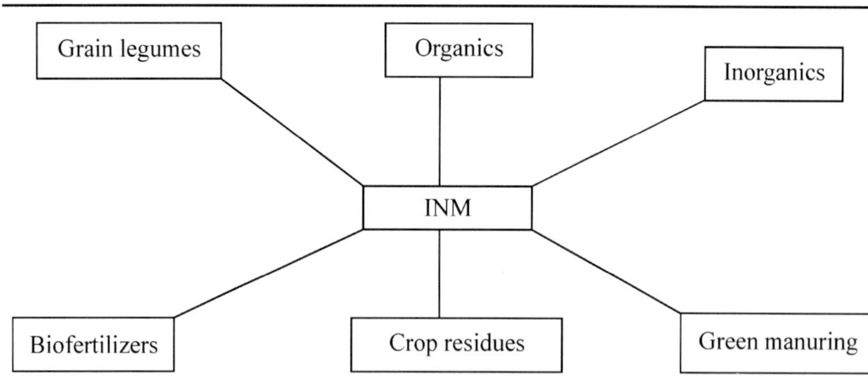

Fig. 1. Components of integrated nutrient management

4. EFFECT OF INM ON SOIL PROPERTIES

4.1 Effect on Soil Chemical and Physical Properties

Soil chemical properties are evaluated under combined use of organic and inorganic fertilizers as their combined effect improves soil organic matter, per centage of organic carbon and total nitrogen. Organic matter in soil is critical for better soil health and higher soil productivity. Addition of organic fertilizer results in increased soil organic carbon (SOC) levels in the soil, while chemical fertilizer result in decreased SOC (Fig. 2) and basic cation contents, and lowering of soil pH. As a result, a positive effect on soil results in modification of soil structure thereby increases the yield in the long term. Organic carbon can be maintained at sustainable manner in 100% NPK + FYM, whereas organic carbon is continuously decreasing in both 100% NPK and control under intensive cultivation (Sarkar, 2000). Along with the evaluation of soil chemical properties, soil physical properties are also enhanced by the use of INM, as it reduces soil erosion and increases cycling of organic residues. Thus it improves both nutrient and water retention capacity of the soil. This management also improves the soil structure, water infiltration, and soil aeration (Das *et al.,* 2014).

4.2 Effect on Soil Biological Properties

The microorganisms perform a key role in nutrient cycling for sustaining the productivity of the soils; because they are the source and sink for mineral nutrition and can carry out biochemical transformations. Moreover, an increase in soil microbial-biomass C and nitrogen (N) is obvious in soils receiving combined application of organic manures and chemical fertilizers compared to soils receiving chemical fertilizers only. Basal and glucose-induced respiration, potentially mineralizable N, and arginine ammonification are higher in soils amended with organic manures with chemical fertilizers, indicating that more active micro-flora is associated with integrated system using organic manures and chemical fertilizers together which are important for nutrient cycling. The use of organic fertilizer together with chemical fertilizers, compared to the addition of organic fertilizers alone, has an higher positive effect on microbial biomass and hence soil health (Hati *et al.*, 2008). The INM changes the chemical and biological properties in soils, it improves the soil organic C, total N, phosphorus (P), and potassium (K) status and microbial biomass (C and N), and soil organic matter (OM) content and long-term soil productivity in the tropics where soil OM content is low. Soil biomass is increased by INM as these amendments supply readily decomposable organic matter in addition to increasing root biomass and root exudates due to greater crop growth (Vineela *et al.,* 2008). As for as the enzyme activities are concerned, the urease and alkaline phosphatase activities of soil increase significantly with a combination of inorganic fertilizers and organic amendments as enhanced enzyme activities are always related to soil organic matter content (Goyal *et al.,* 1999) (Table 2).

Table 2. Amounts of microbial biomass C and activities of dehydrogenase, urease and alkaline phosphatase in soils as affected by different fertilizer treatments (*TPF* triphenyl)

Treatment	Microbial biomass C (mg kg^{-1} soil	Biomass C as % of soil C	Dehydr-ogenase activity (μg TPF g^{-1} soil 24 h^{-1})	Urease activity (μg NH_4^+ -N g^{-1}soil h^{-1})	Alkaline phosphatase activity (μg PNP g^{-1} soil h^{-1})
N_0, P_0	147	3.3	38	42	416
N_{60}, P_{30}	195	3.78	37	55	423
N_{90}, P_{45}	213	3.7	37	60	425
N_{120}, P_{60}	226	3.9	35	64	433
N_{90}, P_{45} + N_{30} (FYM)	273	4.4	39	65	498
N_{90}, P_{45} + N_{30} (wheat straw)	423	6.5	67	88	541
N_{90}, P_{45} + N_{30} (green manure)	317	5.1	34	68	473
LSD (*P* 0.05)	15	0.3	4	5	18

(*Source*: Goyal *et al*., 1999)

4.3 Influence on C : N ratio

Soil organic carbon and nitrogen are the main nutrition used for vegetation growth, and are also used as indexes of soil quality assessment and sustainable land use management. Soil organic carbon and nitrogen not only can reflect the soil fertility level, but can also explain the regional ecological system evolution. The relationship between them can be represented as soil C/N ratio, a sensitive indicator of soil quality and for assessing carbon and nutrient cycling of soils (Zhang *et al*., 2011). Many researchers have shown that soil organic carbon and nitrogen content are not only affected by climate, altitude, terrain, but also land use management. Soil organic C and total N are greater in treatments receiving a combination of inorganic fertilizers and organic amendments compared to soils receiving inorganic fertilizers alone. The greatest amounts of both organic C and total N are observed in soils receiving wheat straw and least organic C and N total were present in unfertilized soils. The C:N ratio of soil decreases with fertilization. The organic amendments show a greater effect in decreasing C:N ratio compared to inorganic fertilizers. Soil organic C and N contents provide a measurement of soil organic matter status. The increase in soil organic matter with the application of inorganic fertilizers is because of greater input of root biomass due to better crop growth. The FYM and crop residue having higher C: N ratio is less resistant to decomposition, addition of inorganic N (50% of recommended dose) along with these materials reduced the C:N ratio and enhanced its decomposability (Nayak *et al.,* 2012). The C:N ratio decrease or increase with organic amendments indicates the build-up of N pool in the soil (Goyal *et al.,* 1999).

5. EFFECT ON YIELD

Targeting high yield with a high cropping intensity is the most logical way to raise the total production from the country's limited resources. Since the nutrient turnover in soil plant system is considerably high in intensive farming, neither the chemical fertilizers nor the organic and biological sources alone can achieve production sustainability of rice based cropping system. Even with balanced use of chemical fertilizers, high yield level could not be maintained over the years because of deterioration in soil physical and biological environments due to low organic matter content in soils. In this context and as a further response to economic recession, and also to conserve and improve soil fertility, the concept of INM system has been adopted. At the same time the integrated use of inorganic fertilizer and organic manure results in important yield attributes (Sharma *et al.,* 2007) (Tables 3).

Table 3. Effect of integrated use of organic and inorganic fertilizers on benefit to cost ratio

Treatment		Yield		B:C ratio
Rice	**Wheat**	**Rice (q/ha)**	**Wheat (q/ha)**	
Control	Control	18.49	12.79	1.07
50% NPK	75% NPK	36.96	30.64	1.98
100% NPK	100% NPK	42.52	35.89	2.10
50% NPK+ 50% N (FYM)	100% NPK	47.05	37.22	2.20
75% NPK+ 25% N (FYM)	75% NPK	43.08	33.22	2.16
75% NPK+ 50% N (PS)	100% NPK	45.65	34.00	1.92
75% NPK+ 25% N (PS)	75% NPK	41.98	32.82	2.04
75% NPK+ 25% N (GLM)	75% NPK	40.17	33.29	2.06
Farmer's practice	Farmer's practice	28.23	25.40	1.68
CD(*P*=0.05)		1.86	1.61	0.08

(*Source*: Sharma *et al.*, 2007)

6. EFFECT ON ENVIRONMENT

Soil fertility and environment are closely interlinked. Depletion of soil fertility means degradation of the environment and likewise, its improvement also leads to the better environment. The depletion of nutrients (fertility erosion) is wide spread on the earth including in developing countries. The major forms and causes of nutrient depletion include excessive crop nutrient uptake and removal, leaching, gaseous loss, fixation in the soil and immobilization in the trunks and branches of the tree crops. Balance sheet for rice based cropping system indicates that net loss of nitrogen in rice-rice-sesame and rice-rice-okra cropping system occurred, whereas nitrogen gains under rice-rice-groundnut and rice-rice-cowpea (Table 4). Therefore, rice based cropping should be taken with pulse crops as second or third crops for maintaining the good soil environment for crop growth (Pillai *et al.*, 2007). Moreover, if an unbalanced organic fertilizer is supplied at levels that satisfy the demands of the limiting nutrient (here S or P), a large surplus of other nutrients, especially nitrogen, would be supplied, with a probable risk for leaching (nitrate, NO_3^-) or volatile losses (ammonia, NH_3), and thereby negative effects on the environment (Heeb *et al.,* 2005). INM practices reduces the emission of greenhouse gases (nitrous and nitric oxides) excessive applications of nitrogen fertilizer results in increased leaching of nitrates into ground water, increasing health risks to new borne infant flows cycled through the return of organic residues as compost, manure and/or mulch have significant implications for conserving soil fauna biodiversity (Prasad, 2008).

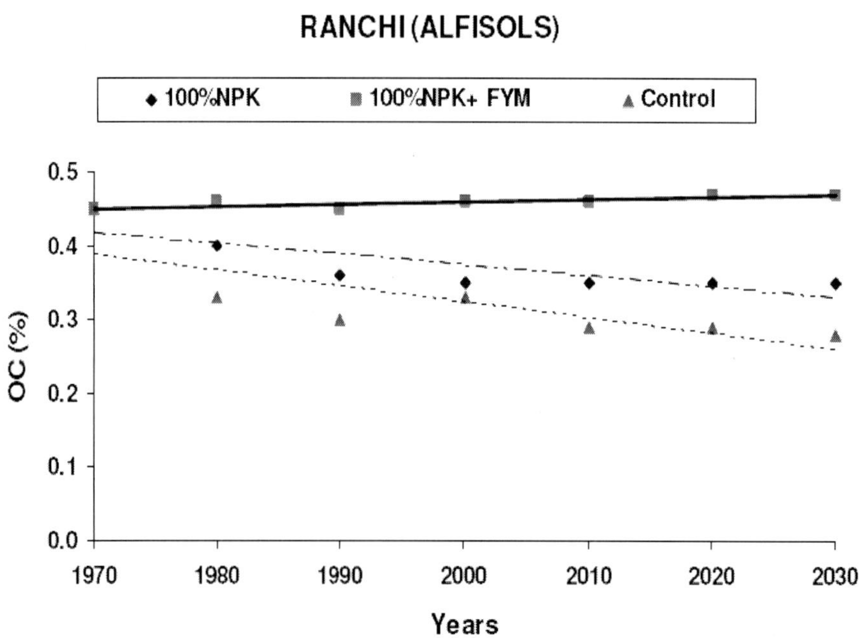

Fig. 2. Long term trend of soil organic carbon status in Alfisol (Ranchi)

Table 4. Balance sheet of nitrogen in soil as influence by INM under different cropping system

Cropping system	Net loss or gain in N (kg ha^{-1})	
	I Year	II Year
Rice-Rice-Sesame	-29.7	-9.4
Rice-Rice-Groundnut	+72.2	+86.7
Rice-Rice-Cowpea	+48.3	+22.7
Rice-Rice-*Okra*	-12.6	-28.9

(*Source*: Pillai *et al.*, 2007)

7. SOIL QUALITY EVALUATION UNDER INM

Soil quality is broadly defined as "the capacity of a living soil to function, within natural or managed ecosystem boundaries, to sustain plant and animal productivity, maintain or enhance water and air quality, and promote plant and animal health" (Doran, 2002). Doran and Parkin (1996) proposed a minimum data set called indicators for characterizing and monitoring soil quality. The minimum data set includes soil attributes and properties such as, texture, soil and rooting depth, bulk density, infiltration, water retention characteristics, soil organic matter, electrical conductivity, extractable N,P and K, microbial biomass and soil respiration. The methodology for evaluation of soil quality is described below:

Quantitative evaluation of changes in soil quality: By introducing the concept (Karlen and Stott, 1994) of Relative soil quality index (RSQI), the 9 indicators were combined into an RSQI. The equation for calculating RSQI value is given below:

$$\text{Relative soil quality index (RSQI)} : (SQI/SQI_m) \times 100$$

Where,

SQI = Soil quality index

SQI_m = Maximum value for soil quality index

$SQI = \sum W_i I_i$

W_i = Weights of the indicators

I_i = Marks of the indicators classes

According to the RSQI values, soils were classified into 5 classes from best to worst, represented as follows by I, II, III, IV and V, respectively.

Soil quality increased under INM by 12–19 units as compare to 7–9 unit of farmer's practice (Table 5). In 2001, there was no class I, II and III type soil, and experimental plots were with class IV type soil. After INM practices for 3 years, areas of class IV were improved to class III. This indicates that INM efforts improved low quality soils, but some high quality soils were not maintained. The reason is the INM study only focused on surface soil properties and the soil functions which support plant growth.

Table 5. General soil quality changes in Farmer's Practices and INM

Parameter	Initial			Farmer's practice			INM		
	2001-2002	2002-2003	2003-2004	2001-2002	2002-2003	2003-2004	2001-2002	2002-2003	2003-2004
SQI	225	221	243	263	244	275	275	283	320
RSQI	57	55	61	66	61	68	69	71	80
Class	IV	V	IV	IV	IV	IV	III	III	III
Δ RSQI		N.A.		9	6	7	12	16	19
Remark		N.A.			Moderate increase			Great increase	

(*Source*: Singh, 2007)

Table 6. Soil quality change (as % over fallow) under different nutrient management practices and cropping systems

Treatment/Cropping system	Rice-wheat	Rice-lentil	Jute-rice-wheat
Control	- 56.0	- 8.0	- 49.0
N only	-	-11.7	- 35.0
NPK only	-10.8	-9.7	19.0
NPK + FYM	18.7	8.6	45.1

(*Source*: Mandal, 2005)

Cultivation without any fertilization (control), only with N caused net degradation of soil quality. Moreover, cultivation even with application of balanced NPK could hardly maintain such quality at the level where no cultivation was practiced. Only integrated use of organic and inorganic sources of nutrient could aggraded the system (Mandal, 2005) (Table 6).

8. ECONOMIC PERFORMANCE

The highest gross return (benefit cost ratio) can be obtained if integrated dose of inorganic fertilizer is applied with organic manure (Table 3). Significantly higher B: C ratio observed under 50 per cent recommended dose of NPK+50% N substitute by FYM among all the treatments (Sharma *et al.*, 2007; Gill *et al.*, 2008).

9. COMMON CONSTRAINTS IN ADOPTION OF INM

As per the feedback of farmers in training, farmers obtained more agricultural knowledge and experience from their neighbours than from the agricultural technology extension technicians. Lack of extension services also contributed to the low adoption of technology. The small-holding farms restricted nutrient management technologies. Small number of staff engaged in soil and fertilizer management in the extension system could not meet the technological demands of farmers. Low levels of education and insufficient trainings to improve agricultural knowledge of farmers in developing countries are another constraint for extension of nutrient management technology.

10. CONCLUSION

On the basis of above chapter on the use of integrated nutrition management (INM) of organic and inorganic fertilizers in agriculture, it can be concluded that it is extremely important combination for sustainable agriculture. It improves the soil organic matter, adds soil macro- and micronutrients, improves soil physical and chemical properties, rejuvenate soil health, and stimulate soil biological and enzyme activities. It has very valuable effect on tomato yield and quality parameters, gives maximum economic benefit and most imperatively, it is environmentally friendly. But if doses of organic and inorganic fertilizers are applied more than the recommended doses according to particular soil type it may affect adversely on the soil and crop. At the same time it is concluded that nitrate fertilizers are better than ammonium fertilizers and poultry manure, FYM or wheat straw are better than any other manure when are combined with inorganic fertilizers. Moreover, governments should take practical steps to promote the implication of INM to enhance the yield of rice based cropping system.

REFERENCES

Das B, Chakraborty D, Singh VK, Aggarwal P, Singh R, Dwivedi BS and Mishra RP (2014) Effect of integrated nutrient management practice on soil aggregate properties, its stability and aggregate-associated carbon content in an intensive rice–wheat system. *Soil and Tillage* **136:** 9–18.

Doran JW (2002) Soil health and global sustainability: translating science into practice. *Agriculture, Ecosystem and Environment* **88:** 119–127.

Doran JW and Parkin TB (1996) Quantitative indicators of soil quality: A minimum data set. In: *Methods for Assessment of Soil Quality* (JW Doran and AJ Jones, Eds). SSSA Special Publication No. 49, Soil Science Society of America, Madison, WI.

Gill MS, Pal SS and Ahlawat IPS (2008) Approaches for sustainability of rice (*Oryza sativa*)-wheat (*Triticum aestivum*) cropping system in Indo-Gangetic plains of India – A review. *Indian Journal of Agronomy* **53:**81–96.

Goswami NN (1998) Some thoughts on the concept, relevance and feasibility of IPNS under Indian conditions. *Bulletin of Indian Institute of Soli Science* **2:** 3–9.

Goyal S, Chander K, Mundra MC and Kapoor KK (1999) Influence of inorganic fertilizers and organic amendments on soil organic matter and soil microbial properties under tropical conditions. *Journal of Biology and Fertility* **29:** 196–200.

Hati Kuntal M, Swarup Anand, Mishra B, Manna MC, Wanjari RH, Mandal KG and Misra AK (2008) Impact of long-term application of fertilizer, manure and lime under intensive cropping on physical properties and organic carbon content of an Alfisol. *Geoderma* **148:**173–179.

Heeb A, Lundegardh B, Ericsson T and Savage GP (2005) Effects of nitrate, ammonium and organic nitrogen based fertilizers on growth and yield of Tomatoes. *Journal of Plant Nutrition and Soil Science* **168:** 123–129.

John PS, George M and Jacob R (2001). Nutrient mining in agro-climatic zones of Kerala. *Fertiliser News* **46:** 45–52, 55–57.

Karlen DL and Stott (1994) A framework for evaluating physical and chemical indicators of soil quality. In: *Defining Soil Quality for a Sustainable Environment* (JW Doran, DC Coleman, DF Bezdicek, and BA Stewart, eds). pp 53–72. SSSA Special Publication. No. 35, Soil Science Society of America, American Society of Agronomy, Madison, WI.

Mandal B (2005) Assessment and improvement of soil quality and resilience for rainfed production system. Completion Report, National Agricultural Technology Project, Indian Council of Agricultural Research, New Delhi, pp. 30.

Nayak AK, Gangwar B, Shukla AK, Mazumdar SP, Kumar A, Raja R, Kumar A, Kumar V, Rai PK and Mohana U (2012) Long-term effect of different integrated nutrient management on soil organic carbon and its fractions and sustainability of rice wheat system in Indo Gangetic Plains of India. *Field Crops Research* **127:**129–139.

Pillai SP, Geetakumari VL and Issac SR (2007) Balance-sheet of soil nitrogen in rice (*Oryza sativa*) - based cropping systems under integrated nutrient management. *Indian Journal of Agronomy* **52:**16–20.

Prasad R (2008) Rice-wheat cropping systems. *Advance in Agronomy* **86:** 255–321.

Sarkar AK (2000) Integrated nutrient management for better livelihood of farmers in rainfed areas. *Journal of the Indian Society of Soil Science* **48:** 690–696.

Sharma AK, Thakur NP, Kour M and Sharma P (2007) Effect of integrated nutrient management on productivity, energy use efficiency and economics of rice-wheat cropping system. *Journal of Farming System Research & Development* **13:** 209–213.

Singh AK (2007) Evaluation of soil quality under integrated nutrient management. *Journal of the Indian Society of Soil Science* **55:** 58–61.

Timsina J and Connor DJ (2001) The productivity and sustainability of rice-wheat production systems: Issues and challenges. *Field Crops Research* **69(2):** 93–132.

Vineela C, Wani SP, Srinivasarao C, Padmaja B and Vittal KPR (2008) Microbial properties of soils as affected by cropping and nutrient management practices in several long-term manurial experiments in the semi-arid tropics of India. *Applied Soil Ecology* **40:** 165–173.

Zhang CH, Wang Z M, Ju WM and Ren CY (2011) Spatial and Temporal Variability of Soil C/N Ratio in Songnen Plain Maize Belt. *Environmental Science* **32:** 1407–1414

■■■■

2

Biopesticides: An Integral Partner of Sustainable Agriculture

***Amit Yadav*[1], *Pushpa Singh*[2], *Veer Singh*[3] and *Abhishek Yadav*[4]**

[1]*Department of Entomology and Agricultural Zoology, Institute of Agricultural Sciences,Banaras Hindu University, Varanasi – 221005 India*
[2]*Rajendra Agricultural University, Pusa Bhiar, Samastipur – 848101 India*
[3]*Department of Entomology, College of Agriculture, Swami Keshwanand Rajasthan Agriculture University, Bikaner –334006 India*
[4]*Department of Entomology, College of Agriculture, Swami Keshwanand Rajasthan Agriculture University, Bikaner –334006 India*

ABSTRACT

Increasing global population along with demand of safe, nutritious and quality food have made it necessary to increase and sustain food production with relatively less adverse impact on human and environment. With the increased awareness of ecological and human health the focus has shifted on eco-friendly pest management tactics. Extensive and injudicious use of conventional pesticides have led to development of insecticide resistance, resurgence of secondary pests, pesticidal residues and harm to natural enemies. Biopesticides which are the integral part of integrated pest management are means for management of insect and diseases as well as reducing the pesticidal load in environment. It includes a broad array plant derived biochemicals, micro-organism, microbial pesticides as well as genes encoding insecticidal toxins. Biopesticides are considered relatively safe as being target specific they are considered relatively safe to non–target organisms and humans and ecology as whole. Some of the biopesticides like neem based insecticides, *Trichoderma*, Bt, NPV are slowly becoming popular among farmers. But, there is need to increase its production, distribution and marketing as there is large potential of its spread and popularity in agriculture.

Keywords: Biopesticides, Eco-friendly, Microbial pesticides, Microorganism

1. INTRODUCTION

Sustainable agriculture depends on the effective management of resources of agriculture along with satisfying the constant changing human needs while conserving the natural resources and maintaining the environmental quality. The major restrain and limiting factor in crop production are the herbivorous insects pests diseases and weeds. Reducing the incidence of insect, pest and diseases to an extent that they do not cause serious damage to crop without upsetting the balance of nature is also part of sustainable agriculture. Integrated Pest Management (IPM) includes the integration of biological, cultural, physical and chemical controls, host plant resistance, and decision support tools, which may play a significant role in sustainable agriculture. Integrated pest management encourages most compatible and ecologically sound combination of available pest suppression techniques to keep pest populations below economically damaging levels. The IPM concept is based on the principle that it is not necessary to eliminate all pests, but to suppress pest population to a level at which these pests do not cause significant losses.

It can be defined as the rational selection, integration, implementation of available compatible methods, such as biological, cultural, mechanical, botanical and chemical pest control methods based on predicated economic, ecological, geographical and sociological consequences. The first line of defence against pests is prevention through the use of good agronomic practices or cultural methods which are unfavourable for the development of pest problems. Regular monitoring of pest activity is essential for decision making in IPM. Selected management measures to check pest are to be taken at economic threshold level (ETL) a level of pest density at which control measures should be initiated to prevent the increasing pest population from reaching the economic injury level and it justifies the application of control measures. Primary objective of IPM is to ensure minimal and judicious use of chemical pesticides so that there is least disruption to bio-control agents and natural enemies, leading to the conservation of biodiversity. Further, this method of pest control avoids insecticide resistance, insect resurgence, secondary pest attack and evolution of minor pests as major pests. Chemical pesticides have made great contribution against the destructive insects and diseases affecting the production and productivity of agriculture. It is one the most important weapon in our fight for food and nutrition security. Since Second World War, the production as well as consumption of synthetic organic pesticides has greatly increased. It is one of the major contributors providing umbrella protection against pests bringing about the green revolution. Classical organic pesticides like DDT (dichloro-diphenyl-trichloroethane) introduced in 1940 and others pesticides like organophosphates and carbamates in 1960s have been the most common method for pest control. Although very popular method it raised many issues like soil, air and water contamination, resistance development of insects against chemical group, negative effect on non-target organisms including humans (Casida and Quistad, 1998; Fitches *et al*., 2009; Al-Zaidi *et al*., 2011). With time, the availability and use of many broad spectrum pesticides declined due to resistance development as well as legislation (Chandler *et al.*, 2011; Ishtiaq *et al.*, 2012) making way for safer alternative methods (O'Callaghan and Brownbridge, 2009). The burden of over indulgence and indiscriminate use have made extensive damage to environment, human health, development of insecticidal resistance, lethal effect on non-target insects, pest resurgence (Abudulai *et al.*, 2001).

It is in this perspective that there is need for development of biopesticides for preventing environmental pollution as well controlling the menace of agricultural pests without causing harm to ecology. Organisation for Economic Co-operation and Development (2009) defined biopesticides as a mass produced agent manufactured from living microorganism or a natural product and sold for the control of plant pests. The natural resource from where biopesticides are derived are classified as Microbial pesticides, botanicals, Zooid pesticides and genetically modified plants (Chandler *et al.*, 2011). Most of the biopesticides are steadily becoming popular as they have the advantage of higher selectivity and safety to non-target living. They are characterized by low residue toxicity, less toxic side effects along with high performance. Bacterial pesticides control nearly 74 per cent of market share, 10 per cent by fungal biopesticides, viral biopesticides 5 per cent and rest by other biopesticides (Thakore, 2006). Across the world biopesticides like viruses, bacteria, fungi, insect specific toxins, pheromones, hormones, botanicals and genetically modified organisms have been used and developed (Nicholson, 2007; Erlandson, 2008; Mazhabi *et al.*, 2011).

2. BIOPESTICIDES

Naturally occurring substances that control pests by nontoxic mechanism that may be living organisms (natural enemies) or their products (phytochemicals, microbial products) or byproducts (semiochemicals) which are used to manage injurious pests are biopesticides, they are integral part of bio-intensive IPM programme. They offer ecologically sound and target specific solution to pest problem with less impact on environment and human health. These biopesticides include bio-insecticides like *Bacillus thuringiensis*, bio-herbicide like *Phytopthora* and bio-fungicides like *Trichoderma* offering a range of benefits like:

i. Less environmental load and harmfulness

ii. Toxic to target organism only with high specificness

iii. Biologically degradable in short period of time

iv. Very effective as a component of IPM programme.

Major categories of biopesticides are:

A. *Microbial biopesticides*: This includes microorganisms like bacterium, fungi, protozoa, algae and viruses as active ingredient with specific mode of action against the target pests. The most popular and common microbial pesticide is *Bacillus thuringiensis* or Bt which is very effective against lepidopteron as well as coleopterans and is used worldwide to manage insects of potato, cabbage, corn etc.

B. *Plant –incorporated protectant (PIPs):* These are pesticidal substances produced from genetic material and incorporated into plants as the case of Bt protein gene introduced in plant genetic material thereby the plant produces the pesticidal protein and destroys the pest.

C. *Biochemical Pesticides*: Naturally occurring substances which may be plant extracts, pheromones, fatty acids, Hormones etc. which have the ability of controlling insects by non-toxic mechanism either by interfering with their growth, mating, attractants or repellant comes under this category.

2.1 Microbial Biopesticides

They are naturally occurring bacteria, fungi, protozoans, viruses or algae which manage the pest specific toxins, by causing disease, preventing establish-ment of other microorganism by competition or by any other mode of action (Clemson HGIC, 2007). Presently there are nearly 73 microbial active ingredients registered by USEPA including 35 bacterial, 15 fungi, 6 viruses, 1 each yeast and protozoa (Steinwand, 2008).

2.1.1 *Bacterial biopesticides*

They are the most common bio-pesticides generally used to control insects besides bacteria, fungi and viruses. The first record of Bacillus was in 1901 as the cause of “sotto” disease of silkworm in Japan (Ishikawa, 1936). *Bacillus thuringiensis* was isolated by Berliner (1915) from Mediterranean flour moth, *Ephestia kuehniella* and was re-isolated in 1927 by Mattes, both reported high pathogenicity. This bacterium occurs naturally in in soil worldwide with 1000 isolates reported from soils of five continents. It was recognized as living insecticides by Steinhaus (1954) with bio-potency against insect pests of order Lepidoptera, Diptera, Hymenoptera, Orthoptera and Coleoptera. With a wide host range especially against lepidopteron pests it is regarded as novel insecticide (Estruch *et al.*, 1996). They are needed to be effectively ingested by the target pest.

Entomopathogenic bacteria are divided into four groups on the basis of their pathogenicity *i.e.*, obligate (*B. popilliae*), Crystalliferous spore formers *(B. thuringiensis*), facultative pathogens (*B. spharicus*) and potential pathogens (*Serratia marcescens*). The strains and subspecies of *B. thuringiensis* is the most widely used microbial pesticide accounting for 90 per cent of the biopesticides market (Chattopadhyay *et al.*, 2004; Romeis *et al.*, 2006). About 6,000 isolates are stored in various repositories of world with each strain producing different types of protein which specifically kills one or few related insect species. It produces three types of insecticidal crystal protein α exotoxin, β exotoxin and δ endotoxin (Heimpel, 1967; Chattopadhyay *et al.*, 2004).

B. thuringiensis is a gram positive spore-forming, host specific bacterium that produces crystalline proteins δ endotoxins released to the environment after lysis of the cell wall at the end of sporulation and causing death within 24 hrs (Jisha *et al.*, 2013). There are different commercial *B. thuringiensis* products which have been developed for management of insect pests of crops as well as mosquito species. The spore crystal formulations are obtained from different strains like *B. thuringiensis* var. *kurstaki* (Btk)-isolate HD1 which includes Cry1Aa, Cry1Ab, Cry1Ac, and Cry2Aa proteins, *B. thuringiensis* var. *kurstaki* (Btk)-isolate HD73 containing Cry1Ac), *B. thuringiensis* var. *aizawai-isolate* HD137 including Cry1Aa, Cry1B, Cry1Ca, and Cry1Da, *B. thuringiensis* var. San Diego and *B. thuringiensis* var. *tenebrionis* having Cry3Aa and *B. thuringiensis* var. *israelensis* containing Cry4A, Cry4B, Cry11A, and Cyt1Aa toxins (Schünemann *et al.*, 2014).

After ingestion, the *B. thuringiensis* crystal protein is solubilized in the alkaline midgut of insect and proteolytically converted into toxic fragment, during which peptides from the N terminus and C terminus are cleaved from the full protein and activated toxin binds to receptors located on the apical microvillus membranes of epithelial midgut cells and inserts into the cell membrane. All this leads to oligomerization and the oligomer forms pore or ion channel within the functional receptors of the brush border membrane thereby disrupting the membrane transport and lysis of cell ultimately resulting in death of insect (Jisha *et al.*, 2013; Schünemann *et al.*, 2014). The families belonging to Lepidoptera are agriculturally important Noctuidae Cossidae, Gelechiidae, Lymantriidae, Pyralidae Pieridae, Thaumetopoetidae, Tortricidae and Yponomeutidae (Gathmann and Priesnitz, 2014). Tephritidae, Culicidae, Muscidae, Simuliidae and Tipulidae are the Dipteran families susceptible to *B. thuringiensis* (Lysyk, 2006). Chrysomelidae, Curculionidae, Tenebrionidae and Scarabeidae are Coleopteran families susceptible to δ endotoxin toxin (Gathmann and Priesnitz, 2014). The cry protein of *B. thuringiensis*are also reported to be toxic against diverse insect order like Hymenoptera, Hemiptera, Orthptera, Acaria and Phthiraptera (Mallophaga) (Eswarapriya *et al.*, 2010). Several workers have reported entomocidal activity of *B. thuringiensis* endotoxins against a number of insects like Locusts (Orthoptera : Acrididae): *Locusta migratoria manilensis* (Wu *et al.*, 2011), the cotton aphids *Aphis gossypii* (Hemiptera: Aphididae) and white-flies *Bemisia tabaci* (Hemiptera: Aleyrodidae) (Malik and Riazuddin, 2006).

Vegetative Insecticidal Proteins (VIP) is another protein derived from *B. thuringiensis,* which includes the binary toxin Vip1 and Vip2 with Coleopteran specificity and Vip3 with a wide activity spectrum against Lepidoptera (Yu *et al.*, 2011). The insecticidal potency of Vip1/Vip2 against the Coleopteran store grain pest, *Sitophilus zeamais* (Curculionidae family) with 60 per cent mortality has been reported by Shingote *et al.* (2013). Vip3Ac1 have been reported to show high insecticidal activity against the Lepidoptera larvae, the fall armyworm *Spodoptera frugiperda* (Noctuidae) and the cotton bollworm *Helicoverpa zea* (Noctuidae) but very low activity against the silk worm *Bombyx mori* (Bombycidae) (Fang *et al.*, 2007).

2.1.2 *Fungal biopesticides*

Entomopathogenic fungi which are over 750 species and 100 genera are distributed throughout the fungal kingdom, majority belonging to sub-division Deuteromycotina (imperfect fungi) and Zygomycotina (Benjamin *et al.*, 2002; Zimmerman, 2007). Most of the entomopathogenic fungi are obligate, facultative or symbiotic and high humidity with ambient temperature are necessary for sporulation and spore germination. They may be ectoparasite or endoparasite and insects are usually infected by spores or conida (Zygomycotina), conidia (Deuteromycotina), zoospores (Mastigomycotina) and ascospores (Ascomycotina). *B. bassiana* and *M. anisopliea* infect over 100 different insect species of different insect orders (Keller *et al.*, 2003), although, their isolates have found to have high level of specificity also (Benjamin *et al.*, 2002), which may be related to insect maturation and host plant, the characteristic of insect integument and also to the nutrition requirement of the fungus as well as the cellular defence of host insect (Kerwin and Washino 1986a,c; Scholte *et al.*, 2004).

Entomopathogenic fungi invade their hosts by direct penetration in the host exoskeleton or cuticle or through natural body opening, the conidia germinate on the insect cuticle and often form an appressorium (Brey *et al.*, 1986). The hyphae penetrates into the cuticle of infected insect, into the haemocoel, proliferate in the insect's body, producing toxins and removing nutrients thereby causing insect death which is caused by tissue destruction as well as toxins produced by the fungus which finally emerges from insect bodies to produce spores which spreads by wind, rain or other insects, therby spreading infection. A group of cyclic peptides containing ester linkages destruxins (DTXs) produced by *M. anisopliae*, are responsible for insect mortality by the fungus (Suzuki *et al.*, 1970) and they are insecticidal by injection (Fargues *et al.*, 1986) or when ingested by mouth with toxicity most acute to Lepidopteran larvae and adult Diptera including mosquito larvae.

B. bassiana, a parasite on different insect species, grows naturally in soils throughout the world, causing white muscardine disease (Shah and Pell 2003). It is one of the most common entomopathogenic fungal species that shows strong pathogenicity to Lepidoptera, Hymenoptera, and Coleoptera (Harris *et al.*, 2000; Fernandez *et al.*, 2001; Mannion *et al.*, 2001; Phoofolo *et al.*, 2001) and is used to control pest like termites, whiteflies, beetles and mosquitos (Donald and Neil 2005). The microscopic spores of the fungus come into contact with insect host germinates, penetrates the cuticle, and grows inside, thereby killing the insect after which a white mold emerges from the cadaver and produces new spores.

2.1.3 *Viral biopesticides*

More than 20 groups of viruses of 12 viral families and over 1100 hosts, including a few mites and ticks are known to be insect pathogens (Martignoni and Iwai 1986). Baculoviruses are the most common and widely studied group of viruses pathogenic for insects (Martignoni 1984) with subgroups, Nuclear

Polyhedrosis Viruses (NPV) and the Granulosis Viruses (GV) having different disease symptoms. Both groups have an enveloped rod shaped virion (approx. 50 × 250 nm) containing a circular double stranded DNA genome with molecular weight ranging from 50 to 100 million Dalton (Bilimoria 1986). Both NPV and GV are in a polyhedral proteinaceous inclusion body (IB) having diameter of 40–80 μm and length greater about 300–420 μm. Granulosis Viruses (GV) are more host specific than Nuclear Polyhedrosis Viruses. The period of ingestion of GV which is transmitted orally or by egg and death of host is between 4 to 25 days with the larvae developing light color and liquefaction if epidermis is involved (Martignoni and Iwai, 1986).

Viruses enter the insect's body via the gut, multiplies in tissues, disrupting components of an insect's physiology as well as interfering with feeding, egg laying and movement. Infectious virus particles when ingested by host insect liberate and become active internally. In the larval gut, the virus's protein overcoat disintegrates, and the viral DNA infects the digestive cells. Within a few days, the host larvae become unable to digest food, weaken and die (Thakore, 2006b). A browninsh fluid oozes from the infected larvae and it hangs upside down from leaves and twigs. This fluid is highly infective and spreads amongst the healthy insect population. The Baculo viruses are highly host specific, the killed insect have shiny-oily appearance, very fragile to touch and ruptures to release infective virus fluid particle. They have tendency to remain attached to foliage and rupture.

NPVs are highly specific and are safe to parasites and predators as well as to fishes, birds and mammals. They are very effective against the lepidopterous pests of vegetables, and can also be used in a variety of field crops, including sorghum, chickpea, cotton and maize. It has been found to be very effective and highly specific against *H. armigera, S. litura* and *P. xylostella.*

Applications of *Ha*NPV @ 450LE (Larval equivalent) ha^{-1} at 10 days interval was found to be most effective on chickpea in pod damage reduction. Two spray of NPV @ 500 LE ha^{-1} on chickpea against *H. armigera,* recorded, lowest pod damage of 8.6 and 7.9 per cent when 7 and 14 days after treatment. The highest reduction in larval population of 79.9 and 81.5 per cent was recorded 7 and 14 days after NPV spray with highest yield of 17.9 q ha^{-1} (Rai *et al.*, 2001).

NPV is found to be very effective against larval population of *H. armigera* alone as well as with endosulfan (Dahiya and Chauhan, 2001). Singh *et al.* (2001) recorded HaNPV @ 1.5 × 1012 POB ha^{-1} spray 3 times at 10 days interval with 3 sprays of endosulfan (700 g a.i. ha^{-1}) at 15 days interval in comparison to control against *H. armigera* on tomato. Similarly, when five biorational insecticides were evaluated against the major insect pest of pigeonpea *viz.*, *Bt* (biolep, dipel and halt), NPV, *B. bassiana* (dispel), NSKE and Juvenile hormone (RH-2485), NPV was found best (Saxena, 2001). The specificity of NPVs is very useful in preservation of natural predators in Integrated Pest Management systems and can also cause sudden and severe outbreaks within the host population for complete control (Sylvar, 2008). The limitation with baculo virus is that it needs to be ingested which reduces its efficacy as well as its high cost of production.

2.1.4 *Entomopathogenic nematodes*

Five orders and eleven families of nematodes are known to contain entomopathogenic nematodes (Tanada and Kaya, 1993) where, insects serve as vectors or intermediate hosts for a number of parasitic nematodes of vertebrates (Schmidt and Roberts, 1989; Grewal *et al.,* 2001). R.W. Glaser (1929) found the nematode, *Steinernema* (syn. *Neoaplectana*) *glaseri* infecting the Japanese beetle, *P. japonica*, and cultured for the first time this parasitic nematode on artificial media and uses it in field tests against the beetle (Glaser and Farrell, 1935; Grewal *et al*., 2005).

Parasitic nematodes parasitize their insects by directly penetrating through the cuticle into the hemocoel or by entering through natural openings spiracles, mouth, and anus (Sankar, 2009). Species in the order Tylenchida possess a stylet and in the order Mermithida possess a spear, which is technically called an odontostyle. *Heterorhabditis* has an interior tooth that is used to scrape and rupture the host's cuticle (Bedding and Molyneux, 1982). Passive infection of insect occurs when a mermithid deposits its eggs on the hosts food which are ingested by an insect, and the nematode, bore through the midgut and enters the hemocoel (Rishipal *et al*., 2008). Active infection of insect takes place when the nematodes search their hosts and penetrate directly through the integument into the haemocoel. Which takes from 10 min to 2 hours, the adhesive substance seals the wound after the nematode has entered the insect. Insects infected with nematodes have shown abnormal behaviour compared which may be as a result of other anomalies which is usually observed late. Many physiological changes from a number of insects groups have been recorded due to nematode infection (Sankar, 2009). Steinernematid nematodes are known to carry a bacterial symbiont in their digestive system, in infective juvenile stage, called a dauer larva, penetrates the host cuticle entering the haemocoel, and release sing the bacteria *Xenorhabdus* which kill the insect host within 48-hrs and the juvenile nematodes feed on the bacterial and bacterial byproducts in insect body. The nematodes develop to adult stage, lay eggs inside the host and become infective juvenile daver larvae, which exit the dead host only to infest other host insects. Steinernematid nematodes are not host specific and have the ability to kill their host within few days.

2.2 Plant-Incorporated-Protectants (PIPs)

PIPs are pesticidal substances that plants produce from genetic material that has been added to the plant. For example, scientists can take the gene for the *Bt* pesticidal protein, and introduce the gene into the plants own genetic material. Then the plant, instead of the *Bt* bacterium manufactures the substance that destroys the pest. Both the protein and its genetic material are regulated by EPA;

the plant itself is not regulated. Genetically modified plants express genes encoding insecticidal toxins and the adoption of genetically modified (GM) crops has increased dramatically in the last few years. Transgenic plants express insecticidal δ-endotoxins derived from the soil bacterium *Bacillus thuringiensis* (*Bt* plants) were first commercialized in the US in 1996. The expression of these toxins gives protection against insect crop destruction. The toxicity of *Bt* endotoxins is dependent on the alkalinity of the insect gut which also makes its safer to human and other vertebrates. The δ-endotoxins proteins are commercially produced and introduced against the highly damaging pests of cotton, tobacco, tomato, potato, maize and paddy. More than 60 Cry proteins in different strains of *Bt*, have been identified. *Bt* maize expresses the Cry1Ab protein, and a few express the Cry1Ac or the Cry9C protein targeting against the European corn borer (*Ostrinia nubilalis* Hubner) (Lepidoptera), a major pest of maize in North America and Europe. Some maize hybrids express the Cry3Bb1 protein against the corn rootworm complex (*Diabrotica*) (Coleoptera). Cry1Ac protein in cotton targets against the cotton bollworm (*Helicoverpa zea)* Boddie) (Lepidoptera) a major pest of cotton, potato the Cry3A or Cry3C i against the Colorado potato beetle (*Leptinotarsa decemlineata* Say) (Coleoptera) and Cry4 proteins target against Diptera flies like *Lycoriella castanescens* Lengersdorf and mosquitoes *Culex pipiens.*

2.3 Biochemical Pesticides

Biochemical pesticides are characterized by their non-toxic mode of action toward target organisms and natural occurring (Steinwand, 2008). The active ingredients which may be a single or a mixture of molecules, like a plant essential oil, or isomers as in the case of insect pheromones. While, active ingredients of biochemical pesticides occur in nature, or they may be a synthetic analog. These are categorized into biologically functional classes, like semio-chemicals, plant extracts, natural plant growth regulators, and natural insect growth regulators. Presently 122 biochemical pesticide active ingredients registered that include 18 floral attractants, 20 plant growth regulators, 6 insect growth regulators, 19 repellents, and 36 pheromones.

2.3.1 *Plant extracts and oils*

Plant extract or Phytochemicals show wide range of activity including hormonal, neurological, nutritional, and enzymatic activity. This category of biopesticides is diverse in compositions and specific mode of action which target the pest. They are mostly used as insecticides, product to product variation in mode of action. Their use for insecticidal purpose in storage pest control has also been well documented. The mode of action varies greatly from product to product.

Sex pheromones interrupt the reproductive cycle of insects but plant extracts and oils acts less directly and specifically, botanical extracts such as floral essence attracts insects to traps, caynene is deterrent. Extracts of *Hoodia gardoni* deterred the feeding and oviposition of *Trichoplusia ni* (Akhtar *et al.*, 2009). Similarly, extract of *Humulus lupulus* is known to manage the larval stage of Colorado potato beetle. Plants synthesize two types of metabolites, primary and secondary. The primary plant metabolites are the starting materials for the biosynthesis of specific and genetically controlled, and enzymatically catalyzed complex compounds known as secondary metabolites. Secondary plant metabolites play an important role in the defense against insect herbivores as they act as insect repellents, feeding inhibitors, and toxins. Terpens, the allied terpenoid compounds and the steroids are largest and most structurally diversified class of naturally occurring organic products of secondary plant metabolites.

Plant products which are not considered biopesticides include nicotine extracted from tobacco (highly toxic to bees), and rotenone extracted from roots of the legume family (Ware and Whitacre, 2004). Tannins coagulate muco-proteins in insect oral cavity and affected its digestion. Coumarins which are ovicidal to Colorado potato beetles are located in roots and seed coats. Thymol control the Varroa mite (*Varroa destructor*) which is parasitic to bees is a chemical constituent of thyme essential oil obtained from *Thymus vulgaris*. It is volatile in nature and permeates into hives, irritating the mites and causing them to withdraw from the bees and die from starvation.

2.3.2 *Insect growth regulators*

Insect growth regulators are naturally found chemical compounds that affect the growth and development of insects (Siddall, 1976). Juvenile hormones (JH) disrupt immature development and prevent the emergence of an adult. Bowers *et al.* (1976) isolated and identified anti-JH substances from the plants which provide protection against insects.

Slama and Williams (1965) found that the "Paper Factor" was responsible for supernumerary larva in *Pyrrhocoris apterus*. Precocene I and II, and their analogs isolated from plant, *Ageratum houstoniatum*. Interfere with the JH activity causing precocious metamorphosis and sterilization in Heteroptera, Homoptera and Orthoptera insects (Bowers *et al.*, 1976; Bowers, 1991; Grafton-Cardewell *et al.*, 2006). Chitin synthesis inhibitors is the third group of IGR which inhibits the insect to produce a new exoskeleton after molting and also affects the elasticity and firmness of endocuticle (Ishaya and Horowitz, 1999; Schneider *et al.*, 2003). So the insect unprotected from the cuticle and from the prey, its chances of survival gets reduced. A number of chitin synthesis inhibitors *i.e.*, benzoylureas, buprofezin and cyromazine act as conventional pesticides.

Azadirachtin from the neem tree *Azadirachta indica* (Meliaceae) is the botanical pesticide greatest potential and is effective at doses as low as 0.1 ppm, acts as a feeding deterrent to more than 100 species of herbivorous insects (Saxena, 1989). Saponins act as toxins and feeding deterrents to a number of species of mites, lepidopteron, beetles and many other insects (Ishaaya *et al.*,

1969), whose toxicity is due to the binding of saponins to free sterols in the gut of insect so it reduces the rate of sterol uptake into the haemolymph and interfering with the insect molting process (Chaieb, 2010). As maximum registered insect growth regulators are structurally similar to juvenile hormone also known as juvenoids which can be naturally occurring or synthetic. Phytojuvenoids (sesamin, sesamolin, juvacimene, juvadecene, and others) showing modest to very high JH activity has been isolated from several plant families (Bowers, 1991). Juvocimenes from sweet Basil, *Ocimum basilicum* disrupts the insect hormonal process leading to the development of second generation of JH-active commercial products that is fenoxycarb. These hormone-based insect growth regulators are effective against several field crop pests (Lopez *et al.*, 2005). Insect Juvenile hormones JH I-III, JH-0, and iso-JH-0 are synthesized and secreted from the corpora allata (Miyamoto *et al.,* 1993). These JHs control a number of processes, such as embryogenesis, molting, metamorphosis, reproduction, diapause, communication, migration/dispersal, caste differentiation, pigmentation, silk production, phase transformation and any disturbance in normal hormone balance affects growth and development of the insects. As these JH are very specific, but due to their instability difficulty in synthesis not used in pest control, but their mimics or analogs (JHAs) are easy to synthesize, the first compound being methoprene a terpenoid compound used against household pests due to its low activity against agricultural pests and low residual on plants under field conditions (Smith, 1995).

The 'ecdysteroides' or 'phytoecdysones' are a group of terpenoids discovered in plants (Eswaran, 2009). It is known as ecdysone or molting hormone is a group of steroids α- and β-ecdysone isolated from insects by Butenandt and Karlson (1954) and affect ecdysis or molting in insects. Ecdysones resemble plant cholesterol in structure and its high concentrations in plants suggest a defensive function against non-adapted insect herbivores.

Phytoecdysones and several synthetic ecdysone analogs inhibit growth, development, and reproduction when fed to several species of insects (Singh *et al.,* 1982). These are effective in very small quantity and not species specific. Iridoid glucosides present in *Plantago lanceolata* act or toxic agent to a variety of generalist insect herbivores and have been shown to act as anti-feedants to grasshoppers and lepidopteron larvae (Bowers and Puttick, 1988).

The feeding inhibition activity of sesqueterpenes against some lepidopteron larvae may be due to blocking of the stimulatory effects of glucose, sucrose and inositol on chemosensory receptor cells located in the insect mouth parts (Frazier 1986). Sesquiterpene lactones are poisonous to several lepidopteron, flour beetles and grasshoppers (Isman and Rodriguez, 1983; Eswaran, 2009). Gossypol is toxic to a number of herbivorous insects, causing an antibiosis effect on many insect pests of cotton including the tobacco bud worm, *Heliothes virescens* and cotton leaf worm, *Spodoptera littoralis* (Gang *et al.*, 2009).

2.3.2.1 Attractants

Attractants use in pest management as precise, specific, and ecologically sound has been described as "new, imaginative, and creative approaches to the problems of sharing earth with other creatures" (Carson, 1962). Pheromones are semiochemicals that act intra-specifically in chemical communication between individuals of the same species (Karlson and Butenandt, 1959), behavioural message between individuals of different species are called allomones if they favour the producer and kairomones if they favour the receiver (Brown *et al.*, 1970). This complex, biochemical and biophysical process are the reason of precise chemical communication system whereby chemically specific pheromones or kairomones induces directed insect behaviour. Structurally optimized synthetic molecules which have similar steriochemical structures can also induce similar response as that of natural pheromones or kairomones. Such parapheromones or parakairomones besides being cheaper are also less volatile or are more persistant.

Peter Karlson and Martin Luscher in 1959 introduced the term pheromones which are the secretions of exocrine glands of insects causing a specific reaction sexual attraction and mating behaviour, alarm, aggregation, trail marking, or specific changes in physiological development, such as maturation or sexual determination in the insect of the same species, (Karlson and Butenandt, 1959). 3000 insect species, pheromones including 1700 that of lepidopteron have been identified (Klowden, 2007).

2.3.2.2 Sex pheromones

Sex pheromones are distributed in the class insect of 10 orders (Jurenka, 2004; Reddy and Guerrero, 2004; El-Sayed, 2005; Gould *et al.*, 2009). They have been identified as specific chemical structures in hundreds of insects (Tamaki, 1985; Mayer and McLaughlin, 1990). In Lepidoptera they are volatile esters, alcohols, aldehydes, and ketones produced in the last abdominal segments of the females. A large number of sex pheromones have been isolated and characterized as simple straight chain aliphatic compounds (El-Sayed, 2005; Gould *et al.*, 2009).

In pest management synthetic sex pheromones are used to attract insect to a trap which contains lethal pesticide or to disrupt mating. They are also be used to monitor pest populations to determine right timing and application of pesticides. These insect pheromones form a large per centage of the biochemical pesticides market. They can be successfully used to manage insects when pest pressure is moderate to low as well as to get information about population levels, to monitor control or to forewarn about new pest introductions (Ware and Whitacre, 2004). These insect pheromones are highly species specificity and have low toxicity making them one of the most targeted pest management strategies (Goldansaz *et al.*, 2004). This specificity maintains an ecological balance by leaving undisturbed populations of other insect species and non-target organisms (Ware and Whitacre, 2004), but the limitation is that they have to be used in combination with other steps of pest management to achieve desirable control especially when pest pressure is very high.

Pheromones of many insect species are now synthetically produced and used in in insect pest management programs (Reddy and Guerrero, 2004). Some pheromones are specific to one insect, whereas some like clearwing borer *Podosesia syringae* lure attract several related species. Lures of Japanese beetle *P. japonica* may also include floral or food analogues that attract females as well as males.

(i) Alarm pheromones

These pheromones are released by social insects to warn other colony members of danger and to be ready for colony defense (Klowden, 2007). It induces for a collective response to some traumatic stimulus rather than an individual response (Fujiwara-Tsujii *et al.*, 2006). The alarm pheromones are produced in abundance and appear to be the least specific of all pheromones (Witte *et al.*, 2007).

(ii) Aggregation pheromones

Aggregation pheromones unlike sex pheromone which acts on only one sex induces group formation by bringing many individuals of both sexes together.

(iii) Trail-marking pheromones

Social insects like termites and ants that forage on the ground, but a few nonsocial insects mainly produce these pheromones to communicate. When a worker locates a resource it lays down a trail, while returning to the colony so that other worker can use to find the resource (Klowden, 2007).

These pheromones are used in various ways:

a. Mating disruption

Pheromones, *i.e.* the chemicals produced by an insect which evoke a specific response in the other individuals of the same species are used as mating disrupters and is based on the principle that when a specific pheromone is released in the air in high quantity, the males are unable to orient to natural sources of pheromone and fail to locate the calling female and the reproduction is prevented (Witzgall *et al.*, 2010). Olfactory receptors for detection of pheromone in moths located on the antennae, when exposed to a constant stimulus of high and uniform concentration of the pheromone shuts down the ability of sensory organs to detect the pheromone (Witzgall *et al.*, 2010). The exposure of insect to high concentration of pheromone results in slow decline of behavioural response lasting several minutes or few hours affecting the central nervous system and is called habituation and nerves do not recover in the normal manner playing an important role in suppressing normal male responsiveness in the mating disruption.

b. False trails

The time and energy of male moths of *Cydia pomonella* in search of females when pheromones are placed at different parts of field is critically affected (Pfeiffer *et al.,* 1993a, b). So male of *C. pomonella* finding a calling female would diminish, with all pheromone sources releasing nearly same amount of pheromone as a calling female.

c. Masking

Here background level of the pheromone is high enough to mask the odor trail from a calling female and although, the sensory system of the male moths is functioning, but due to high background level of the pheromone the relatively low concentration trails emitted by females cannot be perceived by male moth. It is effective against pink bollworm, *Pectinophora gossypeilla* by permeating the air with synthetic pheromone, gossyplure (Lykouressis *et al.,* 2004) and the natural sex attractants emanating from the female is masked causing disruption in mating and ultimate result in suppression (Witzgall *et al.,* 2008).

d. Volatile kairomones as lure for insects

Volatile plant kairomones which are semiochemicals synthesized by plants are used as lure to attract insects to traps for monitoring populations, for trapping, for control operations timing, and as a lure in poison baits (Suckling, 2000; Tooker *et al.,* 2005). The discovery of methyl eugenol or 3,4-dimethoxy-1allylbenzene, from lemon grass oil, *Cymbopogon nardus* by Howlett's (1915) which was found to be highly attractive to the male fruit flies, *Dacus diversus* and *D. zonatus* and at least 60 other closely related *Dacus* Sp. Methyl eugenol attracts fruit flies from a distance of 800 m has both olfactory as well as phagosti-mulatory action (Roomi *et al.,* 1993; Wee *et al.,* 2007). Male annihilation technique uses methyl eugenol, along with an insecticide impregnated into a suitable substrate has been successfully used for management and control of *Bacterocera* species (Cunningham, 1989; Wee *et al.,* 2002). This discovery initiated the beginning of the modern chemical ecology of insects (Metcalf, 1990) as well as chemical characterization of plant kairomone for insect control (Cornelius, 2000),

Semiochemical attractants has been most extensively used and are an integral part of many IPM programs (Cornelius, 2000; Suckling, 2000).

2.3.4 *Neem as biopesticide*

Azadirachtin (Tetranortriterpenoids) a derivative of neem is the dominant active insecticidal component present in seeds and leaves of neem is strong anti-feedant growth inhibitor of many insects (Butterworth and Morgan, 1968; Al-Fifi, 2009). Their formulations are environmentally safe and low toxic to non-target natural enemies (Saxena, 1989). It interferes with the life processing behaviour of insects without killing it, but the insects cannot feed, breed or undergo metamorphosis

(Elahi, 2008), disrupting molting by inhibiting biosynthesis or metabolism of ecdysone and juvenile hormone (Ware and Whitacre, 2004). The mode of action is as below:

a. Insect growth regulator

The neem derivatives act as insect growth regulator on nymphs of heteroptera (Leuschner, 1972) and larvae of Lepidoptera (Ruscoe, 1972). Its action on growth of larval form of an insect depends on an enzyme, ecdysone is suppressed and the larva fails to molt, remaining in the larval stage and ultimately dies (Rembold *et al.,* 1984; Akhtar and Ismam, 2004; Al-Fifi, 2009). When the concentration of azadirachtin was low, the larvas enter the pupal stage but dies and if still less the emerging adults from pupae are 100 per cent malformed, sterile and without any capability of reproduction (Schmutterer, 1987; Lowry and Isman, 1996; Di-Ilio *et al.,* 1999). The survival, fecundity, development and oviposition of *P. xylostella* were adversely affected by the extracts of *Melia azadirach* and *A. indica*. The larval growth of *Helicoverpa armigera* was inhibited by sublethal dose of azadirachtin of 1 μg/insect (Gujar, 1997). Morphological deformity like larval pupal intermediate, larva forms pupa and deformed adult of *Helicoverpa armigera* was caused by neem formulations like bioneem, neemazol, econeem, achook, nimata and nimbicidine @0.5 and 1% (Srivastava and Mathur, 1998).

b. Feeding deterrent

Feeding deterrence is an important characteristic of neem (Al-Fifi, 2009). Due to presence of azadirachtin, salanin and melandriol on neem treated plants anti-peristalitic wave in the alimentary canal producing vomiting sensation in the insects due to which insects stop feeding and its ability to swallow is also blocked (Ayyangar and Rao, 1989, 1990).

c. Oviposition deterrent

Neem based pesticides do not cause direct mortality, but can cause substantial reduction in the fecundity of insects (Al-Fifi, 2009), resulting in population below economic threshold level in subsequent generations. It also reduces pests by not allowing the females to deposit eggs. Injection of azadirachtin inhibited the juvenile hormone production thereby preventing vitellogenin synthesis and egg production (Rembold *et al.,* 1984). The sterilizing effect of azadirachtin is due to inhibition of oogenesis and vitellogenesis (Schmutterer, 1987). Egg laying is prevented by coating of seeds in storage with neem kernal powder and neem oil (Dorn *et al.*, 1987). The fecundity of cabbage aphid *Brevicoryne brassicae* was reduced by Azadirachtin @ 30–60 ppm (Koul, 1998). Similarly, the ovicidal effect on the eggs of *Leucinodes orbonelis* was seen by use of Neem T/S (1% Azadirachtin) and neem azal-F (5% azadirachtin) (Kumar and Babu, 1998). Other pesticidal action by neem includes inhibition of chitin formation (exoskeleton), disruption of sexual and mating communication, repelling larvae and adults of, sterilization of adults and poisoning of adults and larvae.

d. Control of store grain insects

Loss due to storage pests is nearly 9 per cent in developed countries to 20 per cent or more in developing countries (Phillips and Throne, 2010). Neem is effective safe, natural and inexpensive alternative to chemical pesticide (Umoetok and Gerard, 2003). In rural India, it was an age old practice to store grains mixed with dried neem, protecting and preventing spread of storage pests if present. Neem oil or kernel powder coating protects stored products up to 20 months without any impact on its palatability. Due to anti feeding property insects do not feed and oviposition deterrent characteristic prevents the female insect from egg laying.

3. BIOPESTICIDES: NEW CHALLENGES, NEW OPPORTUNITIES

Biopesticides offer powerful tools to create a new generation of Sustainable Agriculture by having a powerful tool in the form of biopesticides, an effective alternative to chemical pesticide which has been under severe scrutiny and criticism, offering solution to ever increasing problems of pest resistance, resurgence as well concern of ecology and non-target pests.

The limited spectrum of activity against the different pest complexes especially those of viruses and fungi in comparison to chemical insecticides is one of major constraint in biopesticide use. The efficacy of biopesticide is the result of interaction between the pathogen, host as well as the temperature and environmental condition (Blanford *et al.,* 1998; Blanford and Thomas, 2000). Generally, the biopesticide takes 4–7 days to kill the host in temperature range of 25–30°C, but temperature in field conditions do not always remain constant and there can be huge fluctuation in 24 h period. Microbial insecticides as compared to chemical insecticides are not harmful to non-target species and have limited potential for environmental degradation so there should be increased efforts to improve their efficacy and to increase their usefulness in IPM programs (Cuperus *et al.,* 2004). Plant based many compounds have been isolated, characterized, and evaluated for its anti-insect activity from plants, but there is no standardization of plant based anti-insect preparations which is its major constraint hindering their potential marketability as compared to conventional pesticides. The concept of phytosynergistic strategy and assessing the toxicity of co-occurring toxins has the potential leading to further advancement in the development of biopesticides (Koul and Dhaliwal, 2001).

Since biopesticides are not designed to be one-for-one replacements of broad-spectrum chemical pesticides, it is best to be used as integral part of integrated pest management with more skill and understanding.

4. BIOPESTICIDE REGULATIONS

Bio-intensive Integrated Pest Management concept emerged by the advent of biologically based alternatives, which are environment friendly ensuring food safety and security, as well as maintaining the ecobalance. All pesticides in India are regulated by the Insecticide Act, 1968 with mandatory registration from the Central Insecticide Board and Registration Committee.

5. MARKETING AVENUES FOR BIOPESTICIDES

Despite active promotion and campaigning of IPM programme, the awareness and use of biopesticides among farmers is poor, although the market's prospects are very high in light of increased awareness to food safety standards in domestic as well as international markets. Chemical pesticides give stiff competition to biopesticides which are trying to establish identity for themselves with preconc-eived notion, lower performance and efficacy. The cooperation of government agencies, private sectors and farmers to develop and promote the number of model projects and success stories showing success in 'cash-crops' like oil palm and cocoa, orchards, cotton and other commercial and field crop will encourage and form new markets. World wild biopesticides are used for controlling insect pests and diseases. They are increasingly growing market segments and are expected to only increase in future.

There are 175 registered biopesticide active-ingredients and 700 products available in the market globally and its market was valued at US $1.3 billion in 2011, and it is expected to reach US $3.2 billion by 2017. The increase awareness and popularity of residue free organic food products are increasing its demand. North America dominates the world biopesticide market and accounts for about 40 per cent of the global biopesticide demand in 2011. It is valued at around $205 million and is expected to increase to approximately $300 million by 2020. European market is nearing $200 million and is one of the fastest growing markets.

6. CONCLUSION

The motive of a pest management program is to protect the crop most economically with minimum impact on agro-ecosystem. So every attempt to prevent losses caused by pests can be done by implementing application of biopesticides. Insecticides should be used only when the mode does not keep pests below damaging levels and supplement integrated pest management program. The use of pest monitoring and forecasting strategies as well as economic thresholds will ensure that pest management strategies can be employed appropriately. There has been increasing concern about the degrading quality of land as well as high ecological price that we are paying by practicing unsustainable conventional agriculture and need is to change our methodology before it is too late and sustainable productive agriculture. The challenge before present agriculture is to make more productive agricultural resources which is not only economically viable and productive, but also safer to environment as a whole. The use of chemical pesticides to fulfill the increasing demand of growing population for food and fibre pest management cannot be denied, but utilized in integrated pest management programme and avert economic losses. Biopesticides with all its benefits, safety, their increasing availability and popularity can be utilized for safer productive food production programme.

REFERENCES

Abudulai M, Shepard BM and Mitchell PL (2001) Parasitism and predation on eggs of Leptoglossus phyllopus (L.) (Hemiptera: Coreidae) in Cowpea: impact of endosulfan sprays. *Journal of Agicultural Urban Entomology* **18:** 105–115.

Akhtar Y, Shikano I and Isman MB (2009) Topical application of a plant extract to different life stages of *Trichoplusia ni* fails to influence feeding or oviposition behaviour. *Entomology Experimental Applied* **132:** 275–282.

Al-Fifi Zia (2009) Effect of different neem products on the mortality and fitness of adult *Schistocerca gregaria* (Forskal). *JKAU* **21(2):** 299–315.

Al-Zaidi AA, Elhag EA, Al-Otaibi SH and Baig MB (2011) Negative effects of pesticides on the environment and the farmers awareness in Saudi Arabia: a case study. *Journal of Animal and Plant Sciences* **21(3):** 605–611.

Ayyangar GSG and Rao PJ (1989) Azadirachtin effects on consumption and utilization of food and midgut enzymes of *Spodoptera litura* (Fabr.). *Indian Journal of Entomology* **51:** 373–376.

Ayyangar GSG and Rao PJ (1990) Changes in haemolymph constituents of *Spodoptera litura* under the influence of azadirachtin. *Indian Journal of Entomology* **52:** 69–83.

Bedding RA and Molyneux AS (1982) Penetration of insect cuticle by infected juveniles of *Heterorhabditis* spp (Heterorhabtidae, Nematoda) *Nematologica* **28:** 354–359.

Benjamin MA, Zhioua E and Ostfeld RS (2002) Laboratory and field evaluation of the entomopathogenic fungus, *Metarhizium anisopliae* (Deuteromycetes) for controlling questing adult, *Ixodes scapularis* (Acari: Ixodidae). *J Med Entomol.* **39(5):** 723–728.

Berliner E (1915) Ueber die schlaffsucht der Ephestia kuhniella und *Bac. thuringiensis n.* sp. *Z Angew Entomol* **2:** 21–56.

Bilimoria SL (1986) Taxonomy and identification of baculoviruses. In: Granados RR, Federici BA (eds) In the biology of baculoviruses, vol 1. CRC Press, Boca Raton, Fl, pp 37–59.

Blanford S and Thomas MB (2000) Thermoregulation in two acridid species: effects of habitat and season and implications for biological control using pathogens. *Environ Entomol.* **29(5):** 1060–1069.

Blanford S, Thomas MB and Langewald J (1998) Behavioural fever in a population of the Senegalese grasshopper, *Oedaleus senegalensis*, and its implications for biological control using pathogens. *Ecol Entomol.* **23:** 9–14.

Bowers MD and Puttick GM (1988) Response of generalist and specialist insects to qualitative allelochemicals variations. *J Chem Ecol.* **14:** 319–334.

Bowers WS (1991) Insect hormones and antihormones in plants. In: Rosenthal GA, Berenbaum MR (eds) Herbivores: their interactions with secondary plant metabolites, vol 1. Acad Press, New York, pp 431–456.

Bowers WS, Ohta T, Cleere JS and Marsella PA (1976) Discovery of insect anti-juvenile hormones in plants. *Science* **193:** 542–547.

Brey PT, Latge JP and Prevost MC (1986) Integument penetration of the pea aphid, *Acrythosiphon pisum* by *Conidiobolus obscurus* (Entomophthoraceae). *J Inverteb Pathol* **48:** 34–41.

Brown WL, Jr Eisner T and Whitlake RH (1970) Allomones and kairomones: transpacific chemical messenger. *Bioscience* **20:** 21–22.

Butenandt A and Karlson P (1954) Uber die Isolierung cines metamorphosehormons der insektan in kristalissierter Form. *Z Naturforsch* **9:** 389–391.

Butterworth JH and Morgan ED (1968) Isolation of a substance that suppresses feeding in locusts. *J Chem Soc Chem Commun.* **1968:** 23–24.

Carson R (1962) Silent spring. Houghton-Mifflin, Boston.

Casida, JE and Quistad GB (1998) Golden age of insecticide research: past, present, or future? *Annual Review of Entomology* **43(1):** 1-16.

Chaieb I (2010) Saponins as Insecticide: a Review. *Tunisian J Plant Protec.* **39(5):** 39–50.

Chandler D, Bailey AS, Tatchell GM, Davidson G, Greaves J and Grant WP (2011) The development, regulation and use of biopesticides for integrated pest management. *Phil. Trans. R.Soc. B: Biol. Sciences* **366(1573):** 1987–1998.

Chattopadhyay A, Bhatnagar N and Bhatnagar R (2004) Bacterial Insecticidal Toxins. *Critical Review of Microbiology* **30:** 33.

Clemson Extension [Clemson HGIC], Home and Garden Information Center (2007) Organic Pesticides and Biopesticides. URL: http://hgic.clemson.edu/factsheets/HGIC2756.htm Clemson (SC): Clemson University.

Cornelius ML, Duan JJ and Messing RH (2000) Volatile host fruit odors as attractants for the Oriental fruit fly (Diptera: Tephritidae) *J Econ Entomol.* **93(1):** 93–100.

Cunningham RT (1989) Male Annihilation. In World Crop Pests. Robinson AS, Hooper G (eds) Elsevier, Amsterdam, Netherlands, pp 78–81.

Cuperus GW, Berberet RC and Noyes RT (2004) The essential role of IPM in promoting sustainability of agricultural production systems for future generations. In: Koul O, Dhaliwlal GS, Cuperus G (eds) Integrated pest management: potential, constraints and challenges. CAB Internat, Wallingford, UK, pp 265–280.

Dahiya B and Chauhan R (2001) Evaluation of insecticides and biopesticides on pod borer *Helicoverpa armigera* Hubn in Chickpea. 'Symp on biocontrol based pest manage-ment for quality plant protection in the current millennium. July 18–19 PAU, Ludhiana.

Di-illio V, Cristofaro M, Marchini D, Paola N and Dallai R (1999) Effects of a neem compound on the fecundity and longevity of *Ceratitis capitata* (diptera : Tephritidae). *J Econ Entomol* **92:** 76–82.

Donald G and McNeil JR (2005) Fungus Fatal to Mosquito May Aid Global War on Malaria, The New York Times, 10 June.

Dorn A, Rademacher JM and Shen E (1987) Effects of azadirachtin on reproductive organs and fertility in the large milkweed bug, *Oncopeltus fasciatus*. In: Schmutterer H, Ascher KRS (eds) Natural products from the neem tree and other tropical plants. Proc 3rd Internat neem Conf.GTZ, Eschborn, Germany, pp 273–288.

Elahi KM (2008) Social Forestry, Exotic Trees and Mango. The Daily Star Published 6 Sept 2008. URL: http://www.thedailystar.net/story.phpnid=53438.

El-Sayed AM (2005) The pherobase: database of insect pheromones and semiochemicals. http:// www.pherobase.com.

Erlandson ME (2008) Insect pest control by viruses In: Mahy, B. W. J. and M. H. V. van Regenmortel (eds) Encyclopedia of virology, 3rd ed. Elsevier, Oxford. pp. 125–133.

Estruch J, Gregoury J, Warren W, Martha MA, Gardon JN, Joyce AC and Michael GK (1996) *Bacillus thuringiensis:* vegetative insecticidal protein with a wide spectrum of activities against lepidopterous insect. *Proc National Academy of Sciences USA* **93(11):** 5389–5394.

Eswaran SV (2009) Why does the grasshopper not eat spinach. *Resonance* **14:** 978–984.

Eswarapriya B, Gopalsamy B, Kameswari B, Meera R and Devi P (2010) Insecticidal activity of *Bacillus thuringiensis* IBT- 15 strain against *Plutella xylostella*. International *Journal of Pharmacy Technological Research.* **2:** 2048-2053.

Fang J, Xu X, Wang P, Zhao JZ, Shelton AM, Cheng J, Feng MG and Shen Z (2007) Characterization of chimeric *Bacillus thuringiensis* Vip3 toxins. *Applied Environmental Microbiology* **73:** 956-961.

Fargues J, Robert PH, Vey A and Pais M (1986) Toxixite relative de la destruxine e pour le Lepidoptera *Gelleria mellonella*. *CR Acad Sci Paris* **303:** 84–86.

Fernandez S, Groden E, Vandenberg JD and Furlong MJ (2001) The effect of mode of exposure to *Beauveria bassiana* on conidia acquisition and host mortality of Colorado potato beetle, *Leptinotarsa decemlineata*. *J Inverteb Pathol* **77:** 217–226.

Fitches, EC, Bell HA, Powell ME, Back E, Sargiotti C, Weaver RJ and Gatehouse JA (2009) Insecticidal activity of scorpion toxin (ButaIT) and snowdrop lectin (GNA) containing fusion proteins towards pest species of different orders. *Pest Management Science* **66(1):** 74-83.

Frazier JL (1986) Perception of allelochemicals that inhibit feeding. In: Brattsten LB, Ahmad S(eds) Molecular aspects of insect-plant associations. Plenum Press, New York, pp 1–42.

Fujiwara-Tsujii N, Yamagata N, Takeda T, Mizunami M and Yamaoka R (2006) Behavioural responses to the alarm pheromone of the ant, *Camponotus obscuripes* (Hymenoptera: Formicidae). *Zool Sci* **23(4):** 353–358.

Gang W, Harris MK, Guo J Y *et al* (2009) Temporal allocation of metabolic tolerance in the body of beet armyworm in response to three gossypol-cotton cultivars. *Sci China Ser C-Life Sci* **52(12):** 1140–1147.

Gathmann A and Priesnitz, KU (2014) How susceptible are different lepidopteran/ coleopteran maize pests to Bt-proteins: a systematic review protocol. *Environment. Evid.* **3:** 12.

Glaser RW and Farrell CC (1935) Field experiments with the Japanese beetle and its nematode parasite. *J N Y Entomol Soc* **43:** 345–371.

Goldansaz SH, Dewhirst S, Birkett MA, Hooper AM, Smiley DWM, Pickett JA, Wadhams L and McNeil JN (2004) Identification of two sex pheromone components of the potato aphid, *Macrosiphum euphorbiae* (Thomas). *J Chem Ecol* **30:** 819–834.

Gould FL, Groot AT, Vasqnez GM and Schal C (2009) Sexual communication in Lepidoptera: a need for wedding genetics, biochemistry and molecular biology. CRC Press, Boca Raton.

Grafton-Cardwell EE, Leea JE, Stewartb JR and Olsenc KD (2006) Role of Two Insect Growth Regulators in Integrated Pest Management of Citrus Scales. *J Econ Entomol* **99(3):** 733–744.

Grewal P, Ehlers RU and Shapiro-Ilan DI (eds) (2005) Nematodes as Biocontrol Agents. CABI Publishing, Wallingford, UK.

Grewal PS, Nardo ED and Aguillera MM (2001) Entomopathogenic nematodes: potential for exploration and use in South America. *Neotrop Entomol* **30(2):** 191–205.

Gujar GT (1997) Biological effects of azadirachtin and plumbagin on *Helicoverpa armigera*. *Ind J Entomol* **59(4):** 415–422.

Hajeck AE and Leger ST (1994) Interactions between fungal pathogens and insect hosts. *Annual Review of Entomology* **39:** 293–322.

Harris RS, Harcourt SJ, Glare TR, Rose EA and Nelson TJ (2000) Susceptibility of *Vespula vulgaris* (Hymenoptera: Vespidae) to generalist entomopathogenic fungi and their potential for wasp control. *J Inverteb Pathol* **75:** 251–258.

Heimpel AM (1967) A critical review of *Bacillus thuringiensis* var. thuringiensis Berliner and other crystalliferous bacteria. *Annual Review of Entomology* **12:** 287–322.

Howlett FM (1915) Chemical reaction of fruit flies. *Bull Entomol Res* **6:** 297–305.

Ishaaya I, Birk Y, Bondi A and Tencer Y (1969) Soybean saponins IX. Studies of their effect on birds, mammals and cold blooded organisms. *J Sci Food Agricul* **20:** 433–436.

Ishaaya I and Horowitz AR (1999) Insecticides with novel modes of action: an overview. In: Ishaaya I, Degheele D (eds) Insecticides with novel modes of action. Springer, Berlin, pp 1–24.

Ishtiaq M, Saleem MA and Razaq M (2012) Monitoring of resistance in *Spodoptera exigua* (Lepidoptera : Noctuidae) from four districts of the Southern Punjab, Pakistan to four conventional and six new chemistry insecticides. *Crop Protection* **33:** 13–20.

Islam MT and Omar DB (2012) Combined effect of *Beuveria bassiana* with neem on virulence of insect in case of two application approaches. *The Journal of Animal. Plant. Science.* **22(1):** 77–82.

Isman MB and Rodriguez E (1983) larval growth inhibitors from species of *Parthenium* (Asteraceae). *Phytochem* **22:** 2709–2713.

Jisha VN, Smitha RB and Benjamin S (2013) An overview on the crystal toxins from *Bacillus thuringiensis*. *Advance Microbiology*. **3:** 462–472.

Jurenka R (2004) Insect pheromone: biosynthesis, chemistry of pheromones and other semiochemical, pp 97–131.

Karlson P and Butenandt (1959) Pheromones (ectohormones) in Insects. *Ann Rev Entomol* **4:** 39–58.

Karlson P and Luscher M (1959) Pheromones: a new term for a class of biologically active substances. *Nature* **183:** 55–56.

Keller S, Kessler P and Schweizer C (2003) Distribution of insect pathogenic soil fungi in Switzerland with special reference to *Beauveria brongniartii* and *Metarhizium anisopliae*. *Bio Cont* **48:** 307–319.

Kerwin JL and Washino RK (1986a) Oosporogenesis by *Lagenidium giganteum*: induction and maturation are regulated by calcium and calmodulin. *Can J Microbiol* **32(8):** 663–672.

Kerwin JL and Washino RK (1986c) Regulation of oosporogenesis by *Lagenidium giganteum*; promotion of sexual reproduction by unsaturated fatty acids and sterol availability. *Can J Microbiol* **32:** 294–300.

Klowden MJ (ed) (2007) Physiological Systems in Insects. 2nd ed. Acad Press, London.

Koul O (1998) Effect of neem extract and *azadirachtin* on fertility and fecundity of cabbage aphid, *Brevicoryne brassicae* (L). *Pestic Res J* **10:** 258–261.

Koul O and Dhaliwal GS (2001) Phytochemical biopesticides. Harwood Acad, Amsterdam, p 223.

Kumar SP and Babu PCS (1998) Ovipositional deterrent and ovicidal effects of neem azal against brinjal shoot and fruit borer and leaf beetle. *Shaspa* **5(2):** 193–196.

Leuschner K (1972) Effect of an unknown plant substance on a shield bug. *Naturuwissenschaften* **59:** 217–218.

Lopez O, Fernandez-Bolanos JG and Gil MV (2005) New trends in pest control: the search for greener insecticides. *Green Chem* **7:** 431–442.

Lowry DT, Isman MB (1996) Inhibition of aphid (Homoptera: Aphididae) reproduction by neem seed oil and azadirachtin. *J Econ Entomol* **89(3):** 602–607.

Lykouressis D, Perdikis D, Michalis C, Fantinou A (2004) Mating disruption of the pink bollworm *Pectinophora gossypiella* (Saund.) (Lepidoptera: Gelechiidae) using gossyplure PB-rope dispensers in cotton fields. *J Pest Sci* **77:** 205–210.

Lysyk TJ (2006) Abundance and species composition of culicoides (Diptera: Ceratopogonidae) at cattle facilities in southern Alberta, *Canada Journal Medicinal Entomology* **43:** 840–849.

Malik K and Riazuddin S (2006) Immunoassay-based approach for detection of novel Bacillus thuringiensis-endotoxins, entomocidal to cotton aphids (*Aphis gossypii*) and whiteflies (*Bemisia tabaci*). *Pakistan Journal of Botany* **38:** 757–765.

Mannion CM, McLane W, Klein MG, Moyseenko J, Oliver JB and Cowan D (2001) Management of early-instar Japanese beetle (Coleoptera: Searabaeidae) in field-grown nursery crops. *J Econ Entomol* **94:**1151–1161.

Martignoni ME (1984) Baculovirus: An attractive biological alternative. In: Garner WY, Harvey J (eds) Chemical and biological control in forestry. ACS Symp Ser No 238, Seattle, Washington, pp 55–67.

Martignoni ME and Iwai PJ (1986) A catalogue of viral disease of insects, mites, and ticks. US Dep Agric Forest Serv Pac NorthWest Res Stn Gen Tech. Rep PNW-195.

Mayer MS and McLaughlin JR (1990) Handbook of insect pheromones and sex attractants. CRC Press, Boca Raton, FL.

Mazhabi M, Nemati H, Rouhani H, Tehranifar A, Moghadam EM, Kaveh H and Rezaee A (2011) The effect of *Trichoderma* on polianthes qualitative and quantitative properties. *Journal of Animal Plant Science.* **21(3):** 617–621.

Metcalf RL (1990) Chemical ecology of Dacinae fruit flies (Diptera: Tephritidae). *Ann Entomol Soc Am* **83:** 1017–1030.

Miyamoto J, Hirano M, Takimoto Y and Hatakoshi M (1993) Insect growth regulators for pest control, with emphasis on juvenile hormone analogs: Present status and future prospects. ACS Symp. Ser ACS, Washington, DC **524:** 144–168.

Nicholson GM (2007) Fighting the global pest problem: Preface to the special Toxicon issue on insecticidal toxins and their potential for insect pest control. *Toxicon* **49:** 413-422.

O'Callaghan M and Brownbridge M (2009) Environmental impacts of microbial control agents used for control of invasive pests. In: Hajek, A.E., T. Glare and M. O'Callaghan (eds) Use of Microbes for Control and Eradication of Invasive Arthropods. Springer, Dordrecht, The Netherlands, pp: 305-32.

Pfeiffer DG, Kaakeh W, Killian JC, Lachance MW and Kirsch P (1993a) Mating disruption to control damage by leaf rollers in Virginia apple orchards. *Entomol Exptl Appl* **67:** 47–56.

Pfeiffer DG, Kaakeh W, Killian JC, Lachance MW and Kirsch P (1993b) Mating disruption for control of damage by codling moth in Virginia apple orchards. *Entomol Exptl Appl* **67:** 57–64.

Phillips TW and Throne JE (2010) Biorational approaches to managing stored-product insects. *Ann Rev Entomol* **55:** 375–397.

Phoofolo MW, Obrycki JJ and Lewis LC (2001) Quantitative assessment of biotic mortality factors of the European corn borer (Lepidoptera: Crambidae) in field corn. *J Econ Entomol* **94:** 617–622.

Rai HS, Verma ML and Gupta MP (2001) Efficacy of foliar application of Nuclear Polyhedrosis Virus against Pod Borer, *Helicoverpa armigera* Hubn in Chickpea. Symp Biocontrol Based Pest Management for Quality Plant Protection in the Current Millennium. PAU, Ludhiana, July 18–19.

Reddy GVP and Guerrero A (2004) Interactions of insect pheromones and plant semiochemicals. *Trends Plant Sci* **9(5):** 253–261.

Rembold H, Forster H, Czoppelt CH, Rao JP and Sieber KP (1984) The azadirachtins: a group of insect growth regulators from the Neem Tree. In: Schmutterer H, Ascher KRS (eds) Natural pesticides from the neem tree and other tropical plants. Proc 3rd Int Neem Conf Rauischholz-Hausen, 1983, Eschborn, GTZ, pp 153–162.

Rishipal M, Hussain A and Prasad CS (2008) Natural occurrence of entomopathogenic nematodes in Meerut district, North India. *Int J Nematol* **18(2):** 198–202.

Romeis J, Meissle M and Bigler F (2006) Transgenic crops expressing *Bacillus thuringiensis* toxins and biological control. *Nature Biotechnology* **24:** 63–71.

Roomi MW, Abbas T, Shah AH, Robina S, Qureshi AA, Sain SS and Nasir KA (1993) Control of fruit flies (*Dacus* sp.) by attractants of plant origin. *Anzeiger fiir Schadlingskunde, Pflanzeschutz, Umwelschutz*, **66:** 155–157.

Ruscoe CNE (1972) Growth distruption effects of an insect antifeedant. *Nature New Biol* **236:** 159–160.

Sankar M (2009) Investigation of indigenous entomopathogenic nematodes, *Heterorhabditis indica* as a potential biocontrol agent of insect pest in rice. PhD thesis Osmania Univ, Hyderabad, India, pp 168–179.

Saxena H (2001) Biorational management of insect pest of pigeonpea. Symp Biocontrol Based Pest Management for Quality Plant Protection in the Current Millennium. PAU, Ludhiana, July 18–19.

Saxena RC (1989) Insecticides from neem. In: Arnason JT, Philogene BJR, Morand P (eds) Insecticides of plant origin. ACS Symp. Series 387. Am Chem Soc, Washington, DC, pp 110–135.

Schmidt GD and Roberts LS (1989) Foundation of parasitology 4th Edn. Times Mirror Mosby College pub, Toronto, pp 243–243.

Schmutterer H (1987). Fecundity-reducing and sterilizing effects of neem seed kernel extracts in the Colorado potato beetle, *Leptinotarsa decemlimeat.* In: Schmutterer H, Ascher (eds) Natural pesticides from the neem tree and other tropical plants. KRS Proc. 3rd Internt Neem Conf. Nairobi, 1986, Eschborn, GTZ.

Schneider MS, Smagghe G, Gobbi A and Vinuela E (2003) Toxicity and pharmacokinetics of insect growth regulators and other novel insecticides on pupae of *Hyposoter didymator* (Hymenoptera: Ichneumonidae), a parasitoid of early larval instars of lepidopteran pests. *J Econ Entomol* **96(4):** 1054–1065.

Scholte EJ, Knols BGJ, Samson RA and Takken W (2004) Entomopathogenic fungi for mosquito control: a review. *J Insect Sci* **4:** 19.

Schünemann R, Knaak N and Fiuza LD (2014) Mode of action and specificity of *Bacillus thuringiensis* toxins in the control of caterpillars and stink bugs in soybean culture. ISRN Microbiol. 2014, 135675.

Shah PA and Pell JK (2003) Entomopathogenic fungi as biological control agents. *Appl Microbiol Biotechnol* 413–423.

Shingote PR, Moharil MP, Dhumale DR, Jadhav PV, Satpute NS and Dudhare MS (2013) Screening of vip1/vip2 binary toxin gene and its isolation and cloning from local *Bacillus thuringiensis* isolates. *Science Asia* **39:** 620-624.

Siddall JB (1976) Insect growth regulators and insect control: a critical appraisal. *Environ Health Perspec* **14:** 119–126.

Singh P, Russel GB and Fredericksen S (1982) The dietry effects of ecdysteroids on the development of housefly. *Entomol Exptl Appl* **32:** 7–12.

Singh SP, Murphy ST and Ballal CR (2001) Augmentative Biocontrol- Proc ICAR-CABI Workshop, pp 250.

Slama K and Williams CM (1965) Paper factor as inhibitor for the embryonic development of European bug, *Purrhocoris apterus*. Nature **210:** 329–330.

Smith CA (1995) Searching for safe methods of flea control. *J Am Vet Med Assoc* **206:** 1137–1143.

Srivastava JP and Mathur YK (1998) Insect growth regulatory effect of some neem based formulations against sesamum pod borer. Abs. in ICPPMSA, 11–13 Dec, 1998, CSAU&T, Kanpur, pp 228.

Steinwand (2008) Washington DC: US Environmental Protection Agency. Biopesticide Ombudsman (Personal communication).

Suckling DM (2000) Issues affecting the use of pheromones and other semiochemicals in orchards. Crop Protect **19(8–10):** 677–683.

Suzuki A, Taguchi H and Tamura S (1970) Isolation and structure elucidation of three new insecticidal cyclodepsipeptides, destruxin C and D and dasmethyldestruxin B, produced by *Metarrhizium anisopliae*. *Agric Biol Chem, Tokyo,* **34:** 813–816.

Sylvar Technologies (2008) Research. URL: http://www.sylvar.ca/content/13636 (accessed 19 Aug 2008). Fredericton (NB).

Tamaki Y (1985) Sex pheromones. In: Kerkut GA, Gilbert LS (eds) Comprehensive insect physiology. *Biochem Pharmacol* **9:** 145–191.

Tanada Y and Kaya HK (1993) Insect pathology. Acad Press, San Diego.

Thakore Y (2006b) The new biopesticide market. Report code: CHM029B.

Tooker JF, Crumrin AL and Hanks LM (2005) Plant volatiles are behavioural cues for adult females of the gall wasp, *Antistrophus rufus*. *Chemoecol* **15:** 85–88.

Umoetok SBA and Gerard MB (2003) Comparative efficacy of *Acorus calamus* powder and two synthetic insecticides for control of three major insect pests of stored cereal grain. *Global J Agric Sci* **2(2):** 94–97.

Ware G and Whitacre D (2004) An Introduction to Insecticides. In: Radcliffe E, Hutchison W, Cancelado R, and Radcliffe's IPM World Textbook. URL: http://ipmworld.umn.edu. St. Paul (MN): University of Minnesota.

Wee S, Tan KH and Nishida R (2007) Pharmacophagy of Methyl Eugenol by Males Enhances Sexual Selection of *Bactrocera carambolae*. *J Chem Ecol* **33:** 1272–1282.

Wee SL, Hee AKW and Tan KH (2002) Comparative sensitivity to and consumption of methyl eugenol in three *Bactrocera dorsalis* (Diptera: Tephritidae) complex sibling species. *Chemoecol* **12:**193–197.

Witte V, Abrell L, Attygalle AB, Wu X and Meinwald J (2007) Structure and function of Dufour gland pheromones from the crazy ant, *Paratrechina longicornis*. *Chemoecol* **17(1):** 63–69.

Witzgall P, Kirsch P and Cork A (2010) Sex pheromones and their impact on pest management. *J Chem Ecol* **36:** 80–100.

Witzgall P, Stelinski L, Gut L and Thomson D (2008) Codling moth management and chemical ecology. *Ann Rev Entomol* **53:** 503–522.

Wu Y, Lei CF, Yi D, Liu PM and Gao MY (2011) Novel *Bacillus thuringiensis* dendotoxin active against *Locusta migratoria manilensis*. *Applied Environmental Microbiology* **77:** 3227-3233.

Yu X, Zheng A, Zhu J, Wang S, Wang L, Deng Q, Li S, Liu H and Li P (2011) Characterization of vegetative insecticidal protein vip genes of Bacillus thuringiensis from Sichuan Basin in China. *Current Microbiology* **62:** 752-757.

■■■■

3

Year Round Green Fodder Production and Conservation for Sustainable Dairy Farming in India

Rakesh Kumar

Forage Research and Management Centre,
ICAR-National Dairy Research Institute (Deemed University)
Karnal – 132001 (Haryana)

ABSTRACT

For the sustainability of agriculture in India, integrated farming system based on dairying will play a major role. Farmers in India, are mainly falls under the category of small and marginal holders and their livelihood is mainly depends on livestock as secondary agricultural enterprise. In organised dairy farm out of total expenditure about 65 per cent accounts for feed and fodder. The green fodder is the cheapest source as it is easily digestible, palatable and gives immediate response, in terms of increased milk yield. By the adoption of suitable crop rotation and conservation of green fodder in the form of silage and hey during glut season for further use during lean period, the productivity can be sustained in the dairy farming India.

***Key words*:** Green fodder, Silage, hey and crop rotations.

1. INTRODUCTION

For food security livestock production and agriculture are complementary to each other and both are crucial for farmers with small and marginal holding in India. In mixed crop-livestock production system, dairy production contributes 20 to 50 per cent of family income. The share of livestock for underprivileged marginal and landless livestock owner is as high as 70 to 80 per cent during drought year. India is gifted with the largest livestock population in the world. It accounts for about 57.3 per cent of the world's buffalo population and 14.7 per cent of the cattle population. There are about 71.6 million sheep and 140.5 million goats in the country. Farmers of marginal, small and semi-medium operational holdings (area less than 4 ha) own about 87.7 per cent of the livestock. Hence development of livestock sector would be more inclusive. The livestock sector achieved an average growth rate of 4.8 per cent during the Eleventh Five Year Plan (Economic Survey, 2012–13) and it contributed 3.6 per cent of national GDP during the Eleventh Plan. India, the largest producer of milk in the world, is set to produce over 137 million ton milk during 2012–13. The value of output of milk is Rs. 3,05,484 crore in 2011–12, which is higher than the value of output of paddy and wheat.

The average milk yield of cattle in the world and Europe is about 2040 kg and 4250 kg per lactation, respectively, the average milk yield of Indian cattle is about 1000 kg. While genetic improvement and health care are the prerequisites for sustainability, efficient feeding and marketing will help in increasing the profitability, because 65–70 per cent of the total cost of livestock farming is attributed to feeding. With feeding of good quality forage, particularly leguminous fodder, feeding of concentrate can be reduced significantly. At present, the country faces a net deficit of 35.6 per cent green fodder, 10.95 per cent dry crop residues and 44 per cent concentrate feed ingredients. The demand of green and dry fodder will reach to 1012 and 631 million tonnes of by the year 2050 (Table 1). At the current level of growth in forage resources, there will be 18.4 per cent deficit in green fodder and 13.2 per cent deficit in dry fodder in the year 2050. To meet out the deficit, green forage supply has to grow at 1.69% annually (Anonymous, 2013).

Table 1. Demand and supply estimates* of dry and green forages (million tons)

Year	Demand		Supply		Deficit		Deficit as %	
	Dry	Green	Dry	Green	Dry	Green	Dry	Green
2010	508.9	816.8	453.2	525.5	55.72	291.3	10.95	35.66
2020	530.5	851.3	467.6	590.4	62.85	260.9	11.85	30.65
2030	568.1	911.6	500.0	687.4	68.07	224.2	11.98	24.59
2040	594.9	954.8	524.4	761.7	70.57	193.0	11.86	20.22
2050	631.0	1012.7	547.7	826.0	83.27	186.6	13.20	18.43

(*Source*- Vision, 2050, IGFRI)

The deficit and supply in crude protein (CP) and total digestible nutrients (TDN) are given in Table 2. The situation is further aggravated due to increasing growth of livestock particularly that of genetically upgraded animals. The available forages are poor in quality, being deficient in available energy, protein and minerals. To compensate for the low productivity of the livestock, farmers maintain a large herd of animals, which adds to the pressure on land and fodder resources.

Table 2. Demand and availability estimates of CP and TDN (million tons)

Year	Requirement		Availability		% Deficit	
	CP	TDN	CP	TDN	CP	TDN
2010	60.04	347.8	42.95	271.3	28.47	21.99
2020	62.58	362.5	47.18	290.5	24.60	19.87
2030	67.01	388.2	53.09	320.2	20.78	17.52
2040	70.19	406.6	57.61	342.8	17.92	15.69
2050	74.44	431.2	61.92	364.5	16.81	15.47

(*Source*- Vision, 2050, IGFRI)

Rapid industrialization and mining areas has caused shrinkage of grazing and fodder producing lands. Due to non-availability of quality green fodder throughout the year, milk producers are forced to utilize extra concentrates for optimum milk production. On account of this cost of milk production is higher in such areas. The current milk production level can be sustained and enhanced by year round green fodder production through intensive year round green production rotations and fodder preservation techniques.

YEAR-ROUND FORAGE PRODUCTION THROUGH INTENSIVE YEAR ROUND GREEN PRODUCTION ROTATIONS

Overlapping cropping systems developed to full fill the needs of dairy farmers for green fodder throughout the year and for small farmers requiring maximum forage from a piece of land. It consists of raising berseem, inter-planted with hybrid Napier in spring and intercropping the inter-row spaces of the grass with cowpea during summer after the final harvest of berseem (Table 3). This system was found superior to multiple crop sequences both in terms of production and economic returns. The hybrid Napier could be successfully replaced with relatively soft and palatable perennial grasses like *Setaria* and guinea grass and berseem with lucerne wherever required.

Table 3. Round-the-year fodder production systems

Crop sequence	Green fodder yield (tonnes/ha/year)
Napier x Bajra hybrid + Cowpea - Berseem	260
Maize + Cowpea – MP Chari + Cowpea – Berseem + Japanese rape	197
MP Chari + Cowpea – Berseem + Japanese rape	184
Cowpea – MP Chari + Cowpea – Berseem + Japanese rape	176
Napier x Bajra hybrid + Cowpea – Berseem – Cowpea	255

The intensive cropping systems when managed properly using modern techniques of soil and crop management are able to yield 180–300 tonnes of green fodder (30–55 tonnes dry fodder) per ha/year. Some of the intensive cropping systems have been suggested for different regions.

North Zone

Maize + Cowpea – Sorghum + Cowpea (two cuts) – Berseem + Mustard.

Sudan grass + Cowpea – Maize + Cowpea – Turnip – Oats (two cuts).

Hybrid Napier or *Setaria* inter-planted with cowpea in summer and Berseem in winter (9–10 cuts/year).

Teosinte + Cowpea (two cuts) – Carrot – Oats + Mustard/*Senji* (two cuts).

Western and Central Zone

Bajra + *Guar* (Clusterbean) (two cuts) – Annual Lucerne (6 cuts).

MP Chari + Cowpea (2 cuts) – Maize + Cowpea - Teosinte + Cowpea (2 cuts).

Hybrid Napier or *Guinea* or *Setaria* grass inter-planted with Cowpea in summer + Berseem in winter (8–9 cuts/year).

Hybrid Napier or *Guinea* or *Setaria* grass interplanted with Lucerne (8–9 cuts/ year).

Southern Zone

Sorghum + Cowpea (3 cuts) – Maize + Cowpea – Maize + Cowpea.

Hybrid Napier or *Guinea* or *Setaria* grass inter-planted with Lucerne (8–9 cuts) or Hybrid Napier + *Subabul* / *Sesbania* (9–11 cuts/year).

Sudan grass + Cowpea (3 cuts) – M.P. *Chari* + Cowpea (three cuts).

Para grass + Centro (*Centrosema pubescens*) (9–11 cuts/year).

Eastern Zone

Maize + Cowpea – Teosinte + Rice bean (2 cuts) – Berseem + Mustard (3 cuts).

M.P. Chari + Cowpea – *Dinanath* grass (2 cuts) – Berseem + Mustard (3 cuts).

Para grass + *Centrosema pubescens* (8–9 cuts/year).

Hybrid Napier or Setaria grass inter-planted with *Subabul* or Common Sesban (*Sesbania sesban*) (9–10 cuts/year).

YEAR-ROUND FORAGE AVAILABILITY THROUGH FORAGE PRESERVATION

SILAGE-MAKING

Silage-making is a management tool that allows producers to match feed resources (forages, crop residues, agro-industrial by products, etc.) with feed demand for a dairy herd. The basic function of silage-making is to store and preserve feed for later use with minimal loss of nutritional qualities. Silage making is less dependent than hay-making on good weather conditions and can be extended to a great variety of forage crops (corn, sorghum, immature cereal grains, etc.) and locally available agro-industrial by products (sugar beet pulp, brewers grain, etc.). However, silage making requires considerable capital and labour investments on the farm; it also demands a fairly high level of technical expertise. This knowledge is key to making the best management decisions for minimizing the inevitable losses that occur when fresh feed resources are ensiled and preserved for long periods of time in a silo.

Advantages of silage-making

Silage making is a tool for producers to achieve whole farm management goals. Silage making has some distinct advantages compared to grazing or hay-making:

1. Intensification of forage production (i.e., increased yield of forage per hectare);
2. Minimization of risk factors associated with weather conditions (rainfall losses) when trying to harvest high quality forages. For example, compared to hay-making, silage making shortens the time between cutting wider range of weather conditions, and risk of dry matter losses due to rainfall is minimized;
3. Improvement of the producer's control over cutting dates and optimal stage of maturity at harvest;
4. Minimization of loss of leaves and other small plant parts of high quality in the field;
5. Storage of non-forage feeds that cannot be preserved as hay, such as agro-industrial by products; Stabile composition of the feed (silage) for a longer period (up to 5 years);
6. Plants can be harvested at optimal phase of development and are efficiently used by livestock.
7. Reduction of nutrient loses (below 10%) which in hay may amount to 30 per cent of the dry matter.
8. More economical use of plants with high yield of green mass;

9. Better use of the land with 2–3 crops annually; Requires 10 times less storage space compared to hay; Silage is produced in both cold and cloudy weather;
10. The fermentation in silage reduces harmful nitrates accumulated in plants during droughts and in over-fertilized crops.
11. Allows by-products (from sugar beet processing, maize straw, etc.) to be optimally used;
12. Maize silage has 30–50 per cent higher nutritive value compared to maize grain and maize straw; 2 kg of silage (70% moisture) has the equal nutritive value of 1 kg of hay.

Limitations and drawbacks of silage making

Silage making also has some limitations or drawbacks that need to be taken in mind:

1. Silage making requires high capital investment. Harvesters are needed to chop the forage, tractors or other heavy equipment are used to pack the silage, storage facilities (silo structures) may be expensive, and additional equipment may be required to remove silage from a silo;
2. The management of silos is sometimes difficult on the farm because once a silo is opened, silage should be removed on a daily basis (to minimize loss of nutritive value). Adjusting the number of silos and their dimension to the expected feed out rate for a given herd size is difficult. Usually, only large herds can afford to feed out of different silos of varying forage qualities for different groups of animals on the farm (heifers, dry cows, lactating cows, etc.);
3. Once silage is removed from a silo, it becomes unstable (because of exposure to oxygen) and tends to spoil within a day or two (especially in warm weather conditions with well-preserved silages); Silage cannot be marketed easily (difficult to transport long distances);
4. Loss of nutrients during storage in a silo is unavoidable and may be high if the silage is not prepared properly.

Principles and practices of silage preservation

As a forage crop is cut, harvested and stored, loss of dry matter (quantity) and nutritional quality inevitably occur. These losses are due to enzymes that degrade the plant after it has been cut. Thus, the goal in silage making is to stop enzymatic reactions and minimize loss of energy, protein and other nutrients. Hence silage making can be defined simply as a method of forage preservation in which most of the energy, protein, and other nutrients that were in the original plant remain in a form that can be efficiently utilized by cows. A **more technical definition,** "Silage is a method of feed resource preservation - which is based on the removal of air (oxygen) from a mass of feed to promote the fermentation of sugars into lactic acid by lactic-acid bacteria causing an increase in acidity (a reduction in pH) which inhibits further silage degradation by: plant enzymes (primarily protein degrading enzymes); undesirable bacterial species (enterobacteria, clostridia), yeast and molds; and the lactic acid bacteria themselves".

As silo is filled, each load of freshly chopped forage is packed to expulse as much air as possible from the growing mass of silage. The absence of oxygen allows lactic acid bacteria to grow by converting sugars (simple sugars and starch) into lactic acid, a strong organic acid. As lactic acid bacteria grow, lactic acid accumulates in the ensiled mass and the acidity increases, that is, pH drops. As pH declines, the degrading actions of plant enzymes and undesirable bacteria (clostridia and enterobacteria), yeast and molds are slowed. When pH is sufficiently low (pH of 3.8–4.2 in corn silage and pH of 4.2–4.7 in alfalfa silage) most degrading enzymes are inhibited and the growth of lactic acid bacteria is also inhibited. Thus, the goal in silage-making is to exclude oxygen from the forage mass in order to promote the fermentation of sugars by lactic acid bacteria and drop the pH as rapidly as possible to stop all forms of degradation activity (Brodenic, 1995).

Suitable forage crops to be preserved as silage

The chemical composition of a forage crop or agro-industrial by-product plays an important role in determining the ease with which lactic acid fermentation can take place; and thus the ease with which a particular feed can be preserved as silage. It is easier to ensile forages that have: A high level of fermentable sugar; a low level of protein; a low buffering capacity; ideal dry matter content at ensiling time.

As indicated in Table 4, corn is preserved as silage more easily than grasses or alfalfa because of its high sugar content, low protein content and low buffering capacity. In contrast, alfalfa is more difficult to ensile well because of its low sugar content and high buffering capacity, which is due in part to its high protein content. Thus, the higher the quality of alfalfa, the more challenging it is to ensile successfully.

Table 4. Chemical composition and buffering capacity of typical forage crops

Crop	Water soluble carbohydrate (g/kg DM)	Crude protein (% DM)	Ratio wsc/cp	Buffering capacity (mEq/kg DM)	"Aptitude" for silage preservation
Corn	80 - 300	80 - 100	1.0 – 3.0	150 - 300	High
Grasses	35 - 300	100 - 160	0.4 – 1.8	250 - 550	Intermediate
Alfalfa	20 - 150	140 - 200	0.1 – 0.75	350 - 650	Low

Stage of growth

Crops should be cut at a proper stage of maturity as it is the most important factor for controlling the silage quality. The appropriate stage of growth for cutting different fodder crops for silage making is Sorghum - Flowering to dough stage; Maize - Milk to dough stage; Oat - 50 per cent flowering to dough stage; Grasses - Early flowering stage

The "right" dry matter content

The dry matter content of silage strongly influences the type of fermentation that takes place in a silo. Losses of dry matter in effluents occur when silage dry matter content is less than 25 per cent. On the other hand, too much dry matter makes it difficult to pack and expulse the oxygen from the silage mass. The "ideal" dry matter content of the forage depends on the type of silo (which influences the mode and level of level of compaction that can be achieved). The recommended levels of dry matter in forages at ensiling are: 30 to 40 per cent for bunker silos; 35 to 50 per cent for tower silos; 40 to 50 per cent for wrapped round bales.

Often, wilting a green forage for 24–48 hours to increase dry matter is highly desirable, because wilted silage requires less lactic acid production and will stabilize at a higher pH level than silage that has a lower dry matter content. This is the case because higher dry matter content increases the concentration of soluble dry matter in the silage, and the resulting increase in osmotic pressure inhibits bacterial growth. Thus fermentation stops in wilted silage because of the high acidity, high osmotic pressure combination (rather than high acidity alone in silage with lower dry matter content). Thus wilting has a "sparing" effect on the level of sugar and the level of fermentation needed to stabilize the silage. As a result, wilting is particularly important when the ensiled forage is a legume, which has a relatively low level of fermentable sugars as compared to grass or corn.

Methods of controlling fermentation

There are several ways in which the clostridial type fermentation can be checked. One of the commonly used practices is to increase the dry matter of the herbage. In ensiled crops containing 30 per cent or more dry matter, crop ensile satisfactorily. Promotion of lactic acid fermentation is important pathway for controlling clostridial growth. Other method of controlling fermentation is by addition of preservatives and additives. Carbohydrates in the forages may be naturally occurring or may be added as a separate ingredient, such as molasses obtained as sugar industry byproducts, which acts as a fermentable substrate.

Additives for effective ensiling of nutrients

Various types of additives can be used to improve or inhibit the fermentation or supplement nutrients needed by ruminants to be fed as silage. Propionic acid, formaldehyde, etc. have been used to increase the rate of lactic acid fermentation and produce stable silage. Carbohydrate sources, such as molasses, whey, yeast and other energy-rich ingredients have also been used as additives to increase the fermentation and feeding value of silage. Common additives are as next page:

Molasses: Some green forages, such as legumes and certain grasses are rather low in sugar content. Adding molasses definitely improves the quality of silage by increasing lactic and acetic acid production. It also increases the palatability and nutritive value of the silage. Molasses may be added (3.5 to 4 per cent of the green weight of the forage) in either liquid or dehydrated form. Molasses and starches when added in the form of grains supply the silage bacteria with ample food so that fermentation proceeds normally.

Urea: Adding urea at a level of 0.5 per cent of fresh forages is recommended. The very idea of adding urea is to enrich the silage with nitrogen as cereal forages are mostly deficient in this element.

Limestone: This is calcium carbonate and may be added at a level of 0.5 to 1.0 per cent to maize silage to increase acid production. It neutralizes some of the initial acids as they are formed allowing the lactic acid bacteria to perform longer and to produce more desirable acids.

Sodium metabisulphite: Sulphur dioxide (SO_2 a gas) is a very good antibacterial preservative. It also improves carotene content.

Organic acids: Propionic and formic acids are used for enhancing preservation of forages without the loss of palatability. These are costly and when added, the following guidelines may be observed: Add 1 per cent propionic acid to the forage in the field at the time of harvest or at the chopper. Limit the presence of oxygen by using a sound well built silo. Prevent dilution of organic acid treated silage by rain and cover it with plastic when it is stored outside.

Bacterial cultures: Silage preservatives containing cultures of acid forming bacteria like *Lactobacillus acidophilus, Torulopsis* sp., and *Bacillus subtillis*, are added to silage crops. The basis for including these as a preservative is to provide an inoculum or to increase the number of bacteria for helping rapid fermentation.

Silos

The different types of silos generally used are: (i) pit silo, (ii) tower silo, (iii) trench cum bunker, (iv) trenches, and (v) drum and PVC silo. The silo must provide a solid surface to permit consolidation of the ensiled material and elimination of air. It must protect the silage from water. In India, pit silo is the most common.

Chopping

Crop should be chopped before ensiling. For good silage, the shorter the chop length, the better is the quality. Chopped silage is more palatable to livestock and has little chance of secondary fermentation.

Filling of silo

Silo should be filled rapidly and should not be left open. It should be sealed as soon as possible. Packing is important to create anaerobic conditions. It should be thoroughly pressed so that no air pocket is left in the silo otherwise chances of mould formation will be there which will spoil the silage. After filling, silo should be covered with polythene sheet followed by that of a layer of soil, etc.

Removal of silage

After 45 days of ensilage, the silage can be removed for feeding to animals. Care should be taken in removing the silage from silo. It should not be allowed to deteriorate after the silo is opened for feeding. Covers should be kept firmly in place as long as possible and the minimum face should be exposed at one time. The sugars, proteins and lactic acid present in the silage are subject to attack by mould growth and oxidation as some air is allowed to fermentation, and causes loss of feeding value and intake by the animals.

Silage quality

Silage quality is determined mainly by the odour, physical state, pH, ammonical nitrogen, volatile acids and lactic acid. For desirable fermentation, the forage should be rich in water soluble sugar (more than 5% on dry-matter basis). A good-quality silage should have the following characteristics: (i) pH 4.5–5.0, (ii) ammonical nitrogen of total N – less than 10% of total N, (iii) butyric acid, less than 0.2%, (iv) lactic acid –3 to 12%

Change in Forage Composition in a Silo

Loss of soluble carbohydrates and proteins

As dry matter is lost during silage-making, forage composition also changes. The changes in dry matter composition are due primarily to the fact that the most valuable nutrients (soluble carbohydrates and proteins) are also the first to be lost during respiration and fermentation and in the effluents (juices). fibre on the other hand remains essentially unaffected by the naturally occurring fermentation process in a silo. Thus the overall effect of these losses is to increase the proportion of fibre; the per centage of both the acid detergent fibre (ADF) and neutral detergent fibre (NDF) tend to be higher in silage than in freshly cut forage.

Changes in the protein (nitrogen) fractions

Silage-making change the protein fraction of forages. Respiration is responsible for protein breakdown. As plant cells die after cutting, proteolytic enzymes break down large proteins into smaller soluble compounds including: peptides, amino acids (the building blocks of proteins) and ammonia. In addition, enterobacteria have proteolytic enzymes that remain active even after the pH has dropped to 5.0. Thus most of the protein degradation that takes place in a silo occurs within the first 24 to 72 hours (Phase 1 and Phase 2 of silage fermentation). By the time pH is about 4.0, proteolytic enzymes have lost 65 to 85 per cent of their activity.

Consequently, a rapid drop in pH is desirable to reduce the amount of protein breakdown in a silo. Nevertheless, recent research indicates that as much as half of the total nitrogen in alfalfa silage may be in the form of non-protein nitrogen. The extensive breakdown of alfalfa protein in a silo may be a factor in limiting milk yield in high producing cows (Broderick, 1995). Some scientists have proposed the use of ammonia content as one indicator of adequate silage fermentation. For example, grass silage containing less than 5 per cent of total nitrogen in the form of ammonia might be classified as excellent. In contrast, silage that undergoes clostridial fermentation might contain more than 35 per cent nitrogen in the form of ammonia and thus would be classified as poor (Muck, 1998).

Although limiting protein solubilization and ammonia production in a silo is desirable, one should realize that this process is relatively similar to what happens in the rumen of a cow after ingesting silage. Not all soluble protein and ammonia produced in the silo are necessarily lost. Ammonia and amino acids are needed for microbial synthesis in the rumen. Another nitrogen transformation that may occur during hay or silage-making results from the browning reactions that take place when excess heat is produced. Excess heat cause chemical reactions that combine amino acids with plant sugars (usually derived from hemicellulose) to form a compound resembling lignin. This reaction results in increased levels of acid detergent fibre (ADF) and acid detergent insoluble protein (ADIP). Excess heat is an indication of aerobic degradation due to the presence of air (oxygen) in a silo (Harrison an Fernsen, 1991).

HAY MAKING

Conservation of high-quality forages by drying is termed as hay making. The principle of hay making is to preserve nutritional value of forages through drying it to a level at which the activity of microbial decomposers is inhibited. In India, sunlight is available in abundance, which enables farmers to dry the green forage in open sunlight and thus making hay more economical. The hay making leads to reduction of moisture content to 10–20 per cent, which inhibits the enzyme activity in the plant to be conserved. Thin stemmed cereal crops like sorghum, oat, guinea grass, range grasses, range legumes particularly *Sylosanthes, Siratro,* lablab bean, and all the cultivated legume fodders like berseem, lucerne, and cowpea are suitable for hay making. Leguminous forages have high buffering action and high nitrogen content, and hence are more suitable to be conveniently conserved as hay. Hay making is relatively more convenient and easy for Indian farmers. It can be done by sun drying. During inclement weather, hay making may be done by drying under shade or artificial drying through solar energy. Harvested forage particularly thick stemmed should be chopped and spread over the ground for sun curing and the layers should be changed every day to prevent any sort of fermentation or bacterial growth. After it is well dried (dry-matter content at the time of storing should be around 85–90%), this can be stored for feeding during the lean periods. Thin stemmed crops including legumes can be dried without chopping while thick stemmed fodders like sorghum, maize and *bajra* (pearl millet) require chopping or crushing before they are allowed to dry.

Factors affecting the quality of hay

The following factors affecting the quality of hay are important and should be given due consideration, *i.e.,* plant species, stage of harvest, leaf : stem ratio, chemical composition, physical form, and deterioration during storage.

Hay refers to grasses or legumes that are harvested, dried and stored at 85–90 per cent dry matter. In other words, Hay refers to a forage plant when preserved through reducing the moisture content to the level at which tissues are dead or dormant. High quality hay is green in colour, leafy and pliable and free from mustiness. When harvested in the proper physiological stage of growth and well cured to 15 per cent or less moisture at the time of storing, hay can be utilized as an excellent feed for dairy cattle, particularly when fodder is scarce or pasturage is insufficient (Pitt, 1990).

Principles of hay making

The principle involved in hay making is to reduce the water content of the herbage so that it can safely be stored in mass without undergoing fermentation or becoming mouldy. This must be accomplished in such a manner that the hay is not leached by rain and that the loss of leaves is kept at a minimum (Pitt, 1991).

Requisites of good quality hay

1. Good hay should be leafy. It has been found that leaves are richer in food value compared to other parts of the plant. The leaves are generally rich in proteins, vitamins and minerals. Loss of leaves in hay making would mean deterioration in feeding value of the ultimate product.
2. It should be prepared out of herbage, cut at a stage nearing maturity, preferably at the flowering stage when it has the maximum of nutrients. Delay in cutting would mean losses of a part of nutrients which would be used up by the plant in seed formation.
3. It should be green in colour. The green colour of leaves indicates the amount of carotene which is a precursor of vitamin A.
4. It should be soft and pliable. It should be free from dust and moulds. It should be free from weeds and stubbles. It should have the smell and aroma characteristic of the crop. The moisture per centage in hay should not exceed 15 per cent.
5. Hay of average quality will usually have 25–30 per cent crude fibre and 45–60 per cent TDN. Hay is primarily feed to cattle, buffalo, horse, sheep and goat. Very little of hay of any kind is ever fed to swine.

Types of hay

The hay depending on the various ways of processing can be divided into the following categories. Jungle hay, Sundried hay, Barn dried hay, Dehydrated fodders etc. Jungle hay and Sundried hay mostly refer to the field curing. Barn refers to a simple roofed farm building for storing hay. Hay is baled in the field and brought to the barn. Driers are also used for dehydration. Two types of driers are available Low temperature driers, 80 to 180°C and high temperature driers, 300 to 600°C (Oelberg *et al.,* 1983).

Legume and non-legume hay

Legume hay : Good legume hay has many characteristics that make it of special value to the dairy cattle. It has a higher per centage of digestible nutrients. It has more of digestible proteins because of the high protein content. Furthermore, the proteins of legumes are of superior quality as compared to proteins from other plants. Well-cured legume hays are higher in vitamin contents. They are particularly rich in carotene and may even contain vitamin D. They are also a rich source of vitamin E. The legume hays are particularly rich in calcium and are generally palatable. Among various leguminous fodder crops lucerne, berseem, cowpea and soybean hays are considered first.

Non-legume hay: Non-legume hays made from grasses are inferior to legume hays. They are, as a rule, less palatable and contain less proteins, minerals and vitamins than the legume hays. Non-legume hays have the advantage over legume hays because their output per hectare is more than that of legume hays and the former can be grown easily.

Mixed hay: Hay prepared from mixed crops of legumes and non-legume is known as mixed hay. The composition of such a kind of hay will depend on the proportion of the different species grown as a mixed crop.

Steps in hay making

1. Selection of crop and suitable stage of harvest:
    1. The quality hay can be prepared from the forages having soft and pliable stem. Oat is the best crop for hay making as the crop has soft and pliable stem. Annual and perennial grasses are also suitable for hay making. Legumes are also used for hay making but the problem of leaf shattering should be taken care of by careful handling. Berseem, lucerne, cowpea, pillipesera etc., can be used for hay making.
    2. The crop should be harvested during the day time after the dew has dried off so that plants when spread over the ground may dry evenly. Another factor which needs attention is that the field should not be wet, otherwise uniform drying will not be effected.

   3. The crop cut early is higher in protein, lower in crude fibre, contains more of vitamins, is more palatable and leaf shedding will be less. The best time for cutting a crop for hay making is when it is one-third to a half in blossom.

2. Selection of suitable method of drying:
   1. Quantity of crop available (For a small quantity forage, skilled operation and costly structures would be unsuitable).
   2. Duration for which forage is available for hay making (If forage supply is continuous for the hay making at least 300 days in a year, then the dehydrators may be worthwhile).
   3. Relative humidity (Field curing is not suitable for hay making if the relative humidity is high which may cause moulds/fungi).
   4. Intensity of rains (Rain causes loss of nutrients due to leaching. Continous rains combined with high relative humidity pose a serious problem for hay making).
   5. Atmospheric temperature (The intensity and duration of bright sunshine hours decide the atmospheric temperature. During the hot weather period of March–April to May–June, hay making through sun drying is very convenient and efficient).
   6. Cost of installation (The choice of barns or dehydrators depend on the cost of it. Simple field curing during summer months is cheap and best for the poor farmers with few animals).
   7. In curing, it is necessary that the herbage should be saved from bleaching by the sun and as far as possible, leaves preserved from shattering. The maximum quantity of moisture should be evaporated, so that it can be stored without generation of heat and consequent loss of nutrients.
3. For reasonably rapid curing and production of high quality hay, it is best to let the herbage lie in the field for a few hours until it is well wilted or about one-fourth to one-third cured. It should be raked into small loose heaps known as windrows. If good weather continues, the hay is cured in windrows alone.
4. When the weather is such that the hay cures slowly, turning may also be necessary. Besides field curing, hay can be cured by hanging the herbage on tripods, and on farm fences.
5. In artificial curing, the material is placed in a suitable chamber where it comes in contact with heated air and exposure is regulated depending on the material and the temperature.

Losses of nutrients in hay making

Some nutrients are always lost in field curing of hay, but under favourable conditions this loss is not too much. Drying of green forage at ordinary temperature reduces its digestibility. If the plants are dried without fermentation or bleaching, they contain a high per centage of nutrients.

1. *Due to Late cutting:* Late cutting means greater lignification and lower carbohydrate and protein digestibility. One disadvantage of early cutting is that we get low yield and high moisture content of the forages meant for hay making.
2. *By shattering:* The loss due to shattering of leaves and finer parts in hay making is of importance, especially in the case of legumes. The leaves are much richer in digestible nutrients than the stem and hence losses by shattering materially decrease the nutritive value of hay. To avoid these losses, hay should never be overdried or handled during warm periods of the day.
3. *Loss of Vitamins:* In the process of drying, much of the green colouring matter containing carotene, a precursor of vitamin A is lost with bleaching. In general, the carotene content of freshly cured hay is proportional to the greenness. With severe bleaching, more than 90 per cent of carotene may be destroyed.
4. *Losses in Fermentation:* In fermentation of hay, some of the organic nutrients like starch and sugars are oxidized into CO_2 and water. If drying is prolonged because of unfavourable weather conditions, changes brought about by the activity of bacteria and fungi may occur. Mouldy hay is not only unpalatable but also may be harmful for animals as well as for persons handling the hays due to the presence of mycotoxins. Very often such hays contain actinomycetes, responsible for the allergic condition in man known as "Farmer's Lung". One of the ways to prevent the development of mould growth is to spray propionic acid uniformly on entire hay. In general, it is not uncommon to find patches of mouldy hay in a stack resulting from uneven drying.
5. *By Leaching:* If hay is almost cured and is exposed to heavy and prolonged rains, especially when it is in the field, severe losses may occur through leaching. Unless the rain is so heavy as to soak the material, losses by leaching will not occur. For this reason losses will be much less even in heavy rain if the hay is in good sized windrows.

Storage of hay

Hay is usually stored in stacks in this country. Care should be taken that the hay is in a good and dry condition before it is stored. It should be stacked in a shady place where there is no danger of fire. The stacks should be made at an elevated place. Machines are also available for baling the hay. Baled (a large package or bundle) materials occupy less space.

Brown hay

Sometimes because of very unfavourable weather conditions, good hay cannot be obtained by the ordinary method of curing. Under such circumstances, hay is allowed to dry until about 50 per cent moisture has been removed and then it is packed in stacks or piles. Fermentation takes place and the hay may become very hot, the temperature however, should not be allowed to exceed 80°C . There are great losses in the nutritive value on account of fermentation. These losses range from 30 to 40 per cent. Such hay is often quite palatable.

BAILING AND DENSIFICATION

The transport of dry grass occupies voluminous space and takes lot of time to transport resulting into higher costs. Bailing and densification of dry grass helps in reducing the volume and could be transported economically and efficiently. By using these techniques, the hay could be transported from excess producing areas to deficit areas especially during the calamities of drought. Wheat straw *(bhusa)* could be added with 20 per cent molasses and then densified and bales can be prepared to increase the digestibility of feed intake. Densified block of wheat straw, molasses and urea could be developed through high density bailing machine. The average density of wheat straw and stubble block obtained thus would be 398 kg/m^3 and 355.0 kg/m^3, respectively. Moisture level of admixture was maintained at 20% for densification. The optimum ratio of physical composition of straw block is 78:20:2 (wheat straw: molasses: urea) (Satter *et al.*, 1998).

PRODUCTION OF COMPLETE FEED BLOCKS

Complete feed blocks could also be prepared by densifying machine. Nutritive value of forage is enhanced through mixing with molasses and blending with leguminous fodder, concentrate mixture, minerals and vitamin additives. The composition of complete feed blocks included wheat straw (40%), molasses (20%) dry leaves of berseem (20%), concentrate mixture (19%) and mineral mixture and vitamin additives (1%).

PRESERVATION IN THE FORM OF LEAF MEAL

Preparation of leaf meal out of top feeds and leguminous forages as an animal feed stuff because of high concentration of protein of high biological value and other nutrients such as carotene and minerals. There exists a big deficit of concentrate in the country to the tune of 60 per cent and this deficit can be partially bridged by replacing the concentrate feeds by leaf meals. Leaf meal production technology is simple as well as profitable enterprise for the farmers. Crude protein content (% DM basis) in leaf meals of important forage crops are as follows – *Leucaena leucocephala* (18–21%), *Sesbania sesban* (18%), lucerne (20–21%), *Stylosanthes* sp. (12–18%), *Ziziphus nummularia* (13–16%). The leaf meal of leguminous forages are also known to have rich content of essential amino acids such as lysine, leucine, isoleucine, threonine, methionine, cysteine, valine, histidine, and arginine. A lot of scope exists in establishing a production, processing and marketing chain for its popularization in different parts of the country. One lucerne leaf meal production plant started in 1977 at Udmalpet near Coimbatore produces more than 15 tonnes of leaf meal per month. The setup consists of a chaff cutter feeding sun dried lucerne to a powdering mill for grinding and collection of sieved leaf meal. Farmers around Palladam (Tamil Nadu) have organized through a co-operative to grow and supply lucerne to the leaf meal plant. The arrangement assures regular cash income for small holder farmers through a well developed production, processing and marketing chain. Furthermore, utilization of such meals has grain-saving effects

REFERENCES

Anonymous (2013) Economic Survey, 2012–13.

Broderick GA (1995) Performance of lactating dairy cows fed either alfalfa silage or alfalfa hay as the sole forage. *J. Dairy Sci.* **78:** 320–329.

Harrison JH and Fransen S (1991) Silage management in North America. In Field Guide for Hay and Silage Management in North America, pg. 33. K.K. Bolsen, ed. Natl. Feed Ingredients Assoc. West Des Moines, Iowa.

IGFRI (2014) Vision, 2050 document of ICAR - Indian Grassland and Forage Research Institute, Jhanshi (Uttar Pradesh).

Muck RE (1989) Initial bacterial numbers on Lucerne prior to ensiling. *Grass and Forage Sci.* **4:** 19–25.

Oelberg TJ, Clark AK, McGuffey RK and Schingoethe DJ (1983) Evaluation of covering, dry matter and preservative at ensiling of alfalfa in bunker silos. *J. Dairy Sci.* **66:** 1057–1068.

Pitt RE (1990) Silage and hay preservation. North Regional Agricultural Engineering Service, Cooperative Extension (NRAES-5), 152 Riley-Robb Hall, Ithaca, New York 14853.

Pitt RE (1991) Hay preservation and hay additive products. In Field Guide for Hay and Silage Management in North America, pp. 127. K.K. Bolsen, ed. Natl. Feed Ingredients Assoc. West Des Moines, Iowa.

Satter LD, Muck RE, Woodford JA, Jones BA and Wacek CM (1988) Proc. of the Wisconsin Forage Council's 12th forage production and use symposium.

4

Green House Gas Emissions in Rice and its Mitigation Options for Sustainability

B. Lal[1*], *Priyanka Gautam*[1], *Teekam Singh*[2], *B.P. Meena*[3] *and Rachna Rana*[4]

[1]*Division of Crop Production, ICAR-National Rice Research Institute, Cuttack – 753 006, Odisha, India*
[2]*Regional Rainfed Lowland Rice Research Station, Kamrup – 781 102, Assam, India*
[3]*Division of Soil Chemistry and Fertility Agronomy, ICAR-Indian Institute of Soil Science, Nabibagh – 462 038, Bhopal, India*
[4]*Division of Agronomy, CSK, Himachal Pradesh Krishi Vishvavidyalaya, Palampur – 176 061, H.P., India*
**Corresponding author: E-mail: blalsgnr@gmail.com*

ABSTRACT

A sustainable food future will require reductions in greenhouse gas emissions from agriculture even as the world produces substantially more food. The increasing demand of the growing population requires enhancement in the production of rice. The production of rice, the staple crop for the majority of the world's population, emits large quantities of methane, a potent greenhouse gas and this has a direct bearing on the global environment. Methane in its current concentration of 1.72 ppm V in the atmosphere accounts for 15 per cent of enhanced greenhouse effect. The atmospheric concentration of CH_4 is increasing at 0.3 per cent per year and lowland is considered to be the major contributor. The residence time of this gas in atmosphere is relatively short (10 y) as compared to CO_2 (100y) and N_2O (170y). There are uncertainties in estimation of GHG emissions from Indian rice fields because of its diverse soil and climate conditions and socio-economic status of the farmers. Moreover, various crop management practices, water and fertilizer management, for example, play a major role in the emission. Mitigation options studied for curtailing the methane emissions include rice production management, use of low methane emitting varieties of rice, water management and fertilizer amendment. In baseline practices with high organic amendments, use of compost (58–63%), biogas residues (10–16%), and direct wet seeding (16–22%) should be considered mitigation options. Use of prilled urea as sole N source, use of ammonium sulfate could reduce CH_4 emission by 10–67 per cent. In all rice ecosystems, CH_4 emissions can be reduced by fallow incorporation (11%) and mulching (11%) of rice straw as well as addition of phosphogypsum (9–73%). However, in rainfed and deep water rice, mitigation options are very limited in both number and potential gains.

Key words: Carbon dioxide, Emission, Methane, Nitrous oxide, Rice

1. RICE SCENARIO

Rice is the staple crop for the majority of the world's population (FAO 2014), and rice production is a major source of employment and income, especially in developing countries. In 2013, farmers harvested rice on 165 million hectares worldwide, and flooded ("paddy") rice achieved a global average yield of 4.5 tons per hectare (FAO, 2014). Although more than 100 countries grow rice, 90 per cent of global production occurs in Asia. Of this production, irrigated lowland rice occurs on about 80 million hectares and produces 75 per cent of the world's rice. Roughly 20 million of those hectares in Asia produce two or three crops of rice per year (Fischer *et al.*, 2014). India contains about one-quarter of the world's rice harvested area and produces around one-fifth of the world's rice (FAO 2014). There are mainly four major rice ecosystems in India (1) irrigated lowland, (2) rainfed lowland, (3) rainfed upland and (4) deepwater covering an area of 42.25 million ha. Half of the area (21.41 million ha) is under irrigated lowland and 14.45 million ha is under rainfed lowland rice ecosystems. Upland rice is grown in 4.25 m ha of land, while deepwater rice occupies an area of 2.22 million ha. The irrigated rice area was further sub-divided into (i) continuously flooded (26.9%), (ii) single aeration (35.7%), and (iii) multiple aerations (37.4%) (Gupta *et al.* 2009; Joshi *et al.* 2013). The rainfed area was also sub-divided into flood-prone (27.1%) and drought-prone (72.9%) based on Huke and Huke (1997). In lowland ecosystems, rice seedlings are transplanted in puddled condition and the fields are kept either in continuous submergence or intermittently flooded depending on soil texture, rainfall and availability of irrigation water. Lowland rice fields in north India are generally intermittently flooded, while those from east and south India are flooded continuously. In the case of upland rice, the seeds are directly sown on pulverized seedbed and fields are never flooded. Deepwater rice is grown in low-lying high rainfall areas, where fields are inundated with water. In these areas rice is either direct seeded or transplanted depending upon the onset of monsoon.

2. GAS EMISSION FROM AGRICULTURE

Today, agricultural GHG emissions and associated land use change probably generate around one-quarter of annual global GHG emissions. Such emissions from agriculture alone would almost certainly make it impossible to hold total emissions within the acceptable global limit because they would leave too little room for the much larger and still growing emissions from energy use in non-agricultural sectors such as industry and transport. For the past some decades, the gaseous composition of earth's atmosphere is undergoing a significant change, largely through increased emissions from energy, industry and agriculture sectors; widespread deforestation as well as fast changes in land use and land management practices. These anthropogenic activities are resulting in an increased emission of radiatively active gases, *viz.* carbon dioxide (CO_2), methane (CH_4) and nitrous oxide (N_2O), popularly known as the 'greenhouse gases' (GHGs). The two major greenhouse substances, water vapour, and water and ice clouds, collectively are responsible for over 90 per cent of the greenhouse effect that keeps the earth about 33°C warmer than it would be in the absence of this effect. These substances have short lifetimes and, therefore, abundances that adjust fairly quickly to the longer-term climate forcing represented by changes in the so-called long-lived greenhouse gases. Carbon dioxide, methane and nitrous oxide are the key greenhouse gases (GHG) that contribute towards the global warming at 60, 15 and 5 per cent, respectively (Watson *et al.* 1996). Concentrations of these gases in the atmosphere are increasing at 0.4, 3.0 and 0.22 per cent per year, respectively (Battle *et al.* 1996). Apart from causing global warming, N_2O is also responsible for the destruction of the stratospheric ozone. These GHGs trap the outgoing infrared radiations from the earth's surface and thus raise the temperature of the atmosphere. The global mean annual temperature at the end of the 20th century, as a result of GHG accumulation in the atmosphere, has increased by 0.4–0.7°C, above that recorded at the end of the 19th century. These gases, although responsible for only a small per centage of the direct greenhouse warming, have a great indirect influence. The increase in temperatures that results from their presence leads to increases in water vapour and clouds, which, in net, accelerate the warming.

Increase in temperature can reduce crop productivity in two ways, through more frequent acute heat stress events or through chronic effects of higher average temperatures. One or both types of stress can occur in any season or location. Acute heat-stress injury generally has a threshold response with damage increasing rapidly over a small range of temperature change. Most chronic effects have temperature-response curves that may be modeled as a product of an exponentially increasing rate function and an exponential decay function (Conn and Stumpf 1972). Since acute and chronic heat stresses affect crop productivity through different processes that must be considered separately. Temperatures up to 41°C have little direct impact on leaf CO_2 assimilation. Instead, high temperature may indirectly benefit leaf CO_2 assimilation (Egeh *et al.* 1994). If high temperature reduce the filled spikelet fraction, then plants are sink-limited and panicles remobilize, and translocate less N and C from leaves to growing grains. Therefore, sink limitations caused by high temperatures at flowering may allow leaves to retain N and C, with a resultant delay in canopy senescence and prolonged CO_2 assimilation. Elevated atmospheric CO_2 increases the growth rate and yield of rice through enhanced photosynthesis (Imai *et al.* 1985). Unfortunately, positive responses of rice growth and yield to elevated atmospheric CO_2 diminish above 500 μ l/l (Baker *et al.* 1990). For the Indica rices, panicle-emergence duration generally increased with temperature at ambient CO_2, but decreased with temperature at elevated CO_2. Indica rice are generally considered to be photoperiod-insensitive. They may, in fact, have short-day photoperiod sensitivity with a very long critical day length (Horie and Nakagawa, 1990).

Carbon dioxide, increasing at 0.3–0.4 per cent per year, is added to the air through the combustion of fossil fuels and deforestation, and removed through net uptake by land ecosystems and the oceans. Methane, rising on the average over the past 20 years at 0.8 per cent per year, is produced, in poorly understood proportions, by the gas and coal industries, rice fields, cattle, and naturally in all wetlands. It is destroyed by chemical reactions that are driven by ultraviolet sunlight and involve the very reactive hydroxyl radical (OH). Methane was rising at 1.2 per cent per year in the late 1970s but is now increasing at less than 0.3 per cent per year. Either global methane emissions are decreasing or the methane destruction rate is increasing (or both). Nitrous oxide, increasing at 0.2–0.3 per cent and CFCs were increasing at 5 per cent per year in the 1980s, but they are now increasing at less than 3 per cent per year as a result of the Montreal Protocol. It is important to note that the potency of both nitrous oxide and the CFCs as greenhouse gases is offset significantly by the stratospheric ozone that they destroy.

Among different sources, agricultural soil is the major contributor to greenhouse effect. Globally, agriculture accounts for 54 per cent of anthropogenic methane and 58 per cent of nitrous oxide emissions. In soils, methane is produced during microbial decomposition of organic matter under anaerobic conditions. Rice fields submerged with water, therefore, are the potential source of methane. Continuous submergence, higher organic C content and use of organic manure in puddled soils enhance methane emission. Burning of crop residues also contributes to the global methane budget. Enteric fermentation in ruminants is another major source of methane.

Nitrogenous fertilizers are the source of N_2O in fertilized soils, whereas the indigenous N contributes to its release in unfertilized soil. Soil water content and the availability of carbon enhance the production of N_2O, provided a suitable nitrate source is available. Generally, increase in N_2O emission is observed following irrigation and precipitation. Burning of crop residues also contributes to the global N_2O budget. Agriculture is a source of carbon dioxide also. Soil management practices, such as tillage trigger carbon dioxide emission through biological decomposition of soil organic matter. Tillage breaks the soil aggregates, increases oxygen supply and exposes surface area of organic material promoting the decomposition of organic matter. Fuel-use for various agricultural operations and burning of crop residues are the other sources of carbon dioxide emissions. An off-site source of CO_2 is the manufacturing of fertilizers and pesticides. These emissions of GHGs also occur during production and consumption of food commodities. Comprehensive estimates of GHGs emission from food systems (including production, processing, marketing, etc.) in India are required for evaluating the economic potential of different mitigation strategies.

2.1 Indian Context of GHGs Emission

As per Indian Network for Climate Change Assessment (INCCA) Report (2010), the net GHGs emissions were 1727.7 million tons (mt) of CO_2 equivalent from India in 2007. The main source was the energy sector (57%), followed by industrial (21.7%), agricultural (17.6%) and wastes (3.0%) sectors. In the agricultural sector with a total emission of 334.4 Mt CO_2 equivalent, the major sources are enteric fermentation (63.4%), rice cultivation (20.9%), agricultural soils (13.0%), manure management (2.4%) and on-field burning of crop residues (2.0%). Thus, the crop production sector (rice cultivation, soils, and field burning of crop residues) contributes 35.9 per cent of the total emissions from Indian agriculture (INCCA 2010). Emission of GHGs from agriculture in different states has shown that Punjab, Haryana, Uttar Pradesh and Andhra Pradesh emit higher amount of N_2O because of higher amount of N-fertilizer use. On the other hand, West Bengal, Andhra Pradesh, Orissa, Bihar, Jharkhand and North-Eastern states emit higher amount of methane per ha of rice cultivation.

2.2 GHG Emission from Rice Field

Rice produces roughly four times the GHG emissions per ton of crop as wheat or maize (Linquist *et al.* 2012), mostly in the form of methane and nitrous oxide. Both CH_4 and N_2O are produced in soils under anaerobic conditions by numerous microbial processes. Flooded rice fields, which constitute about 9 per cent of all cultivated land, are thought to be significant sources of trace gases, especially CH_4 (Rennenberg *et al.*, 1992). Most of the world's rice grows in inundated conditions, which leads to both high methane emissions and large demands for irrigation water. Many studies have shown that changes in water management could reduce GHG emissions from rice substantially. Estimates of greenhouse gas emissions from rice production vary, but they all agree that rice production is a significant contributor to overall emissions. As in wetlands generally, flooding rice fields blocks oxygen penetration into the soil, which allows bacteria that produce methane to thrive. According to most estimates today, paddy rice methane generates roughly 500 million tons of emissions of carbon dioxide equivalent per year. Using figures for the warming potency of methane established by the Intergovernmental Panel on Climate Change (IPCC) in 2006, it can be estimated that this number is actually closer to 600 million tons.

The CH_4 emission has been found to vary with different rice ecosystems. The highest emission has been from irrigated continuously-flooded rice (34%), followed by rainfed flood-prone rice (21%). The rainfed drought-prone, single aeration, deep water and irrigated multiple-aeration rice ecosystems have depicted contributions of 17, 16, 8 and 4 per cent of CH_4, respectively. Emission of methane from Indian rice fields has remained almost similar during the period

1995 to 2007 though the rice production has increased from 115 mt to 128 mt. It is because of area under rice remaining almost constant (43–44 mha) and following of similar water and crop management practices by the farmers over the years.

The modeled results indicated that total CH_4 flux from the simulated 42.25 million ha of rice in India ranged from 1.07 to 1.10 Tg C per year under continuous flooding conditions. Earlier several attempts have been made to estimate CH_4 emission from Indian rice fields (Parashar *et al.* 1991; Matthews *et al.* 2000; Yan *et al.* 2003; Bhatia *et al.* 2004). However, only a few studies (Cao *et al.* 1996; Matthews *et al.* 2000) have attempted to calculate detailed regional CH_4 emissions using simulation modelling. The emission estimates vary greatly with the methodology adopted and assumptions made on the importance of different factors affecting CH_4 emission. Ahuja (1990) gave an estimate of 37.8 Tg yr^{-1} CH_4 emission from Indian paddies, which was based on emission data of European and American paddy fields and extrapolated to the Indian region. Later on a value of 3.0 Tg yr^{-1} was estimated on the basis of measurements done up to 1990 at various rice growing regions in the country (Parashar *et al.*, 1991). Gupta *et al.* (2002) using average emission factors for all paddy water regimes, which included harvested areas having soils with high organic carbon and organic amendments, estimated a budget of 5.0 Tg yr^{-1}. Recently, Yan *et al.* (2003) using region specific emission factors estimated India's CH_4 emission to be 5.9, Tg yr^{-1}. Methane emission from the rice fields is estimated to be globally 37 Tg yr^{-1} by IPCC while N_2O emission is much lower as the total N_2O emission from overall cultivated area was put at 1.8–5.3 Tg yr^{-1} (IPCC 1994). Extent of N_2O emission from the rice field is much lower than CH_4 since N_2O is unstable in anaerobic environment of wet-land rice soils (Cai *et al.* 1997; Bronson *et al.* 1997). N_2O is trapped in the soil long enough to get denitrified to N_2 (Granli and Bockman 1994)), but during intermittent drying of the rice fields, N_2O emission increases considerably (Majumdar *et al.* 2000; Shao 1993). In contrast, CH_4 emission is pronounced under anaerobic condition, while intermittent drainage appreciably reduces its emission (Yagi and Minami 1990). Around 80 per cent of CH_4 produced in the soil can be converted to CO_2 in the oxidized zone in the surface soil of rice fields where CH_4 reaches by diffusion (Conrad and Rothfuss 1991).

Most of the N_2O production is believed to result from denitrification of nitrate derived from N fertilizers. Estimates of the production of N_2O-N from rice fields range from 0.01 to 1 per cent of the fertilizer N applied (Rennenberg *et al.* 1992; Bronson and Singh 1994). The direct and indirect N_2O emissions from Indian agricultural soils were estimated to be 186.4 thousand tons (55.3 mt CO_2 equivalent) and 30.61 thousand tons (5.8 mt CO_2 equivalent), respectively. In spite of an increase in N-fertilizer consumption, N_2O emissions from agricultural soils reduced in 2007 with respect to 1994. This is mainly due to the use of India specific emission factors that are lower by almost 38 per cent than the IPCC default values. The revised emission factor is 0.62 kg N_2O-N kg^{-1} N applied for direct emission (Pathak *et al.* 2004; Bhatia *et al.* 2005) and 0.50 kg N_2O-N kg^{-1} N applied for indirect emission. Application of N-fertilizer is the major source of N_2O emission contributing 70 per cent, followed by crop residues and mineralization of organic-N in soil. Emissions of N_2O from rice fields seem to be more important during fallow periods, no matter how brief (Bronson and Singh, 1994).

One mole of ammonia fixed through the Haber-Bosch process produces 1.436 moles of CO_2 (Schlesinger, 1999). Application of 120 kg N ha^{-1} as in the chemical fertilizer translates into off-site emissions of 541.5 kg of CO_2. Using a conversion factor of 21 for the global warming potential of CH_4 in comparison with CO_2, (IPCC 1995), this off-site emission of CO_2 corresponds to the radiative forcing of 25.8 kg CH_4 ha^{-1}.

Emissions of CH_4 resulting from rice straw burning are in the range of 0.43–0.90 per cent of the carbon content, which is similar to the range through straw application into the soil (Miura and Kanna 1997). Moreover, straw burning emits significant quantities of other greenhouse gases such as CO and N_2O (Miura and Kanna 1997) and adversely affects local air quality.

The database on rice cultivars affecting CH_4 emissions is still inconsistent. The two important traits that determine the CH_4 emission potential of rice cultivars are (a) root exudation and (b) gas transfer through the aerenchyma (Butterbach-Bahl *et al.* 1997). However, the CH_4 emission potential of a given cultivar exhibits enormous variation when grown under different greenhouse and field conditions (Wassmann and Aulakh 2000). These variations complicate determination of cultivar-specific emission potentials (Aulakh *et al.* 2000). Therefore, at this point, it is difficult to recommend preferable rice cultivars for mitigating CH_4 emissions.

Overall, paddy rice methane contributes at least 10 per cent (and possibly more) of emissions from global agricultural production, and 1 per cent or more of total human-generated GHG emissions. For most rice growing countries in Southeast Asia, rice contributes around 50 per cent of agricultural emissions and from 2.5 per cent to more than 20 per cent of total emissions.

3. MITIGATION OPTIONS

The strategies for mitigating methane emission from rice cultivation could be alteration in water management, particularly promoting mid-season aeration by short-term drainage; improving organic matter management by promoting aerobic degradation through composting or incorporating it into soil during off-season drained period; use of rice cultivars with few unproductive tillers, high root oxidative activity and high harvest index; and application of fermented manures like biogas slurry in place of unfermented farmyard manure (Pathak and Wassmann 2007). The most efficient management practice to reduce nitrous oxide emission is site-specific, efficient nutrient management (Pathak 2010). The emission could also be reduced by nitrification inhibitors, such as nitrapyrin and dicyandiamide (DCD). There are some plant-derived organics such as neem oil, neem cake and karanja seed extract which can also act as nitrification inhibitors.

Three principal approaches exist for mitigating GHG emissions from rice. The first is to increase rice yields. Increasing yields avoid increases in emissions from land use change and from rice area expansion. If they are high enough, yield increases could even lead to reductions in rice area, reducing methane emissions as well as emissions from land-use change. Second, better management of rice straw, the non-grain portion of rice plants, can hold down emissions. Methane emissions increase when fresh (non-composted) rice straw is added to flooded fields, particularly if not ploughed in until just before planting. Yet rice straw burning, which occurs in some regions, also creates methane and other greenhouse gases, as well as local air pollution. Strategies that reduce emissions include incorporating rice straw into fields well before new production seasons, and removing rice straw from fields to use for other productive purposes, such as growing mushrooms, energy, or biochar.

Third and most important approach for mitigating GHG emissions from rice is reducing or interrupting periods of flooding. It could reduce emissions most dramatically. In addition, the challenges of improving water management, illustrate the challenges of reducing agricultural GHG emissions more broadly. The longer rice is flooded, the more methane-producing bacteria grow and the more they generate methane. Therefore, Decreasing the duration of flooding reduces methane production and emissions (Setyanto *et al.* 2000). The drawdown of water is accomplished by temporarily halting irrigation, allowing water levels to subside through evapotranspiration, percolation, and seepage. Interrupting flooding even with occasional drawdowns has a dual effect: it quickly drives down the populations of methane-producing bacteria, and it stimulates the breakdown of methane by other bacteria. Although the drop in methane emissions is not necessarily proportional to the duration of the drawdown, studies have found that almost any means of reducing or interrupting this flooding reduces methane emissions (IPCC 2006).

With the intermittent flooding scenario, the national CH_4 flux from rice fields was reduced to 0.12–0.13 Tg C per year implying that the water management changes in India, drastically reduced CH_4 fluxes. The intermittent flooding approach has been applied in many Asian countries, such as India (Adhya *et al.*, 2000), Philippines (Corton *et al.* 2000), China (Li *et al.*, 2002), and Japan (Yagi *et al.*, 1996) to reduce CH_4 emissions. With continuous flooding N_2O emission ranged from 0.04 to 0.05 Tg N yr^{-1}. Shifting the water management from continuous flooding to intermittent flooding increased N_2O fluxes to 0.05–0.06 Tg N yr^{-1}, but like CH_4 emission, emission of CO_2 reduced with intermittent flooding.

Mitigation of CO_2 emission from agriculture can be achieved by increasing carbon sequestration in soil through manipulation of soil moisture and temperature, setting aside surplus agricultural land, and restoration of soil carbon on degraded lands. Soil management practices, such as reduced tillage, manuring, residue incorporation, improving soil biodiversity, microaggregation, and mulching can play important roles in sequestering carbon in soil. Some technologies such as intermittent drying, site-specific N management, etc. can be easily adopted by the farmers without additional investment, whereas other technologies need economic incentives and policy support (Wassmann and Pathak, 2007).

Methane emitted from paddy soils can be controlled by various management practices, such as reducing the number of irrigations, multiple drainage system during the crop cycle, direct seeding, alternate wetting and drying, DSR on raised beds or zero tillage, azolla application, semi-dry cultivation, aerobic rice production, arbuscular mycorrhiza and methanotrophs application (Zhao *et al.*, 2006).

- **Dry seeding:** Most paddy rice production in Asia follows the traditional pattern of transplanting seedlings grown in nursery areas into already flooded paddies. But direct seeding of rice is growing in Asia and probably now accounts for a quarter of all rice production there (Kumar and Ladha, 2011). Farmers in the United States use direct seeding because it requires less labour (Kumar and Ladha, 2011). Direct seeding can occur in flooded fields or through drilling seeds into dry fields. If it occurs in flooded fields (wet seeding), it is unlikely to reduce methane emissions, but if it occurs in dry fields (dry seeding), it reduces emissions because it shortens the flooding period by roughly a month (Pittelkow *et al.*, 2014).
- **Single mid-season drawdown:** Studies have shown that a single drawdown during the crop production season, sufficient to allow oxygen to penetrate the soils, substantially lowers GHG emissions. Typically, this kind of drawdown must occur for 5–10 days to generate methane benefits (Itoh *et al.*, 2011). Most farmers in China, Japan, and South Korea already practice this drawdown to increase yields.
- **Alternate wetting and drying (AWD):** This practice involves repeatedly flooding a farm field, typically to a water depth of around 5 centimeters, then allowing field to dry until the upper soil layer starts to dry out (typically when the water level drops to around 15 centimeters below the soil surface), and then rewatering the field. This cycle can continue from 20 days after sowing until 2 weeks before flowering (Siopongco *et al.* 2013). This approach is also known as "controlled irrigation" or "multiple irrigation," depending on country and research context. Since each drying cycle sets back the generation of methane-producing bacteria, AWD achieves even larger reductions in methane than only one drawdown. AWD can be practiced along a continuum, with the frequency of drawdowns ranging from more to less frequent, although the level of methane reductions will depend on how stringently it is practiced.

- **Aerobic rice production:** Like AWD, this system involves adding irrigation water only when needed. It avoids standing water, aiming instead to keep soils moist. This system can drastically reduce or nearly eliminate methane production. In general, however, aerobic rice production has lower yields than rice produced through traditional methods or the three methods listed above. Still, as our case study from China shows, some farmers are maintaining high yields by constructing raised beds and ditches, which limit standing water to furrows (Adhya *et al.*, 2014).
- Application of mid-season drainage which does not coincide with high ammonium in soil, application of urea and NH_4 -based fertilizers in splits with nitrification inhibitors to increase N use efficiency,
- Replacement of ammonium sulphate with other sulphate sources to minimize CH_4 and N_2O emissions,
- Replacement of N broadcasting by foliar-N spray application,
- Sub-surface application of urea supergranules,
- Incorporation or deep placement of prilled urea instead of surface application,
- Application of well-composted organic matter in place of fresh organic matter and green manure, use of single superphosphate (SSP) basally, which in addition to supplying phosphorus, could mitigate CH_4 production by supplying sulphur to soil and
- Cultivation of rice varieties with low gas transport capacities and low exudate formation. These practices can be taken up without much difficulty in irrigated rice fields and can reduce CH_4 and N_2O emission simultaneously (Majumdar *et al.* 2003)

3.1 Comparative emission of greenhouse gases (GHGs) under different crop establishment practices

Flooded rice culture with puddling and transplanting is considered one of the major sources of methane (CH_4) emissions and accounts for 10–20% (50–100 Tg year^{-1}) of total global annual CH_4 emissions (Reiner and Aulakh, 2000). Due to individual or combined effects of various factors as soil characteristics, climatic conditions, and management, such as soil pH, redox potential, soil texture, soil salinity, temperature, rainfall, and water management, amount of CH_4 emission varies between different crop establishment techniques (Harada *et al.* 2007). Methane emission starts at redox potential of soil below –150 mV and is stimulated at less than –200 mV (Jugsujinda *et al.*, 1996). Direct seeding has the potential to decrease CH_4 emissions (Wassmann *et al.*, 2004). Studies comparing CH_4 emissions from different tillage and crop establishment methods under similar water management (continuous flooding/mid-season drainage/intermittent irrigation) in rice revealed that CH_4 emissions were lower in DSR than with TPR

(Gupta *et al.*, 2002; Tyagi *et al.*, 2010). In Wet-DSR, the reduction in CH_4 emission increased from 16 to 22 per cent under continuous flooding to 82 to 92 per cent under mid-season drainage or intermittent irrigation as compared with TPR under continuous flooding (Corton *et al.*, 2000). Methane gas emission and global warming potential was maximum under TPR and emission of N_2O was maximum under DSR crop with conservation practice of brown manuring as the addition of organic matter to soil increased the decomposition rate, which resulted in higher emission of GHGs (Bhatia *et al.* 2011). In a field experiment in the Philippines, DSR reduced CH_4 emissions by 18 per cent as compared with TPR (Corton *et al.*, 2000). Results of yet another study showed that, just by changing puddling to zero tillage, global warming potential (GWP) declined by 42 per cent in Japan (Harada *et al.*, 2007). Dry-DSR on raised beds or zero tillage showed to have potential to reduce CO_2 equivalent per hectare by 40–44 per cent compared with TPR (Pathak *et al.* 2009). Methane emissions may be suppressed by up to 50 per cent if DSR fields are drained mid-season (Wassmann *et al.*, 2004). Although water-saving technologies including Dry-DSR can reduce CH_4 emissions, relatively more soil aerobic states can also increase N_2O emissions slightly. Nitrous oxide production increases at redox potentials above 250 mV (Hou *et al.* 2000). Aerobic environment and high moisture content under zero tilled direct seeded rice results in nitrogen losses as N_2O gas and contribute to global warming. This tradeoff between CH_4 and N_2O emission is a major hurdle in addressing global warming risks and so, strategies must be devised to reduce emissions of both CH_4 and N_2O simultaneously. Developing water management practices in such a way that soil redox potential can be kept at an intermediate range (–100 to +200 mV) to minimize emissions of both CH_4 and N_2O (Hou *et al.* 2000).

CONCLUSION

In agriculture, rice fields are major sources of CH_4 and N_2O. Continuous flooding of rice fields resulted in annual net emissions of 1.07–1.10, 0.038–0.048 and 21.16–60.96 Tg of CH_4, N_2O and CO_2, respectively, with a cumulated global warming potential (GWP) of 130.93–272.83 Tg CO_2 equivalent. A number of practices have been suggested to minimize the emission of either of these gases but mitigating CH_4 emission may increase N_2O emissions and vice versa. Reducing their emission and making the cumulative radiative forcing a minimum is a priority. The strategies should be effective, applicable on a large scale, technically feasible, economic, less time consuming, environment friendly and should be easily acceptable. A reduction in the application of N fertilizers and organic fertilizers would possibly reduce methane and N_2O emissions from rice fields, but would also reduce total rice production unless the area under rice cultivation is increased. Reducing the period of inundation, growing alternative crops, aeration of water and alternating rice crops with other crops in the dry season are suggested methods of reducing methane emission. Dry seeding in place of transplanting is also suggested, for CH_4 mitigation. Intermittent flooding of rice fields reduced annual net emissions to 0.12–0.13 Tg CH_4 15 and 16.6–48.8 Tg CO_2 while N_2O emission increased to 0.056–0.060 Tg N_2O. The GWP, however, reduced to 91.7–211.8 Tg CO_2 equivalents. Dry seeding is getting increasingly popular among farmers due to labour savings, although it has a lower yield potential than transplanting. But this practice has to be tested for its effects on N_2O emissions. To implement a rice cultivation system based on the proposed practices, institutional and political support is needed which is virtually absent in many countries.

REFERENCES

Adhya TK, Bharati K, Mohanty SR, Mishra SR, Ramakrishnan B, Rao VR, Sethunathan N and Wassmann R (2000) Methane emissions from rice fields at Cuttack (India). *Nutr Cycling Agroecosyst* **58(1):** 95–105

Adhya TK, Linquist B, Searchinger T, Wassmann R and Yan X (2014) Wetting and Drying: Reducing Greenhouse Gas Emissions and Saving Water from Rice Production. Working Paper, Installment 8 of Creating a Sustainable Food Future. Washington, DC: World Resources Institute. Available online at http://www.worldresourcesreport.org.

Ahuja DR (1990) Estimating Regional Anthropogenic Emissions of Greenhouse Gases, US-EPA Climate Change Technical Series, Washington DC, 20

Aulakh MS, Bodenbender J, Wassmann R and Rennenberg H (2000) Methane transport capacity of rice plants. I. Influence of methane concentration and growth stage analyzed with an automated measuring system. *Nutr Cycl Agroecosys* **58:** 357–366

Baker JT, Allen LH Jr and Boote KJ (1990) Growth and yield responses of rice to subambient, ambient, and superambient carbon dioxide concentrations. *J Agric Sci* **115:** 313–320

Battle M, Bender M, Sowers T, Tans PP, Butler JH, Elkins JW, Ellis J T, Conway T, Zhang N, Lang P and Clarke AD (1996) Atmospheric gas concentrations over the past century measured in air from fin at the south pole. *Nature* **383:** 231–235

Bhatia A, Ghosh A, Kumar V, Tomer R, Singh SD and Pathak H (2011) Effect of elevated tropospheric ozone on methane and nitrous oxide emission from rice soil in north India. *Agric Ecosys Environ.* **144:** 21-28.

Bhatia A, Pathak H, Jain N, Singh PK and Singh AK (2005) Global warming potential of manure amended soils under rice-wheat system in the Indo-Gangetic Plains. *Atmos Environ* **39:** 976–984

Bhatia A, Pathak H and Aggarwal PK (2004) Inventory of methane and nitrous oxide emissions from agricultural soils of India and their global warming potential. *Curr Sci* **87(3):** 317–324

Bronson KF, Neue H, Singh U and Abao Jr, E (1997) Automated chamber measurements of methane and nitrous oxide flux in a flooded rice soil: II. Fallow period emissions *Soil Sci Soc Am J* **61:** 988–993

Bronson KF and Singh U (1994) Nitrous oxide emissions from flooded rice. In: Peng *et al.* (eds) Climate change and rice. International Rice Research Institute, Los Baños, Philippines

Butterbach-Bahl K, Papen H and Rennenberg H (1997) Impact of gas transport through rice cultivars on methane emission from rice paddy fields. *Plant Cell Environ* **20:** 1175–1183

Cai ZC, Xu H, Lu WS, Liao ZW, Zhang JG, Wei CF and Xie DT (1998) Effects of water managements in winter on CH4 emission from rice fields. *Chinese J App Ecol* (in Chinese with English abstract) **2:** 171–175

Cai ZC, Xing GX, Yan XY, Xu H, Tsuruta H, Yagi K and Minami K (1997) Methane and nitrous oxide emissions from rice paddy fields as affected by nitrogen fertilizers and water management. *Plant Soil* **196:** 7–14

Cao M, Gregson K, Marshall S, Dent JB and Heal O (1996) Global methane emissions from rice paddies. *Chemosphere* **55(5):** 879–897

Conn EE and Stumpf PK (1972) Outlines of biochemistry. John Wiley, New York

Conrad R and Rothfuss F (1991) Methane oxidation in the soil surface layer of a flooded rice field and the effect of ammonium. *Biol Fert Soils* **12:** 28–32

Corton TM, Bajita JB, Grospe FS, Pamplona RR, Asis CA, Wassmann R, Lantin RS and Buendia LV (2000) Methane emission from irrigated and intensively managed rice fields in Central Luzon (Philippines). *Nutr Cycl Agroecosys* **58:** 37–53

Egeh AO, Ingram KT, Zamora OB (1994) High temperature effects on leaf exchange. Phil J Crop Sci 17:21–26

FAO (Food and Agriculture Organization of the United Nations) (2014) AQUASTAT—FAO's Information System on Water and Agriculture. Rome: FAO. Accessible at: <http://www.fao.org/nr/water/aquastat/main/index.stm>

Fischer T, Byerlee D and Edmeades G (2014) Crop Yields and Global Food Security: Will Yield Increases Continue to Feed the World? ACIAR Monograph No. 58. Canberra, Australia: Australian Center for International Agricultural Research

Granli T and Bockman OC (1994) Nitrous oxide from agriculture. *Nor J Agric Sci* 128

Gupta PK, Gupta V, Sharma C, Das SN, Purkait N, Adhya TK, Pathak H, Ramesh R, Baruah KK, Venkatratnam L, Singh G and Iyer CSP (2009) Development of methane emission factors for Indian paddy fields and estimation of national methane budget. *Chemosphere* **74:** 590–598

Gupta PK, Sharma C, Bhattacharya S and Mitra AP (2002) Scientific basis for establishing country greenhouse gas estimates for rice-based agriculture: An Indian case study. *Nutr Cycling Agroecosys* **64:** 19–31

Harada H, Hitomi K and Hayato S (2007) Reduction in greenhouse gas emissions by no-tilling rice cultivation in Hachirogata polder, northern Japan: Life-cycle inventory analysis. *Soil Sci. Plant Nutr.* **53:** 668–677

Horie T and Nakagawa H (1990) Modelling and prediction of developmental processes in rice. I. Structure and method of parameter estimation of a model for simulating developmental processes toward heading. *Jpn J Crop Sci* **59:** 687–695

Hou AX, Chen GX, Wang ZP, Van Cleemput O and Patrick WHJr (2000) Methane and nitrous oxide emissions form a rice field in relation to soil redox and microbiological processes. *Soil Sci Soc Am J.* **64:** 2180–2186.

Huke RE and Huke EH (1997) Rice Area by Type of Culture: South, Southeast, and East Asia. International Rice Research Institute, Los Baños, Philippines, p 59

Imai K, Colman DF and Yanagisawa T (1985) Increase of atmospheric partial pressure of carbon dioxide and growth and yield of rice (*Oryza sativa* L.). *Jpn J Crop Sci* **54:** 413–418

INCCA (Indian Network for Climate Change Assessment) (2010) Assessment of the Greenhouse Gas Emission: 2007, The Ministry of Environment & Forests, Govt. of India, New Delhi

IPCC – Intergovernmental Panel on Climate Change (1995) Climate Change 1995. The Science of Climate Change. Cambridge University Press Cambridge (UK), 572 p

IPCC – Intergovernmental Panel on Climate Change, Climate Change (1994) The supplementary report to the IPCC scientific assessment.

IPCC (Intergovernmental Panel on Climate Change). 2006. 2006 IPCC Guidelines for National Greenhouse Gas Inventories, Volume 4: Agriculture, Forestry and Other Land Use. Hayama, Japan: Institute for Global Environmental Strategies.

Itoh M, Sudo S, Mori S, Saito H, Yoshida T, Shiratori Y, Suga S, Yoshikawa N, Suzue Y, Mizukami H, Mochida T and Yagi K (2011) Mitigation of methane emissions from paddy fields by prolonging midseason drainage. *Agric Ecosyst Environ* **141:** 359–372

Joshi E, Kumar D, Lal B, Nepalia V, Gautam Priyanka and Vyas AK (2013) Management of direct seeded rice for enhanced resource-use efficiency. *Plant Knowledg J* **2:** 119–134

Jugsujinda A, Delaune RD and Lindau CW (1996) Factors controlling carbon dioxide and methane production in acid sulfate soils. *Water Air Soil Pollut* **87:** 345–355

Kumar V and Ladha JK (2011) Direct seeded rice: Recent development & future research needs. *Advances in Agronomy* **111:** 297–413

Li C, Qiu J, Frolking S, Xiao X, Salas W, Moore B, Boles S, Huang Y and Sass R (2002) Reduced methane emissions from large-scale changes in water management of China's rice paddies during 1980–2000. *Geophysic Res Lett* **29 (20):** 1972.

Linquist BA, Groenigen KJ van,. Adviento-Borbe MA, Pittelkow C and van Kessel C (2012) An agronomic assessment of greenhouse gas emissions from major cereal crops. *Global Change Biol* **18:** 194–209

Majumdar D (2003) Methane and nitrous oxide emission from irrigated rice fields: Proposed mitigation strategies. *Curr Sci* **84:** 1317–1326

Majumdar D, Kumar S, Pathak H, Jain MC and Kumar U (2000) Reducing nitrous oxide emission from an irrigated rice field of north India with nitrification inhibitors. *Agric Ecosys Environ* **81:** 163–169

Matthews RB, Wassmann R and Arah J (2000) Using a crop/soil simulation model and GIS techniques to assess methane emsissions from rice fields in Asia, I. Model development, *Nutr Cycling Agroecosyst* **58:** 141–159

Miura Y and Kanno T (1997) Emissions of trace gases (CO_2, CO, CH_4, and N_2O) resulting from rice straw burning. *Soil Sci Plant Nutr* **43:** 849–854

Parashar DC, Rai J, Gupta PK and Singh N (1991) Parameters affecting methane emission from paddy fields, *Indian J Radio Space Phys* **20:** 12–17

Pathak H (2010) Mitigating greenhouse gas and nitrogen loss with improved fertilizer management in rice: Quantification and economic assessment. *Nutr Cycling Agroecosyst* **87:** 443–454

Pathak H and Wassmann R (2007) Introducing greenhouse gas mitigation as a development objective in rice-based agriculture: I. Generation of Technical Coefficients. *Agric Syst* **94:** 807–825

Pathak H, Gupta PK, Bhatia A, Sharma C, Kalra N and Mitra AP (2004) Nitrous oxide emission from soil-plant system. In Climate Change and India, Uncertainty Reduction in Greenhouse Gas Inventory Estimates, Eds: A.P. Mitra, Subodh Sharma, Sumana Bhattacharya, Amit Garg, Sukumar Devotta, Kalyan Sen. Universities Press (India) Pvt. Ltd., Hyderabad, pp. 244–257

Pathak H, Saharawat YS, Gathala MK, Mohanty S and Ladha JK (2009) Simulating environmental impact of resource-conserving technologies in the rice-wheat system of the Indo-Gangetic Plains. In "Integrated Crop and Resource Management in the Rice-Wheat System of South Asia" Ladha JK, Singh Y, Erenstein O and Hardy B (eds.) International Rice Research Institute, Los Banos, Philippines pp. 321–333

Pittelkow CM, Assa Y, Burger M, Mutters RG, Greer CA, Espino LA, Hill JE, Horwath WR, van Kessel C and Linquist BA (2014) Nitrogen Management and Methane Emissions in Direct-Seeded Rice Systems. *Agron J* **106(3):** 968–980

Reiner W and Aulakh MS (2000) The role of rice plants in regulating mechanisms of methane emissions. *Biol Fertil Soils* **31:** 20–29

Rennenberg H, Wassmann R, Paper H and Seiler W (1992) Trace gas exchange in rice cultivation. *Ecol Bull* **42:** 164–173

Schlesinger WH (1999) Carbon sequestration in soils. *Science* **284:** 2095

Setyanto P, Makarim AK, Fagi AM, Wassmann R and Buendia LV (2000) Crop management affecting methane emissions from irrigated and rainfed rice in Central Java (Indonesia). *Nutr Cycling Agroecosyst* **58:** 85–93

Shao KS (1993) Preliminary study on the relationship of agricultural management measures and methane emission flux from rice paddy near Beijing area. Rural Eco-Environment Supp 19–22 (In Chinese)

Siopongco JDLC, Wassmann R and Sander BO (2013) Alternate wetting and drying in Philippine rice production: feasibility study for a Clean Development Mechanism. IRRI Technical Bulletin No. 17. Los Baños (Philippines): International Rice Research Institute

Tyagi L, Kumari B and Singh SN (2010) Water management – A tool for methane mitigation from irrigated paddy fields. *Sci Total Environ.* **408:** 1085–1090.

Wassmann R, Neue HU, Ladha JK and Aulakh MS (2004) Mitigating greenhouse gas emissions from rice–wheat cropping systems in Asia. *Environ Sustain Dev* **6:** 65–90

Wassmann R and Aulakh MS (2000) The role of rice plants in regulating mechanisms of methane emissions. *Biol Fert Soils* **31:** 20–29.

Watson RT, Zinyowera MC, Moss RH and Dokken DJ (1996) Climate Change 1995, impacts, adaptations and mitigation of climate change: Scientific-technical analyses, Intergovernmental Panel on Climate Change, Cambridge University Press, USA, 879

Yagi K and Minami K (1990) Effect of organic matter application on methane emission from some Japanese paddy fields. *Soil Sci Plant Nutr* **36:** 201–217

Yagi K, Tsuruta H, Kanda K and Minami K (1996) Effect of water management on methane emission from a Japanese rice field: Automated methane monitoring, *Global Biogeochem Cycles* **10:** 255–267

Yan X, Ohara T and Akimoto H (2003) Development of region specific emission factors and estimation of methane emission from rice fields in the East, Southeast and South Asian countries. *Global Change Biol* **9:** 1–18

Yoshida S (1981) Fundamentals of rice crop science. International Rice Research Institute, Manila, Philippines, 269 p

Zhao DL, Atlin GN, Bastiaans L and Spiertz JHJ (2006) Comparing rice germplasm groups for growth, grain yield and weed-suppressive ability under aerobic soil conditions. *Weed Res* **46:** 444–452

■■■■

5 Use of Wastewater for Sustainable Agriculture

M.L. Dotaniya[1], V.D. Meena[1] and B.L. Meena[2]*

[1]ICAR-Indian Institute of Soil Science, Bhopal – 462 038, India
[2]ICAR-Central Soil Salinity Research Institute, Karnal – 132 001
**Corresponding mail: mohan30682@gmail.com*

ABSTRACT

Water is an essential for the living existence on earth. Most of the human civilization developed on the bank of water bodies. It requires everywhere smaller to larger scale like industrial level; and its necessity more among all the natural resources. Water is poorly managed across the global world. Per capita water availability decreasing and the groundwater table grows down and many water bodies dry. The human population increasing with speedy rate and to fulfill the need of thirsty and hungry mouth; from limited natural resources. Decreasing the availability of fresh water for agricultural and industrial use. This situation forces with farmers to use poor quality water or wastewater (WW) for agricultural crop production. The WW comprises of sewage, industrial and household effluent. It contains higher amount of organic matter and plant nutrients, which reduced the fertilizer requirement during the crop production. In the other side of the coin it contains low amount of trace harmful heavy metals. Long term use of theses effluents build up significant amount of trace metals in soil and reached to human and animal body *via* consumption of crop plant parts. Use of WW treatment prior to application in agricultural field; use of non-edible crops like flowering plant, jatropha etc., and regular monitoring can make the WW boon otherwise it can be the bane.

***Keywords*:** Heavy metals, Plant nutrient, Sewage Sustainable agriculture, Wastewater

1. INTRODUCTION

Water, food and energy security are the important pillar of human existence on earth in India or across the globe. A lot of the debate going on related issues of above pillars for sustainable use to fulfill the need of increasing population. Among the natural resources, water has a vital role in food and energy production. It is playing a crucial role in all living creatures and in general 50–97 per cent of plant and animal body and about 70 per cent of human body composed of water (Buchholz, 1998), but it is poorly managed natural resource on the earth (Fakayode, 2005). There is a gradual decline in the availability of fresh water in India and scarcity of irrigation can be seen in agricultural and industrial sectors. India support more than 16 per cent of world human population and the availability of fresh water is only 4 per cent (Singh, 2003). Most of the India river basin are closed or facing shortage of water, and thus affecting the agricultural growth, industrialization and also urbanization. The groundwater level decreasing with alarming rate and may not fulfill the demand of agriculture and industrial sectors. As per estimates approximately more than 80 per cent groundwater is utilized by agriculture mostly as irrigation purpose. With increasing global population, it enhanced the gap between demand and supply of fresh water for drinking as well as for agricultural and industrial purpose. In this changing scenario, use of wastewater (WW), sewage or industrial effluent are practicing common in peri urban areas of developing countries like India. This situation increases each year nearby mega city in vegetable production. Here, WW means industrial effluent, household WW and sewage effluent (Fig. 1). This WW is in untreated form and widely used in developed and developing countries like China and Mexico. Apart from irrigation water, this also contains appreciable amount of plant nutrients (Khurana and Singh, 2012). The farmers of peri urban areas saying that the use of sewage or waste water increased the crop yield and reduced the cost of fertilizers (Rattan *et al.,* 2005). Higher amount of organic matter in sewage water enhanced the soil organic matter (SOM) in irrigated agricultural fields. These WW contained significant amounts of harmful trace metals (Gupta *et al.,* 1998; Brar *et al.*, 2000; Yadav *et al.,* 2002), but the concentration is low. Long term use of WW build up the heavy metal concentration in soil, this concentration varies city to city and the mixer of industrial waste (Sharma and Sharma, 2015; Rattan *et al.*, 2002). The increasing concentration of toxic metals i.e. Cadmium (Cd), chromium (Cr), Lead (Pb), Arsenic (Ar), Nickel (Ni), etc. in soil; these metals reached into animal and human body system through plant parts and can cause clinical problems (Tiller, 1986).

So, on one side decline fresh water, posing a threat to human existence in some part of the beautiful earth; on the other side use of poor quality water enhanced the chances of toxic metals in food chain contamination. A big question arose in front of the global scientific community to deal with the shortage of fresh water and sustainable crop production. A range of brilliant minds is engaged in the solution of this problem and new ways of conserving water. This is the time to find out the solution and recycle the poor quality water for the irrigation purpose and fresh water solely for drinking purpose. Use of modern technologies and pre-treatment of WW prior to application in the agricultural sector can reduce the chances of trace toxic metals in the food chain and adverse effect on human and animal body.

Fig. 1 : View of wastewater generation.

2. AVAILABILITY OF FRESH WATER FOR CROP PRODUCTION

The origin of civilization was on the bank of rivers and peoples worship lord water, but across the globe poorly managed natural resource. Indian agriculture is a gamble in the hand of the monsoon; this is true because the uneven rainfall decide the crop yield and also some extent crop quality. Across the globe the availability of fresh water for use is 2.7 per cent, apart from this more than 75.2 per cent frozen in the form of ice caps in polar regions and 22.6 per cent present as groundwater and rest available in lakes, rivers, atmosphere, soil moisture and vegetation. Peoples are ignoring management practices of water during use. Due to lack of awareness, peoples are using over irrigation and wasting water in day to day life use. Most of the newspapers showing drinking water shortage in mega cities in every day. The total serviceable water in India is about 1123 BCM (In which 690 BCM from surface and 433 BCM as a groundwater). The precipitation meets out the water requirement for irrigation only 28 per cent annually. More than 72 per cent water diverted for irrigation, this requirement will increase 910 BMC in 2025 and 1072 BMC in 2050. Likewise, drinking water demand increasing 73, 102 BCM in 2025, 2050 respectively. Whereas, industry and energy sectors projected the fresh water demand almost 3, 9 times from 2025 to 2050, respectively (Fig. 2). Increasing population reduced the per capita water availability which was 5177, 1869 and 1588 m^3 in the year 1951, 2001 and 2010, respectively.

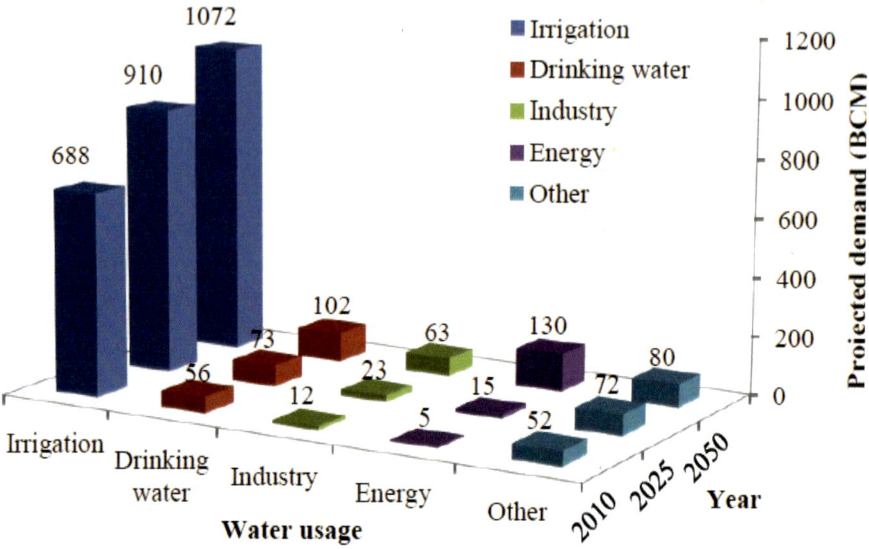

Fig. 2: Projected water demand by different sectors (CWC 2010).

These situations force the farmers to grow the crop with marginal or poor quality of water. Some of the countries are using WW after pre-treatment for industrial purpose.

3. AVAILABILITY AND USE OF WASTEWATER

Large decrease in groundwater levels (up to 200 m in some places) have been observed due to over extraction. The report by the Central Ground Water Board (CGWB) further suggested that, water table is decreasing @ 1 meter/year in the majority of the Indian locations. It is a serious challenge for the researcher and policy maker to reduce this decline rate in coming years. The use of contaminated water by humans being for various activities. As a result of stiff competition from industries for groundwater and surface water resources in agriculture surrounding. That's why many farmers in the areas near to urban/peri-urban localities are even compelled to use WW to irrigate their crops, due to absence of better alternatives (Ghimire, 1994). The use of WW is not our willingness, but it is our limitation or pressure to use for the agricultural purpose. Dryness of natural streams and groundwater forced to use the poor quality of water. The lack of awareness among the farmers, due to poor literacy; they are continuously using the WW. Some of the river tributaries are converted into WW channel and peoples believe that it benefited to crop production. We have surveyed the many WW channels in the adjoining area of peri urban; really surprised that farmers are using WW in mushroom pumsets like extracting fresh water from the well (Fig. 3).

Fig. 3. Use of wastewater for agriculture crop production.

Expansion in domestic water supply, it also increased the WW generation in equal proportion. As per the Central Public Health and Environmental Engineering Organization (CPHEEO) estimates total WW generation is about 70–80% of the water supply for domestic use. These estimates can be more in many Indian cities. The total WW generation by Class I cities (498) was 35,558 MLD and the Class II (410) producing 2,696 MLD in Indian. If see the sewage treatment installation capacity is far behind and it is 11553 and 233 MLD in class I and class II cities (Fig. 4). The mega cities, *i.e.* Maharashtra, Delhi, Uttar Pradesh, West Bengal and Gujarat are the major producer of WW; it is almost 63 per cent of total WW generation in India (CPCB 2007). In India, planned WW treatment capacity is very less, and 234–sewage treatment plants (STPs); which is developed under various river action plans 1978–79 onward. These are located mostly 5 per cent in mega cities or on the bank of major rivers (CPCB 2005). In most of the class I cities, activated sludge process technology working and almost covering 60 per cent of total installed capacity. After this anaerobic sludge blanket technology playing a significant role and covering 26 per cent of total installed capacity. Approximately 28 per cent plants using waste stabilization pond technology. As per the World Bank Report related to WW generation and treatment strongly mentioned in favour of stabilization ponds as the most suitable WW treatment system in most of the developing countries, because land is often available in reasonable opportunity cost and skilled labour and modern technology in short supply (Shuval *et al.* 1986).

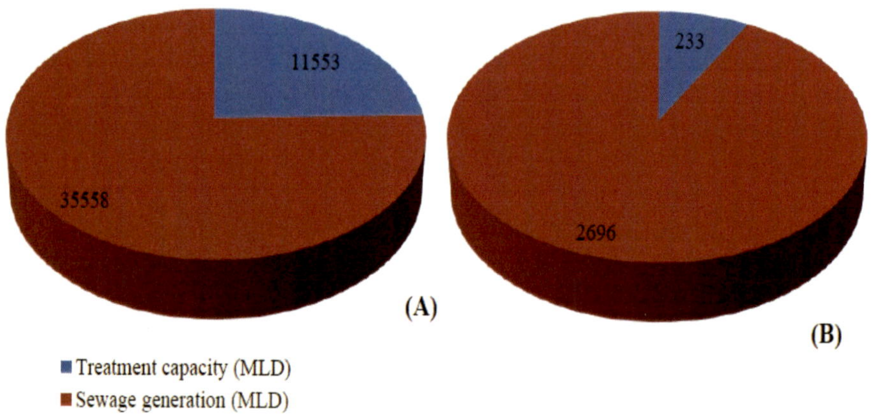

Fig. 4. Sewage generation and treatment capacity in Class I cities (A) and Class II cities (B) in India (CPCB 2009).

4. MANURIAL VALUE OF WASTEWATER

The actual composition of WW varies according to the source of WW generation, but in general it contains organic matter, plant nutrients, inorganic matter, toxic chemicals and pathogens. Sewage water provides the high content of SOM. It provides food material for soil biota, and also determines chemical, physical and biological properties of soil (Datta *et al.*, 2000). The optimum plant nutrient exchange among SOM, soil water, clay and other soil constituents system are essential for soil fertility also for sustainable crop production. Sewage water has been perceived by farmer's liquid fertilizer because of its high nitrogen (N), phosphorus (P) and potassium (K) content (Table 1). Although of wide variation in nutrient concentrations, WW contained 48.3, 7.6, 72.4 and 34.6 mg L^{-1} of N, P, K and sulphur (S), respectively; besides their micro-nutrient manurial value in mg L^{-1} (0.34 for Zn; 10.8 Fe, 0.2 Cu and 0.36 Mn). If farmer apply 5 irrigation of 7.5 cm each with sewage WW, this can add N, P, K and S about 181, 29, 270 and 130 kg ha^{-1}, respectively; which is sufficient to fulfill the crop requirement during crop production. Assuming about 70 per cent potential utilization of WW in agriculture sector shows that sewage farming in the country can annually supply N,P,K and Zn at 380, 60, 520 and 1.4 thousand tons annually, respectively; computed on the basis of average sewage composition which is equivalent to about 1.78 million US$ (Rs. 11.14 crore) worth of plant nutrients.

Table 1. Effect of wastewater irrigation on soil fertility parameters.

Location	Time period	Wastewater type	Parameter	Heavy metal content		Increment (time)	Reference
				WW	GW		
Tamil nadu, India	50 days	Textile waste water	pH	7.92	7.85	1.00	Saravanamoorthy and Rajitha, 2007
			OM (%)	1.29	0.29	4.45	
			N (kg ha^{-1})	118	117	1.00	
			P (kg ha^{-1})	13.45	11.10	1.21	
			K (kg ha^{-1})	56.0	55.0	1.01	
Benin city, Nigeria	8 weeks	Pharmaceutical effluent	pH	7.76	7.23	1.07	Osaigbovo *et al.*, 2006
			OC (%)	1.28	1.00	1.28	
			Ca (cmol kg^{-1})	10.06	0.07	143.7	
			Na (cmol kg^{-1})	1.10	0.05	22.0	
			CEC (cmol kg^{-1})	31.58	0.61	51.8	
			N (g 100 g^{-1})	0.54	0.09	6.0	
			P (g 100 g^{-1})	0.29	0.21	1.38	
Onne, River state, Nigeria	--	Fertilizer plant effluents	pH	8.4	7.9	1.06	Ana and Sridhar, 2002
			K (g 100 g^{-1})	32.3	2.55	12.7	
			Ca (g 100 g^{-1})	0.19	0.86	0.22	
			Na (g 100 g^{-1})	1.50	1.20	1.25	
			Mg (g 100 g^{-1})	0.98	0.68	1.44	
El-Khashb, Egypt	-	Mixed domestic and industrial effluent	pH	6.4	7.2	0.88	El-Hady 2007
			OM (%)	4.07	0.52	7.83	
			TN (mg kg^{-1})	2200	500	4.40	
			NH_4^+-N (mg kg^{-1})	51	20	2.55	
			NO_3^--N (mg kg^{-1})	155	67.5	2.30	
			P (mg kg^{-1})	11.7	2.2	5.30	
			K (mg kg^{-1})	10.2	13.3	0.77	
Khajura, Nepal	-	Distillery effluent	pH	7.7	7.2	1.07	Ale *et al.*, 2008
			OC (%)	0.37	0.19	1.95	
			TN (%)	0.17	0.10	1.70	

Location	Time period	Wastewater type	Parameter	Heavy metal content		Increment (time)	Reference
				WW	GW		
			P_2O_5-P (kg ha^{-1})	183	84	2.18	
			K_2O-K(kg ha^{-1})	590	209	2.82	
Calcutta, India	50-60 years	Sewage Effluent	pH	8.1	8.3	0.98	Morugan *et al.* 2009
			OC (%)	1.73	1.24	1.40	
			TN (%)	0.10	0.06	1.67	
			TP (%)	0.10	0.05	2.00	
			TK (%)	0.13	0.07	1.86	
Kurukshetra, India	25 years	Sewage effluents	OC (%)	1.73	1.24	1.40	Yadav *et al.* 2002
			TN (%)	0.15	0.08	1.89	
			TP (%)	0.88	0.56	1.57	
			TK (%)	0.24	0.18	1.33	

*TN-Total nitrogen; TP-Total phosphorus; TK-total potassium; OM - organic matter; OC - organic carbon, CEC-Cation exchange capacity.

A wheat-soybean based field experiment was conducted during the year 2002–2008 periods on Vertisol of Central India by Dr. J.K. Saha and his team from ICAR-Indian Institute of Soil Science, Bhopal. The use of sewage water in a wheat-soybean cropping system can save chemical fertilizers (60 kg N, 13 kg P and 25 kg K ha^{-1} yr^{-1}) and recover plant nutrients (18 kg N, 7 kg P and 49 kg K ha^{-1} yr^{-1}) through straw biomass for recycling in addition to enhancement of soil fertility. Farmers in the peri urban area usually divert untreated sewage water for irrigation purpose whenever; needed as it is a low cost alternative to the conventional irrigation system.

5. ENVIRONMENTAL HAZARDS RELATED TO WASTEWATER

The use of the WW enriches the agricultural field not only with plant nutrients, but it also supplies the trace toxic metals to the field. The long-term use of tannery effluent for crop production increased 25–30 times more Cr in soil compared to tubewell irrigated fields (Dotaniya *et al.,* 2014a, b). Long-term use of low trace metal WW water build up significant amount of heavy metals like Cd, Cr, Pb, As, Ni, Cu and Zn (Table 2). Some of the metals are also necessary in smaller amount for the smooth physiological functioning of the plants, animal and human being, but a higher amount of the metals, shows toxic symptoms (Rattan and Shukla 1984). The toxicity of metals depends upon the electro-chemical characters, oxidation state, mobility in soil and water system, solubility chemistry in soil, but in a living system, it is governed by age, compounds in body fluids and part of the system. For example, some heavy metals affect crop germination, but have a lesser effect towards a later crop period of maturity. Likewise the intake capacity of toxic metals is also varied by age and the immune system of human being.

Table 2. Total micro-nutrient and heavy-metal concentration (mg kg^{-1}) in soils with wastewater irrigation (WWI) in comparison to groundwater irrigated (GWI) soils.

Location	Years of study	Waste water type	Heavy metal	Heavy metal content		Times	Reference
				WWI	GWI		
Kolkata	50–60	Sewage effluents	Fe	22120	9090	2.43	Gupta and Mitra 2002
			Zn	1210	26	46.5	
			Cu	198	52	3.81	
			Mn	382	446	0.86	
			Cd	3.72	0.04	93.0	
			Pb	385	24.2	15.9	
			Co	46.6	12.0	3.88	
			Ni	61.0	25	2.44	
			Cr	164	24.8	6.61	
			Cu	1.8-6.3	1.15	1.57-5.48	
			Cd	0.12-0.20	0.15	0.80-1.33	
Faridabad	20	Sewage water	Fe	2207	966	2.28	APR 2002-03 CSSRI, Karnal
			Zn	261	53	4.92	
			Cu	60	23	2.61	
			Mn	241	188	1.28	
			Cd	4.2	1.1	3.82	
			Ni	73	19	3.84	
			Cr	79	23	3.43	
Ludhiana	--	Sewage water	Pb	7.12	3.88	1.84	Dheri et al. 2007
			Cr	0.71	0.02	35.5	
			Cd	0.18	0.05	3.60	
			Ni	14.3	1.00	14.3	

Location	Years of study	Waste water type	Heavy metal	Heavy metal content		Times	Reference
				WWI	GWI		
Kurukshetra	25	Sewage effluents	Zn	2.65	0.99	2.68	Yadav et al. 2002
			Cu	2.06	1.45	1.42	
			Fe	22.7	17.8	1.28	
			Mn	7.2	5.8	1.24	
			Pb	1.65	0.99	1.67	
			Ni	3.3	2.4	1.38	
			Cr	6.8	2.2	3.09	
			Cu	17.0	15.5	1.10	
			Pb	16.5	11.0	1.50	
			Cd	40.5	26.5	1.53	
Rohtak	35	Sewage effluents	Fe	11.24	2.48	4.53	Rana et al. 2010
			Mn	2.04	0.46	4.43	
			Ni	4.28	0.31	13.8	
			Cd	0.61	N.D.	-	
			Zn	10.62	2.53	4.19	
			Cu	3.82	0.65	5.87	
			Pb	3.59	1.06	3.38	
Bhopal	5	Sewage effluents	Fe	8.27	7.70	1.07	Saha et al. 2010
			Mn	9.46	8.35	1.13	
			Zn	0.82	0.58	1.41	
			Cu	1.21	1.13	1.07	
Ahmednagar, Maharashtra	-	Sewage effluents	Fe	4.81-7.26	-	-	Kharche *et al.* 2011
			Mn	0.45-1.17	-	-	
			Zn	0.63-2.00	-	-	
			Cu	0.024-1.18	-	-	
Titagarh, WB	-	Waste water	Pb	130.45	-	-	Gupta et al. 2008
			Zn	217.08	-	-	
			Cd	30.72	-	-	
			Cr	148.41	-	-	

Location	Years of study	Waste water type	Heavy metal	Heavy metal content		Times	Reference
				WWI	GWI		
			Cu	89.98	-	-	
			Ni	103.67	-	-	
Zabol, Iran	-	Raw Sewage	Mn	143.92	117.67	1.22	Asgharipour and Azizmoghaddam 2012
			Zn	95.67	86.75	1.10	
			Fe	171.75	191.75	0.90	
			Cu	26.33	17.08	1.54	
			Cd	2.27	N.D.	-	
		50% diluted	Mn	135.42	117.67	1.15	
			Zn	91.50	86.75	1.05	
			Fe	176.92	191.75	0.92	
			Cu	21.33	17.08	1.25	
			Cd	1.00	N.D.	-	
Shandong, China	30	Sewage water	Mn	864.04	452.77	1.91	Zhang *et al.* 2008
			Zn	107.36	41.85	2.57	
			Cu	34.2	16.7	2.05	
			Cd	0.2	0.13	1.54	
			Cr	87.89	45.19	1.94	
			Pb	51.75	24.67	2.10	
			Ni	43.63	11.47	3.80	
			As	6.37	4.21	1.51	
			Co	15.33	9.06	1.69	
			V	85.56	56.85	1.51	

The increasing concentration of heavy metals in soil reduced the soil fertility, by reducing the microbial count and diversity. These HM concentrations affected the mineralization rate of SOM and other nutrients, and also modified the soil physical, chemical and biological properties of soil (Bansal *et al.*, 1992; Rahmani and Rezaer, 2007). Microbial population in soil modifies the plant nutrient availability and also balances the plant nutrient in the soil solution to immobile phase (Wang *et al.*, 2006). Excess concentration of essential plant nutrients N, P and K, invite the excessive vegetative growth and more insect pest attack in agriculture crops. Through the direct use of WW enhanced the pathogen level in the field and reduced the crop productivity. The salinity effect of WW can easily seen in most of the field, they're using as irrigation purpose (Table 3). In Australia, over-abstraction of fresh water for intensive irrigation activities has caused 33 per cent of the land area to be at risk of salination (Gordon, 2003). The harmful impact on crop produce defined on the basis of the economic term like (1) crop yield and income loss; (2) loss of soil health or productivity; (3) poor market value; (4) cost of soil remediation or reclamation; (5) environmental degradation; and (5) health of growing farmers and consumers (Hussain *et al.*, 2002). Few nutrients are leached down into groundwater and caused the severe disease in animal and human. For example, higher concentration of nitrate in groundwater caused nausea, diarrhea, vomiting and also leads methemoglobin (Blue baby syndrome) particularly in infants. Cattle are more susceptible to nitrate poisoning than sheep, and if pregnant, may abort. The world health organization recommended the nitrate concentration in drinking water is 10 mg NO_3^- -N L^{-1} (WHO, 1984; Singh, 2002). The higher concentration of plant nutrients in WW increases the growth of aquatic plants known as eutrophication. The leaching of heavy metals or plant nutrients in groundwater is also affected by soil texture. Sandy soil more permitted than fine textured soil. Availability of heavy metals also affected by soil pH and amount of SOM (Datta *et al.*, 1997; Smith, 1994). The higher concentration of Cd in rice caused itai-itai disease; hair fall due to Cu toxicity; kidney and liver damage due to Cr toxicity; under development of the brain due to Pb; and nervous system badly affected due to As in the human being (Okoro *et al.,* 2012).

Table 3. Pollutant and contaminants in wastewater and their potential impacts through agricultural use (Modified from Hussain *et al.*, 2002).

Pollutant/ Constituent	Parameter	Impacts
Plant nutrients	N, P, K	- Excess N: potential to cause nitrogen injury, excessive vegetative growth, delayed rowing season and maturity, and potential to cause economic loss to farmers - Excessive amounts of N, and P can cause excessive growth of undesirable aquatic species. (eutrophication) - Nitrogen leaching causes groundwater pollution with adverse health and environmental impacts
	Fe, Zn, Ni, Mn	- Higher concentration in soil can affect the availability of other plant nutrients
Suspended solids	Volatile compounds, settleable, suspended and colloidal impurities	- Development of sludge deposits causing anaerobic conditions - Plugging of irrigation equipment and systems such as sprinklers
Pathogens	Viruses, bacteria, helminth eggs, fecal coliforms etc.	- Can cause communicable diseases - Affect the soil fertility - Increased the plant disease and insect pest infestation

Pollutant/ Constituent	Parameter	Impacts
Biodegradable organics	BOD, COD	- Depletion of dissolved oxygen in surface water - Development of septic conditions - Unsuitable habitat and environment - Can inhibit pond-breeding amphibians - Fish mortality - Humus build-up
Stable organics	Phenols, pesticides, chlorinated hydrocarbons	- Persist in the environment for long periods - Toxic to environment - May make wastewater unsuitable for irrigation
Dissolved inorganic substances	TDS, EC, Na, Ca, Mg, Cl, and B	- Cause salinity and associated adverse impacts phytotoxicity - Affect permeability and soil structure
Heavy metals	Cd, Pb, Ni, Zn, As, Hg, etc	-Bioaccumulate in aquatic organisms (fish and planktons) - Accumulate in irrigated soils and the environment - Toxic to plants and animals - Systemic uptake by plants - Subsequent ingestion by humans or animals - Possible health impacts - May make waste water unsuitable for irrigation
Hydrogen ion concentrations	pH	- Especially of concern in industrial waste water - Possible adverse impact on plant growth due to acidity or alkalinity - Impact sometimes beneficial on soil flora and fauna
Residual chlorine in tertiary treated waste water	Both free and combined chlorine	Leaf-tip burn Groundwater, surface water contamination (carcinogenic effects from organochlorides formed when chlorine combines with residual organic compounds) - Greenhouse effect

6. ROLE OF WASTE WATER IN SUSTAINABLE AGRICULTURE

Sustainable agriculture is a practice in which use of ecosystem services for farming to have sustainable agriculture production for satisfying the food and fibre requirement of a growing population. It helps in the most efficient way of non-renewable natural resources and enhances environmental quality and socio-economic viability of farm operation or farmer society. WW also plays the important role in sustainable agriculture in mitigating the water demand for agricultural crop production. The one of the indicators for sustainable water resource development is dependency ratio means dependency of water on neighbouring countries. So, in this regard, marginal water is used for agricultural and industrial sectors and fresh water for drinking purpose. Enhanced the groundwater recharge by use of traditional and modern techniques which are economically feasible and socially accepted at a local or national level. WW carry a huge amount of plant nutrients, which are essential for better crop production and for improving food grain quality (Alam *et al.,* 2003). Higher amount of organic carbon in sewage water improved the soil chemical, physical and biological properties and provide the food materials to soil microorganisms. The soil microorganisms mediated the plant nutrient availability during nutrient transformation. The significant amount of micronutrients in WW water fulfills the crop requirement and farmers are getting good crop yield. The fields situated on the bank of WW channels are full of greenery and farmers are advocating, it is good for their fields. They don't know the harmful effect of untreated WW application in agricultural field; and the toxic effect on human and animal health (Hussain *et al*., 2002). Some time use of WW reduced the potential crop yield due to salinity constitutes. The yield loss can be measured in economic term with the help of productivity loss approaches.

Similarly, the sustainability of WW using the area on soil health or value of soil productivity in term of producing the crop yield and quality. The loss or profitability by the use of WW not only in term of crop yield and quality, but also its effect on human and animal health. In sustainable agriculture, the cost of domestic WW and irrigation cost per capita is important. The annual contribution of WW irrigation in groundwater recharge in volume basis can measure and it can be converted into economic term or in market price value.

7. CONCLUSIONS

Decreasing the availability of fresh water for agriculture crop production is widening with the time period. The increasing population demanding more food and fresh water; which solely depended on limited natural resources. So, use of WW is a common practice in peri-urban areas for agricultural crop production especially vegetable production. These WW supply plant nutrients in higher amount as well as trace toxic metals. Plant nutrients enhanced crop yield, but other side heavy metals reduced the crop quality and soil health. Long term use of WW for crop production reduced the soil biodiversity and sustainable crop yield. Use of sewage water in water scarce areas; can be profitable for growing non-edible food, fiber and oil crop like flowers, castor and jatropha crops. Apart from these, proper treatment of sewage water (through STP) prior to use in agricultural purpose, regular monitoring of sewage irrigated fields, public awareness through mass communication is needed for sustainable use of sewage water.

REFERENCES

Alam MGM, Snow ET and Tanaka A (2003) Arsenic and heavy metal contamination of vegetables grown in Samta village, Bangladesh. *Sci Total Environ* **308:** 83–96.

Ale R, Jha PK and Belbase N (2008) Effects of distillery effluents on some agricultural crops: A case of environmental injustice to local farmers in Khajura VDC, Banke. *Scientific World* **6:** 68–75.

Ana GREE and Sridhar MKC (2002) Soil quality near a chemical fertilizer industry at Port Harcourt, Nigeria. *AJAM/RAGEE* **4:** 50–57.

APR (2002–03) Annual progress report-2002–03. Central Soil Salinity Research Institute, Karnal- 132 001, India.

Asgharipour MR and Azizmoghaddam HR (2012) Effects of raw and diluted municipal sewage effluent with micronutrient foliar sprays on the growth and nutrient concentration of foxtail millet in southeast Iran. *Saudi J Biol Sci.* doi: http://dx.doi.org/10.1016/j.sjbs.2012.06.001.

Bansal RL, Nayyar VK and Takkar PN (1992) Accumulation and bioavailability of Zn, Cu, Mn and Fe in soils polluted with industrial waste water. *J Indian Soc Soil Sci* **40:** 796–799.

Brar MS, Mahli SS, Singh AP, Arora CL and Gill KS (2000) Sewer water irrigation effects on some potentially toxic trace elements in soil and potato plants in northwestern India. *Can J Soil Sci* **80:** 465–471.

Buchholz RA (1998) Principles of environmental management. The Greetings of Business. 2nd Prentice-Hall, London, UK.

CPCB (2009) Status of water supply, wastewater generation and treatment in Class I cities and Class II towns of India. Series: CUPS/70/2009-10. Central Pollution Control Board, India.

CPCB (2005) Parivesh Sewage Pollution – News Letter. Central Pollution Control Board, Ministry of Environment and Forests, Govt. of India, Parivesh Bhawan, East Arjun Nagar, Delhi 110 032 http://cpcbenvis. nic.in/newsletter /sewagepollution /contents ewagepoll-0205.htm.

CWC (2010) Water and related statistics. Water Planning and Project Wing, Central Water Commission, India.

Datta SP, Biswas DR, Saharan N, Ghosh SK and Rattan RK (2000) Effect of long-term application of sewage effluents on organic carbon, bioavailable phosphorus, potassium and heavy metals status of soils and uptake of heavy metals by crops. *J Indian Soc Soil Sci* **48:** 836–839.

Datta SP, Rao, AS and Ganeshamurthy AN (1997) Effect of electrolytes coupled with variable stirring on soil pH. *J Indian Soc Soil Sci* **45:** 185–187.

Dheri GS, Brar MS and Malhi SS (2007) Heavy-metal concentration of sewage-contaminated water and its impact on underground water, soil and crop plants in Alluvial soils of Northwestern India. *Commun Soil Sci Plant Anal* **38:** 1353–1370.

Dotaniya ML, Das H and Meena VD (2014a) Assessment of chromium efficacy on germination, root elongation, and coleoptile growth of wheat (*Triticum aestivum* L.) at different growth periods. *Environ Monit Assess* **186:** 2957–2963. DOI 10.1007/s10661-013-3593-5.

Dotaniya ML, Saha JK, Meena VD, Rajendiran S, Coumar MV, Kundu S and Rao AS (2014b) Impact of tannery effluent irrigation on heavy metal build up in soil and ground water in Kanpur. *Agrotechnol* **2(4):** 77.

El-Hady BAA (2007) Compare the effect of polluted and river Nile irrigation water on contents of heavy-metals of some soils and plants. *Res J Agri Biol* **3:** 287–294.

Fakayode SO (2005) Impact of industrial effluents on water quality of the receiving Alaro River in Ibadan, Nigeria. *Ajeam-Ragee* **10:** 1–13.

Ghimire SK (1994) Evaluation of industrial effluents toxicity in seed germination and seedling growth of some vegetables, M.Sc. Dissertation, Central Department of Botany (1994), Tribhuvan University, Kirtipur, Kathmandu, Nepal.

Gordon LD M (2003) Land cover change and water vapour flows: learning from Australia. *Philosl Trans Royal Soc B: Biol Sci* **358 (1440):** 1973–1984.

Gupta AP, Narwal RP and Antil RS (1998) Sewer water composition and its effect on soil properties. *Bioresour Technol* **65:** 171–173.

Gupta N, Khan DK and Santra SC (2008) An assessment of heavy metal contamination in vegetables grown in wastewater-irrigated areas of Titagarh, West Bengal, India. *Bull Environ Contam Toxicol* **80:** 115–118. Doi: 10.1007/ s00128-007-9327-z.

Gupta SK and Mitra A (2002) In: Advances in Land Resource Management for 21st Century, Soil Conservation Society of India, New Delhi, India, p. 446-460.

Hussain I, Raschid L, Hanjra MA, Marikar F and van der Hoek W (2002) Wastewater use in agriculture: Review of impacts and methodological issues in valuing impacts. Working Paper 37. Colombo, Sri Lanka: International Water Management Institute.

Kharche VK, Desai VN and Pharande AL (2011) Effect of sewage irrigation on soil properties, essential nutrients and pollutant element status of soils and plants in a vegetable growing area around Ahmednagar city in Maharashtra. *J Indian Soc Soil Sci* **59:** 177–184.

Khurana MPS and Singh P (2012) Waste Water Use in Crop Production: A Review. *Res Environ* **2(4):** 116–131.

Morugan A, Gracia-Orenes, Mataix-Solera J, Gomez I, Arcenegui V, Navarro MA and Mataix-Beneyto J (2009) Short-term effects of treated waste water irrigation on soil. Two years of study monitoring a Mediterranean calcareous soil. *Geophysical Res Abstracts* **11:** 1–3.

Okoro HK, Fatoki OS, Adekola FA, Ximba BJ and Snyman RG (2012) A review of sequential extraction procedures for heavy metals speciation in soil and sediments. doi:10.4172/scientificreports.181.

Osaigbovo A, Ulamen O and Ehi-Robert (2006) Influence of pharmaceutical effluent on some soil chemical properties and early growth of maize (*Zea mays* L). *Afr J Biotech* **5:** 612–617.

Rahmani HR and Rezaer R (2007) Effect of industrial effluents of Zob-Han on soil, water and vegetable plants. *J Appl Sci* **17:** 2519–2522.

Rana L, Dhankhar R and Chhikara S (2010) Soil characteristics affected by long term application of sewage wastewater. *Int J Environ Res* **4(3):** 513–518.

Rattan RK, Datta SP, Chandra S and Saharan N (2002) Heavy metals and environmental quality: Indian scenario. *Fertil News* **47 (11)**: 21–40.

Rattan RK, Datta SP, Chhonkar PK, Suribabu K and Singh AK (2005) Long term impact of irrigation with sewage effluents on heavy metal content in soils, crops and groundwater- a case study. *Agric, Ecosys Environ* **109:** 310–322.

Rattan RK and Shukla LM (1984) Critical limits of deficiency and toxicity of zinc in paddy in a Typic Ustipsamment. *Commun Soil Sci Plant Anal* **15:** 1041–1050.

Saha JK, Panwar N, Srivastava A, Biswas AK, Kundu S and Rao AS (2010) Chemical, biochemical, and biological impact of untreated domestic sewage water use on Vertisol and its consequences on wheat (*Triticum aestivum*) productivity. *Environ Monit Assess* **161:** 403–412.

Saravanamoorthy MD and Ranjitha-Kumari BD (2007) Effect of textile water on morphology and yield on two varieties of peanut (*Arachis hypogea* L.). *J Agric Tech* **3:** 335–343.

Sharma UC and Sharma V (2015) Trace elements pollution of soils due to agricultural land irrigation with sewage water in semi-arid environment and its impact on groundwater quality. Intl J Water Res Arid Environ **4(1):** 1–7.

Shuval HI, Adin A, Fattal B, Rawitz E and Yekutiel P (1986) Wastewater Irrigation in Developing Countries: Health effects and Technical Solutions. Technical Paper No. 51. World Bank, Washington DC.

Singh AK (2003) Water resources and their availability. In: Souvenir, National Symposium on Emerging Trends in Agricultural Physics, Indian Society of Agrophysics, New Delhi, 22–24 April 2003, pp. 18–29.

Singh B (2002) Soil pollution and its control. Fundamental of Soil Science, ISSS, New Delhi, p. 499–514.

Smith SR (1994) Effect of soil pH on availability of metals in sewage sludge treated soils. II. Cadmium uptake by crops and implications for human dietary intake. *Environ Pollut* **86:** 5–13.

Tiller KG (1986) Essential and toxic heavy metals in soils and their ecological relevance. Trans. XIII Congr. *Intern Soc Soil Sci* **1:** 29–44.

Wang Y, Shi I, Wang H, Lin Q, Chen X and Chen Y (2006) The influence of soil heavy metals pollution on soil microbial biomass, enzyme activity, and community composition near a copper smelter. *Ecot Environ Saf* **67:** 75–81.

WHO (1984) World Health Organization, Guidelines for water quality. Health and other supporting information. WHO, Geneva (1984) 2, p.100.

Yadav RK, Goyal B, Sharma RK, Dubey SK and Minhas PS (2002) Post-irrigation impact of domestic sewage effluent on composition of soils, crops and ground water—a case study. *Environ Int* **28:** 481–486.

Zhang YL, Dai JL, Wang RQ and Zhang J (2008) Effects of long-term sewage irrigation on agricultural soil microbial structural and functional characterizations in Shandong, China. *Euro J Soil Biol* **44:** 84–91.

■■■■

6

Biochar: An Emerging Technology for Sustainable Agriculture

U.N. Shukla[1], Manju Lata Mishra[2], Ram Swaroop Meena[3], A.K. Pandey[4] and S.K. Verma[3]*

1College of Agriculture, Agriculture University, Jodhpur – 342 304 (Raj.) INDIA
[2]CSK HPKV, Palampur – 176 062 (H.P.) INDIA
[3]Institute of Agricultural Sciences, B.H.U., Varanasi – 221 005 (U.P.) INDIA
[4]NDUAT, Kumarganj, Faizabad – 224 229 (U.P.) INDIA
**E-mail: umanaths7@gmail.com*

1. INTRODUCTION

Global warming is one of the most important challenges facing the modern world. Temperature increases have now been unequivocally proven and are occurring with an unpredicted rate (Prabha *et al.*, 2015). Average global temperature has increased by 0.8°C since industrial era, *i.e.* 1880s and may increase by an additional 3–7°C by 2100 under business as usual scenario (Prabha *et al.*, 2015). The major consequences of agricultural intensification are a transfer of carbon (C) to the atmosphere in the form of carbon dioxide (CO_2), thereby reducing ecosystem of C pools. Agriculture contributes 10–12 per cent of the total global anthropogenic greenhouse gas emissions. To meet the challenges of global climate change, greenhouse-gas emissions must be reduced (Krishnakumar *et al.*, 2014). International efforts aimed to reduce the avoidable greenhouse gas (GHG) emissions or off-setting unavoidable emissions through sequestration of carbon in the environment (Lehmann *et al.*, 2006). It is estimated that use of biochar production method to "tie up" carbon has the potential to reduce current global carbon emissions by 10 per cent (Woolf *et al.*, 2010). The raising concerns about climate change have brought biochar into limelight. Biochar is the carbon-rich product obtained by the heat treatment of biomass under limited or no oxygen (pyrolysis) in a closed container (Mukherjee and Zimmerman, 2013) and it is basically identical to charcoal, but used for different purposes, which is used for non-energy purposes. As a material, the carbon in charcoal is largely inert, showing a relative lack of reactivity both chemically and biologically, and so it is strongly resistant to decomposition. Research scientists have found charcoal particles as old as 400 million years in sediment layers from wildfires that occurred when plant life first began on earth (Venterea and Baker, 2008). Biocharmay also been defined as charcoal that is used for agricultural purposes. It is created by using a pyrolysis process, heating biomass in a low oxygen environment (Prabha *et al.*, 2015 and Lehmann *et al.*, 2006). Once the pyrolysis reaction has begun, it is self-sustaining, requiring no outside energy input. Biochar contains high concen-trations of carbon that can be rather recalcitrant to decomposition, so it may stably sequester carbon (Glaser *et al.*, 2002). By-products of the process include syngas (H_2 + CO), minor quantities of methane (CH_4), tars, organic acids and excess heat. *"Terra-Preta"* an Amazonians' primitive technology for enhancing soil productivity with charred biomass while "Biochar" a viable technique for carbon sequestration in soil (Lehmann *et al.*, 2006). Although, the pyrolysis biochar systems may remove carbon dioxide from the atmosphere, sequestered firstly by plants growth then stored as a stable form of carbon in the soil rather than returning to the atmosphere through decomposition (Prabha *et al.*, 2015).

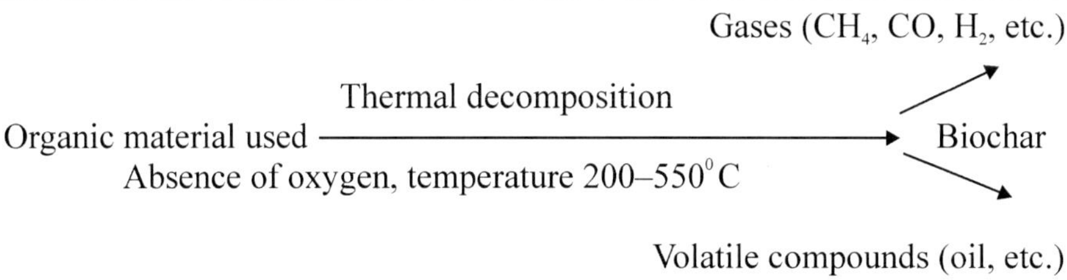

Fig. 1. Schematic sketch of biochar production technique (Jha *et al.*, 2010)

The immediate beneficial effects of biochar additions for nutrient availability are largely due to higher potassium, phosphorus, and zinc availability, and to a lesser extent, calcium and copper (Lehmann *et al.*, 2003a). Once it is produced, biochar is spread on agricultural fields and incorporated into the top layer of soil. Biochar amendments alter soil physical properties; there also could be corresponding impacts on the reliability of flux chamber results from field plots due to differing chamber effects as a consequence of different soil physical properties (Venterea and Baker, 2008). The porous nature of biochar, increase soil water-holding capacity, cation exchange capacity (CEC), surface sorption capacity (Glaser *et al.*, 2002; Keech *et al.*, 2005 and Liang *et al.*, 2008). Thus the high surface area serves as a surface for the sorption of hydrophobic organic compounds. It also helps to prevent fertilizer runoff and leaching, allowing the use of less fertilizers and diminishing agricultural pollution to the surrounding environment. Also it retains moisture, helping plants through periods of drought more easily (Woolf *et al.*, 2010). Most importantly, it replenishes exhausted or marginal soils with organic carbon and fosters the growth of soil microbes essential for nutrient absorption, particularly mycorrhizal fungi.

2. BIOCHAR ACT AS CARBON SEQUESTERED TECHNOLOGY UNDER CLIMATE CHANGE

Biochar not originated as new, but has more innovative concept and technological solution to reduce CO_2 emission and acts as a sequester almost 400 billion tonnes of carbon by 2100 and to lower atmospheric CO_2 concentrations by 37 parts per million (Tim Lenton, 2009). Biochar needs two essential qualities to meet profitable agriculture, firstly adoption of a carbon market and secondly the market price for biochar must be low enough to make farmer friendly (Galinato *et al.*, 2011). The following points to be taken in to consideration to understand the relationship between carbon sequestration and climate change:

- Biochar offers us a golden opportunity to remove excess CO_2 from the atmosphere and sequester it in a virtually permanent and environmentally beneficial way.
- Biochar is one of the only energy production systems that can actually consent more carbon, thereby, creating a "carbon negative and are as economically feasible as pyrolysis or that can simultaneously have so many beneficial effects.
- Basically, biochar production (pyrolysis) effectively locks up atmospheric carbon dioxide (CO_2) by partially combusting plant material (biomass) that are continuously removing carbon dioxide from the environment as they carry out photosynthetic processes necessary for growth.
- Approximately 40 per cent of the carbon content of the plant is locked up as biochar which becomes a safe, stable and beneficial form of soil carbon effectively is able to sequester large quantities of carbon on an annual basis.
- Prior to the industrial revolution there was approximately 280 ppm of carbon in the atmosphere today that number is over 394 ppm. Recent research suggests that the safe upper limit of CO_2 in the atmosphere is no more than 450 ppm. James Hansen, Director of NASA's Goddard Institute of Space Studies, has suggested that the safe upper limit of CO_2 in the atmosphere is likely 350 ppm or below.
- Ability of biochar to sequester carbon with many of these suggesting a range between 0.5 and 3.0 Gigatons (billion metric tons) of carbon per year. Approximately 9.5 billion metric tons of carbon emitted every year from the burning of fossil fuels, cement production, and deforestation, this would equate to a sequestration potential of roughly 5–30 per cent of total global emissions on an annual basis.
- During the production of biochar from the organic residues, it prompted to eliminate pathogens and the speciation of some heavy metal contaminants into forms with reduced levels of toxicity.

- Biochar acts as alternative source as suggested by Day *et al.*, 2005 because, it scrubs CO_2, SOx, and NOx from fossil fuel power plant flue gases, and in the process, creating a slow release fertilizer which sequesters additional CO_2, but the charcoal acts a soil conditioner stressed (Glaser *et al.*, 2002).

Table 1. Some properties of biochar used in different experiments

Materials used for Producing Biochar	pH	Total C (%)	Total N (%)	C: N ratio	Ca (cmol kg^{-1})	Mg (cmol kg^{-1})	P (cmol kg^{-1})	K (cmol kg^{-1})	Cation Exchange Capacity (cmol kg^{-1})	Reference
Papermill waste 1 (waste wood chip)	9.4	50.0.	0.48	104	6.20	1.20	-	0.22	9.0	Zwieten *et al.* 2010
Papermill waste 2 (waste wood chip)	8.2	52.0	0.31	168	11.00	2.60	-	1.0	18.0	Zwieten *et al.* 2010
Greenwaste (grass clippings,cotton trash and plant prunings)	9.4	36.0	0.18	200	0.40	0.56	-	21.0	24.0	Chan *et al.* 2007
Eucalyptus biochar	-	82.4	0.57	145	-	-	1.87	-	4.69	Noguera *et al.* 2010
Cooking biochar	-	72.9	0.76	96	-	-	0.42	-	11.19	Noguera *et al.* 2010
Poultry litter (450°C)	9.9	38.0	2.00	19	-	-	37.42	-	11	-
Poultry litter (550°C)	13.0	33.0	0.85	39	-	-	5.81	-	11	-
Wood biochar	9.2	72.9	0.76	120	0.83	0.20	0.10	1.19	11.90	Major *et al.* 2010
Hardwood sawdust	-	66.5	0.30	221	-	-	-	-	-	Spokas *et al.* 2010

(Source: Jha *et al.*, 2010)

3. STABILITY OF BIOCHAR IN SOILS

The char co-product acts as an energy source and as a soil amendment called biochar. Glaser *et al.* (2001) reported that "black carbon" (analogous to biochar) is very stable due to its polycyclic aromatic carbon structure and able to resist weathering and decomposition by microbial communities, allowing it to persist in soil due to the presence of crystalline morphology, the proportion of which may change with pyrolysis temperature (Cao and Harris, 2010). "Aromacity" refers to a special form of chemical bonding in which a tightly joined ring of atoms form being held together in a series of single and double bonds which freely pass electrons between themselves thereby switching between the two bonding states continuously in a sort of intermediate bond. It is this chemical stability which gives biochar it's highly recalcitrant nature (Glaser *et al.*, 2001). While it is known that charcoal can last in the soil for decades, centuries, millennia and even tens of thousands of years, several studies suggest that there is a certain portion of freshly produced charcoal which is not so resistant to weather and that this could be readily broken down and released as carbon dioxide into the atmosphere. This is referred to as "labile" carbon, the carbon that is easily degraded, as opposed to "recalcitrant" carbon which resists degradation. Generally, the labile portion of biochar is found is approximately 2–10 per cent of the total carbon content of the biochar (Brodowski *et al.*, 2006 and Nguyen *et al.*, 2009).

The cations in the biochar after pyrolysis transformed into oxides, hydroxides, and carbonates (ash) acts as a liming agent when applied to soil. Nguyen *et al.* (2009) found that biochar formed during the conversion of undisturbed land to agricultural land by the burning of the natural vegetation, led to the formation of biochar and irrespective of its origin, the initial biochar content per unit soil mass decreased rapidly by 30 per cent over a period of 30 years. Major *et al.* (2010a) found that approximately 2.2 per cent of carbon being lost from soil respiration, while a much larger per centage of biochar (approx. 50 per cent in their study) is subjected to removal from the field from erosion–large-rainfall events in particular. Research is needed to determine the proportions labile and recalcitrant carbon in fresh biochar. It is likely that this may be influenced by the pyrolysis technique (*e.g.* slow vs. fast pyrolysis) utilized, the feedstock material used, and the temperature and residence times which the biomass is subjected to.

Brodowski *et al.* (2006) found the highest biochar contents are found within the micro-aggregate fraction of the soil, with the macro-aggregates containing lower amounts of biochar. Micro-aggregates play an integral role in reducing biochar decomposition by increasing the encapsulation of the organic fractions. Biochar can initiate increased aggregation when they are broken down to humic acid, aggregation may only play a role in reducing the decomposition rate of the biochar (DeLuca *et al.*, 2006) with its own inherent recalcitrance playing the major role. Biochar addition to soils is a promising alternative to transfer more easily decomposable organic matter (Zech *et al.*, 1990; Fearnside *et al.*, 2001).

4. LARGE SCALE PRODUCTION OF BIOCHAR

4.1 Feedstock

The most suitable materials have high lignin concentration yielding the most biochar (Demirbas, 2004):

- Agricultural residues: paddy straw, wheat straw and maize stalks etc.
- Forestry residues: logging residues, dead wood, excess saplings and pole trees etc.
- Urban wastes: yard trimmings, site clearing, pallets, wood packaging, industry and municipal residues
- Nut shells: groundnut, hazelnut, macadamia nut, walnut, chestnut, coconut
- Bagasse from sugar cane processing,
- Olive or tobacco wastes (Lehmann *et al.*, 2006).

The Carbon Zero Biochar Reactor (Fig. 2a) is specifically designed to produce large quantities of biochar at low cost. In addition, it also produces a clean burning syngas (a combination of hydrogen, carbon monoxide, and methane), and by products of tar and wood vinegar if these are desired. The syngas can be used as a replacement for propane or natural gas, producing heat for poultry sheds or greenhouses for instance. It can also be used to generate mechanical energy for irrigation pumps via a heat exchanger and modern steam engine, or to generate electricity.

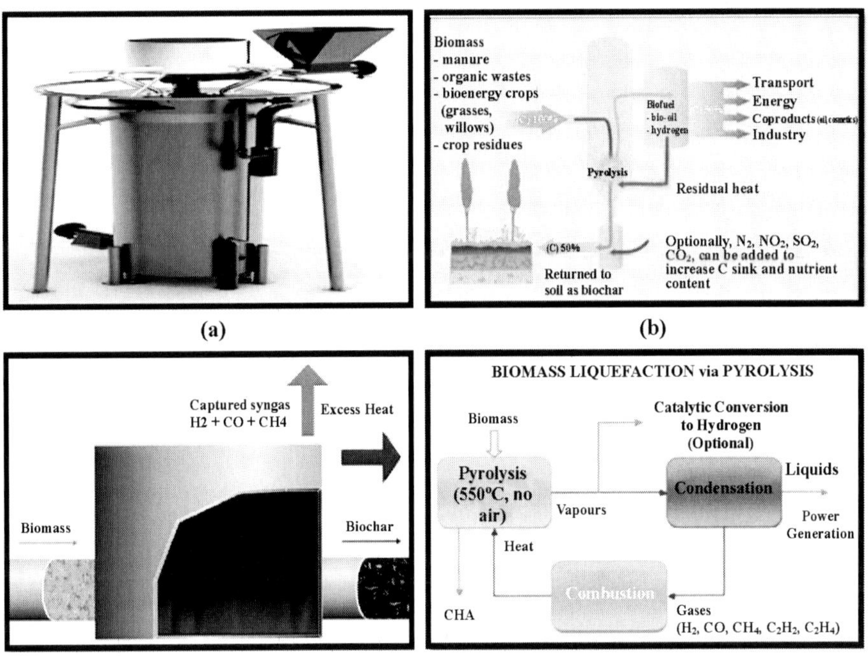

Fig 2. Schematic process of biochar preparation (a-equipment, b-Lehmann process of biochar production (2007), c- simple process of solid production and d-liquid preparation)

Conversion efficiency of woody biomass to biochar is some 30 to 33 per cent, so the unit will produce some 300–330 kilograms of biochar per tonnes of biomass input. Straws and grasses have lower lignin content and hence lower conversion efficiency to biochar, approximately 17 per cent can be expected. Each tonnes of air-dried biomass processed will produce approximately 1.4 MWh to 1.7 MWh of chemical energy in the syngas produced.

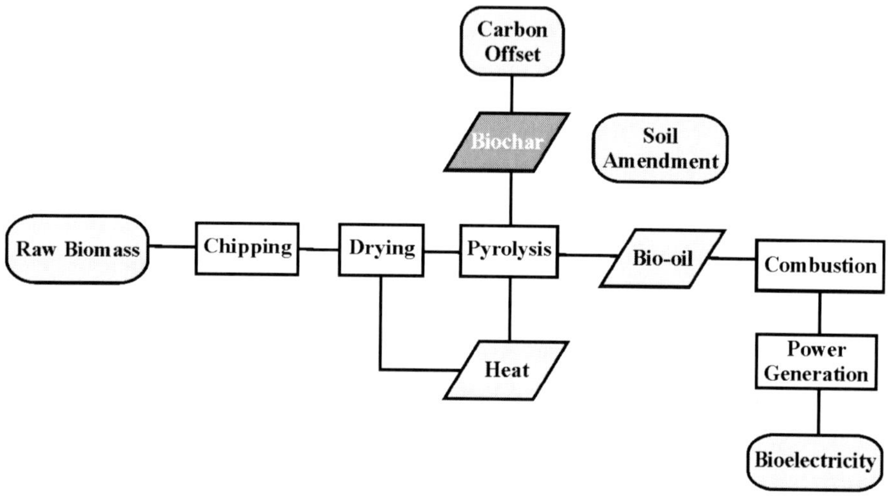

Fig. 3. Steps in biochar production and other diverse commercial uses

4.2 Four types of recognized pyrolysis methods for generating biochar (Fig. 3 and Fig. 4)

- Slow Pyrolysis – traditional (dirty, low char yields) and modern (clean, high char yields)
- Flash Pyrolysis – modern, high pressure, higher char yields
- Fast Pyrolysis – modern, maximizes bio-oil production, low char yields
- Gasification – modern, maximizes bio-gas production, minimizes bio-oil production, low char yields, but highly stable, high ash content.

Table 2. Product yield in different processes

Process	Air filtration	Heat source	Temperature	Time
Slow pyrolysis	None	External	500ºC	30 minutes
Fast pyrolysis	None	External	500ºC	Few seconds
Gasification	20% equivalence ratio	Combustion of infiltrated air	500ºC	Few seconds

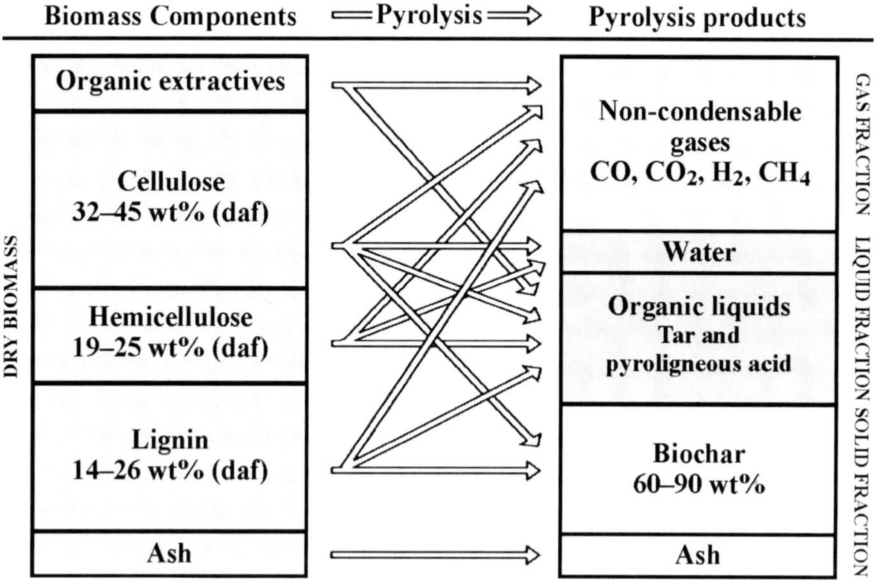

Fig. 4. Conversion of biomass into pyrolysis product (biochar)

5. WHY BIOCHAR ?

5.1 Carbon sink

1. Biochar can store large amounts of greenhouse gases in the ground, potentially reducing or stalling the growth in atmospheric GHG levels; at the same time its presence in the earth can improve water quality, increase soil fertility, raise agricultural productivity and reduce pressure on old-growth forests.
2. Biochar can sequester carbon in the soil for hundreds to thousands of years, like coal and a viable option for mitigation of global warming by greenhouse gas remediation.
3. Biochar can sequester carbon in the soil for hundreds to thousands of years and considered as a potential tool to slow global warming.
4. Annual net emissions of carbon dioxide (CO_2), methane and nitrous oxide could be reduced by a maximum of 1.8∑Pg CO_2-C equivalent (CO_2-C) per year (12% of current anthropogenic CO_2-Ce emissions; 1∑Pg=1∑Gt), and total net emissions over the course of a century by 130∑Pg CO_2-Ce, without endangering food security, habitat or soil conservation.
5. Biochar is a high-carbon, fine-grained residue which today is produced through modern pyrolysis processes. Pyrolysis is the direct thermal

decomposition of biomass in the absence of oxygen to obtain an array of solid (biochar), liquid (bio-oil) and gas (syngas) products.

5.2 Soil amendment

1. For plants that require high potash and elevated pH, biochar can be used as a soil amendment to improve yield.
2. Biochar can improve water quality, reduce soil emissions of greenhouse gases, reduce nutrient leaching, reduce soil acidity, and reduce irrigation and fertilizer requirements.
3. Modest additions of biochar to soil reduce nitrous oxide N_2O emissions by up to 80 per cent and eliminate methane emissions, both are more potent greenhouse gases than CO_2.
4. Biochar reduces leaching of critical nutrients, creates a higher crop uptake of nutrients, and provides greater soil availability of nutrients. At 10 per cent levels biochar reduced contaminant levels in plants by up to 80 per cent, while reducing total chlordane and DDX content in the plants by 68 and 79 per cent, respectively.

5.3 Water retention

1. Biochar is a desirable soil material in many locations due to its ability to attract and retain water, because of its porous structure and high surface area.
2. As a result, nutrients, phosphorus and agrochemicals are retained for the plants benefit. Plants are therefore healthier, and fertilizers leach less into surface or groundwater.

5.4 Energy production: bio-oil and syngas

Bio-oil can be used as a replacement for numerous applications where fuel oil is used, including fuelling space heaters, furnaces, and boilers. Additionally, these biofuels can be used to fuel some combustion turbines and reciprocating engines, and as a source to create several chemicals (Fig. 5).

1. If biochar is used for the production of energy rather than as a soil amendment, it can be directly substituted for any application that uses coal.
2. Pyrolysis also may be the most cost-effective way of producing electrical energy from biomaterial. Syngas can be burned directly, used as a fuel for gas engines and gas turbines, converted to clean diesel fuel through the Fischer–Tropsch process.
3. The greatest potential for bio-oil seems to be its use in a bio-refinery, where compounds that are valuable chemicals, pesticides, pharmaceuticals or food additives are first extracted, and the remainder is either upgraded to fuel or reformed to syngas.

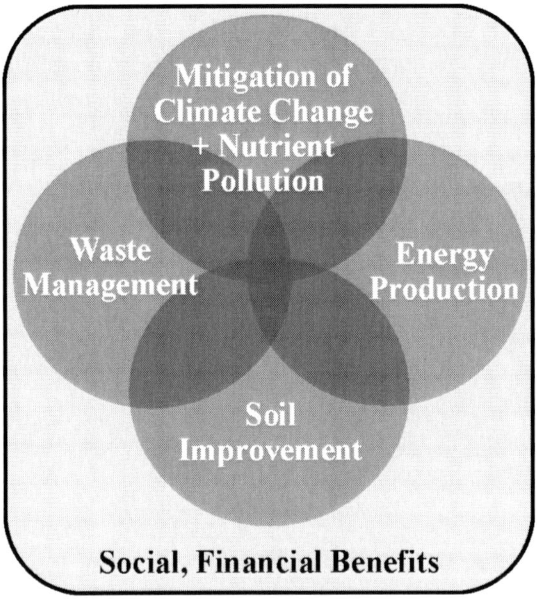

Fig 5. Biochar relationship with environment and agriculture with solution strategies (Adopted from Krishnakumar *et al.*, 2014)

6. BIOCHAR AND AGRICULTURE

Currently, however, very little biochar is utilized in agriculture, in part because its agronomic values in terms of crop response and soil health benefits have yet to be quantified (Prabha *et al.*, 2015 and Krishnakumar *et al.*, 2014):

1. The positive effects of biochar on crop productivity under conditions of extensive agriculture are frequently attributed to direct effects of biochar-supplied nutrients and to several other indirect effects, including increased physico-chemical and biological properties of the soil.
2. Biochar addition to soil alters microbial populations in the rhizosphere, albeit *via* mechanisms not yet understood, and may cause a shift towards beneficial microorganism populations that promote plant growth and resistance to biotic stresses.
3. Moreover, earthworms and biochar increase mineral nutrient availability through an increase in mineralization and nutrient retention, respectively and are likely to interact through various other mechanisms and could thus increase plant growth synergistically.
4. Biochar-induced plant protection against soil borne diseases, the induction of systemic resistance towards several foliar pathogens in many crop systems have been demonstrated.

5. Biochar induces responses along both systemic acquired resistance (SAR) and induced systemic resistance (ISR) pathways, resulting in a broad spectrum controlling capacity in the canopy. Hence, the effects of biochar as soil amendment on the different soil-plant-microbe interactions that may have a role in plant health and improvement of plant responses to disease can be one of the benefits gained from applying biochar to soil.
6. Biochar application also improves the overall sorption capacity of soils and therefore it might influence the toxicity, transport and fate of different heavy metals in the soil.
7. Microbial biomass has been found to increase as a result of biochar additions, with significant changes in microbial community composition and enzyme.
8. Several principal mechanisms have been proposed and partly proven for composts, including:
    i. A direct release of inhibitors of plant pathogens;
    ii. The promotion of microorganisms that acts antagonistic to pathogens, such as parasites, through production of antibiotics, or by successful competition for nutrients;
    iii. Improved plant nutrition and vigour, leading to enhanced disease resistance;
9. Activation of plant defence mechanisms (induced systemic resistance) by enhancing certain microorganisms. Any and all of these four mechanisms may also be applicable to biochar.
10. The known strong sorption of organic compounds onto biochar may modify signalling between plant and pathogens.
11. Affect the mobility and activity of the pathogen.

6.1 Nutrient availability

Biochar application elevates total C, organic C, total N, available P, and exchangeable cations like Ca, Mg, Na, and K increase, and Al decreases in soil, the plant uptakes several of these nutrients after biochar application (Chan *et al.*, 2007; Major *et al.*, 2010a). Major *et al.* (2C10a) reported that nutrient uptake by plants was increased in biochar amended soil, with increased plant yield by greater availability of Ca and Mg in soil. Biochar amendment in soil recycles most of the nutrients that are removed due to the harvest. Because of its high surface area and high surface charge density (Liang *et al.*, 2008), biochar increases the ability of soils to retain nutrients and plant available water and reduces leaching of nutrients and agricultural chemicals (Krishnakumar *et al.*, 2014; Laird *et al.*, 2010; Lehmann *et al.*, 2003 and Glaser at al., 2002). Biochar is a low density material that reduces soil bulk density (Laird *et al.*, 2010) and thereby increases water infiltration, root penetration, and soil aeration, increase soil aggregate stability resulting in to the mineralization of organic derivatives compound to available form, i.e inorganic (Glaser at al., 2002). The biological N fixation (BNF) in plants enhances due to better condition of soil that may be brought up by biochar. This is mainly due to the N availability in soil is lower due to the high C/N ratio of the biochar and the resulting N immobilization (Glaser *et al.*, 2002; Lehmann *et al.*, 2003). Biological nitrogen fixation in boichar amended soils can be preceded due to lower nitrate concentrations and available calcium, phosphorus, and micronutrient concentrations are high, which is ideal for maximum BNF (Lehmann *et al.*, 2003b).

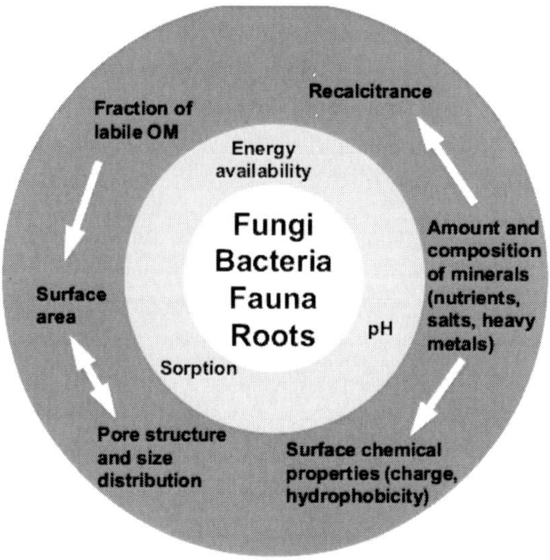

Fig 6. Schematic overview of the connection between primary biochar properties (outer circle), the soil process may influence (intermediate circle) and the soil biota (inner circle), white arrows indicate the influence between biochar properties (Krishnakumar *et al.*, 2014).

It is well corroborated that biomass-derived black carbon (biochar) affects microbial populations and soil biogeochemistry. Both biochar and mycorrhizal association, asymbiotic association in terrestrial ecosystems, are potentially important in various ecosystems (Jha *et al.*, 2010). Biochar and mycorrhizae are two "hot" research challenging areas posed by global warming, alternative energy production and modern non-sustainable agricultural practices (Krishnakumar *et al.*, 2014). Biochar and AMF in soil lead to an altered levels of nutrient availability that affects both plants and mycorrhizal fungi, modifies plant mycorrhizal fungi signalling, serves as a refuge from hyphal grazers and protected from soil predators Fig. 6 (Akiyama *et al.*, 2005). Warnock *et al.* (2007) summarized the effect of biochar on mycorrhizal associations. According to them, mycorrhizal fungi use biochar as a habitat. They further reported that ubiquitous symbioses as association between biochar and mycorrhizal in terrestrial ecosystems, are potentially important in various ecosystem services provided by soils, contributing to sustainable plant production, ecosystem restoration, soil-carbon sequestration and mitigation of global climate changes.

6.2 Soil health

Biochar has many positive properties that make it suitable for use in agriculture or gardening. Due to its porosity increases the soil's ability to retain water and nutrients dissolved in it. Its huge internal surface is a substrate for microbial colonization (Jha *et al.*, 2010). There are no known negative aspects on soil properties if biochar is produced from uncontaminated feedstock and the production conditions are in accordance with the requirements and continuously controlled (Verheijen *et al.*, 2010). The incorporation of biochar into soil can alter soil physical properties, such as structure, pore size distribution and density with logical implications in soil aeration, water holding capacity, plant growth and soil workability (Prabha *et al.*, 2015 and Downie *et al.*, 2009). Biochar has a bulk density much lower than that of mineral soils; therefore, the application of biochar can reduce the overall bulk density of the soil (Gundale and DeLuca 2006) that induces changes in soil chemical properties, *viz.* increasing the pH, total N and available P_2O_5, cation exchange capacity, amounts of exchangeable cations and base saturation, and decrease in the content of exchangeable Al^{3+} (Yamato *et al.*, 2006). Though, significant changes in soil quality, including increase in pH, organic carbon and exchangeable cations as well as reduction in tensile strength were also observed at higher rates of biochar application, *i.e.* > 50 t ha^{-1}. Reduction in tensile strength and increase in field capacity of hard setting soil were the most significant findings (Chan *et al.*, 2007). The buffering capacity of soil can be manage by the application of biochar, but their application rate and method must be taking in to consideration. One of the studies showed that the soils which require more liming, it has been proven from the evidence that biochar may provide similar benefits of improving soil pH balance (Collins, 2008). However, the quantity of biochar that needs to be applied relative to liming may be high. Also, the substitution of biochar for lime can likely provide for net carbon benefit compared to standard liming (Hunt *et al.*, 2010).

Biochar as soil amelioration in degraded landscapes has the potential to increase grassland plant production, enrich soil microbial populations, and stimulate arbuscular mycorrhizal persistence. Biochar addition to soil increases in root colonization of arbuscular mycorrhizae (Matsubara *et al.*, 2002 and Yamato *et al.*, 2006). These qualities in biochar serves as a good sorbent for organic and inorganic pollutants, it can be anticipated that chemical leaching to groundwater and run-off to surface waters will (Steiner *et al.*, 2007; Steiner *et al.*, 2008) be reduced, it can be a tool in the creation of sustainable food and fuel production in areas with severely depleted soils, scarce organic resources, and inadequate water and chemical fertilizer supplies. Biochar is hypothesized to reduce nutrient leaching in well drained soils. Nutrient retention in impoverished post-mine substrates should increase productivity by stimulating biotic-abiotic feedbacks.

However, addition of biochar to the hard-setting soil resulted in significant changes in soil biology in terms of microbial biomass particularly increases in population of earthworm in the soil (Jha *et al.*, 2010).

6.3 Crop yield

Lehmann *et al.*, (2007) reported that increased in yields with increasing biochar applications of up to 140 Mg C ha^{-1} (at which rate, the maximum yield had not yet been reached) on highly weathered soils in the humid tropics. This was not true for all crops however, Rondon *et al.* (2005) found that biomass growth of beans (*Phaseolus vulgaris* L.) rose with biochar applications up to 60 Mg C ha^{-1}, but fell to the same value as for control plots when biochar application was increased to 90 Mg C ha^{-1} (although yield of beans still increased). Lehmann *et al.* (2007) conclude that crops respond positively to bio-char additions up to 50 Mg C ha^{-1} and may show growth reductions only at very high applications. The effect of biochar on crop yield was found positive in different soil. Similarly, rice yield was increased by 294 per cent in *inceptisol* soil when biochar applied at the rate of 30 tonne ha^{-1} without recommended rate of fertilizer application, whereas application of biochar at the rate of 88 tonne ha^{-1} combine with recommended rate of fertilizer, *i.e* N-40, P-20 and K-20 found negative response in *oxisol* soil and yield was declined by 21 per cent due to the negative interaction among fertilizers and biochar, however, under *ferrosol* soil, there was positive interaction reported when biochar applied at the rate 11 tonne ha^{-1} conjunctive with N-30, P-35 and K-50 (Noguera *et al.*, 2010 and Steiner *et al.*, 2008). Zwieten *et al.* (2010) also found the inductive effect of biochar on wheat yield under *ferrosol*.

Table 3. Effect of biochar application on crop yield

Crops	Soil type	Biochar rate (t ha^{-1})	Fertilizer Rate (kg ha^{-1})	Yield/biomass Increase over control[a] (%)	Additional information	Reference
Wheat	Ferrosol	10	1.25 g nutricote per 250 g soil (nutricote contains 15.2% N, 4.7% P and 8.9% K)[b]	+250	Similar response was observed for biomass yield of soybean and radish. Calcarosol (soil type) amended with fertilizer and biochar however gave varied crop responses: Increased soybean biomass, but reduced wheat and radish biomass. No significant effects of biochar were reported in the absence of fertilizer for wheat and soybean, while radish	Zwieten *et al.*, 2010

Crops	Soil type	Biochar rate (t ha^{-1})	Fertilizer Rate (kg ha^{-1})	Yield/biomass Increase over control[a] (%)	Additional information	Reference
					biomass increased significantly.	
Radish	Alfisol	100	N (100)	+266 (biomass)	In the absence of nitrogen fertilizer, application of biochar did not increase the dry matter production of radish even at the highest rate (100 t/ha). Biochar produced from poultry manure increased the dry matter yield of radish, even at the lowest application rate of 10 t/ha, although N fertilizer was not applied.	Chan *et al.*, 2008
Rice	Inceptisol	30	Nil	+294	Sole effect of biochar	Noguera *et al.*, 2010
	Oxisol	88	Nil	+800	Interaction effect of earthworm and biochar	
	Oxisol	88	N (40), P (20), K (20)	-21	Interaction effect of earthworm and biochar	
Maize	Oxisol	20	N (156–170) P (30–43) K (84–138)	+28 (1st year) +30 (2nd year) +140 (4th year)	In the first year after biochar application, no significant effect on crop yield was observed.	Major *et al.*, 2010
Rice	Ferralsol	11	N (30), P (35), K (50)	+29 (stover), +73 (grain)	While charcoal additions alone did not affect crop production, a synergistic effect occurred when both charcoal and inorganic fertilizers were applied.	Steiner *et al.*, 2008

([a]Control means mineral fertilizer alone. + and – signs indicate yield increase and decrease respectively and [b]Rates are in g nutricote/250 g soil)
(Source: Jha *et al.*, 2010)

Table 4. Studies of effect of biochar on crop yield

Study	Results	References
Comparison of maize yields between disused charcoal production sites and adjacent fields, Kotokosu watershed, Ghana	Grain yield 91% higher and biomass yield 44% higher on charcoal site than control.	Oguntunde *et al.*, 2004
Soybean on volcanic ash loam, Japan	0.5 Mg/ha^{-1} char increased yield 151% 5 Mg/ha^{-1} char decreased yield to 63% 15 Mg/ha^{-1} char decreased yield to 29%	Kishimoto & Sugiura, 1985
Sugi trees on clay loam, Japan	0.5 Mg/ha^{-1} wood charcoal increased biomass 249% 0.5 Mg/ha^{-1} bark charcoal increased biomass 324% 0.5 Mg/ha^{-1} activated charcoal increased biomass 244%	Kishimoto & Sugiura,1985
Bauhinia trees onalfisol/ultisol	Charcoal increased biomass by 13% and height by 24%	Chidumayo, 1994
Cowpea on xanthic ferralsol	67 Mg/ha^{-1} char increased biomass 150% 135 Mg/ha^{-1} char increased biomass 200%	Glaser *et al.*, 2002
Pea, India	0.5 Mg/ha^{-1} char increased biomass 160%	Iswaran *et al.*, 1980
Mung bean, India	0.5 Mg/ha^{-1} char increased biomass 122%	Iswaran *et al.*, 1980

7. APPLICATION OF BIOCHAR

Biochar is most commonly incorporated into the soil. First, evenly spread the desired amount onto the soil, then till it in with machinery or by hand. In some cases, such as fruit orchards and other perennial crops where tilling is not an option, biochar can be: (1) applied to the soil surface and, preferably, covered with other organic materials, (2) applied mixed with compost or mulch, or (3) applied as a liquid slurry if finely ground (on a large scale, this could be done with a hydro-mulcher). When planting trees or other potted plants, biochar can be mixed with the backfill material. Deep banding can also be used under appropriate conditions (Hunt *et al.*, 2010). Regardless of the application method, it is important to be cautious when handling dry biochar, which is very dusty and should not be spread in windy conditions. This can be easily remedied by wetting the biochar before application. Respiratory protection (*e.g.*, dust mask) should be worn when handling the dry material (Hunt *et al.*, 2010).

In some soils, 20–25 tonnes biochar/ ha have been applied with beneficial effect. If we assume that the biochar is placed in the top 15 cm of soil and that the soil has a dry specific gravity of 1.2 g/cm^3 and the concentration of biochar applied at the rate of 25 tonne/ha would be 1.4 per cent of soil dry weight, well within the range of this study (0–10%).

8. CONCERNS AND CRITICISM OF BIOCHAR

1. Biochar production could lead to competition with food crops and/or deforestation.
2. Biochar production if scaled up to the extent necessary to sequester massive amounts of carbon dioxide from the atmosphere would lead to the establishment of vast monoculture plantations.
3. Applications and potential carbon dust air pollution. Biochar is very light and easily broken into small particles that can become airborne.
4. Air emission standards from biochar production have not been fully examined and may vary depending on the design of the pyrolysis equipment.
5. Water quality issues related to applied biochar on potentially erodible fields.
6. The production could prevent agricultural wastes from being recycled back into soils.
7. Biochar can lead to faster decomposition rates of native carbon stocks in soils.
8. Potential heavy metal content of biochar (depending on the biomass feedstock) and its effect on human and animal health.
9. Poor biochar production practices could actually lead to greater greenhouse gas emissions and detrimental air quality.
10. There has been little research on biochar and we don't fully know how it will affect soils or crops.

9. FUTURE PROSPECTS

Most biochar researchers agree that the technology needs more study and that the most important thing is to reduce emissions in the first place. "Biochar is not a silver bullet for sequestration," Lehmann says. "We cannot continue the emissions that we generate today and anticipate that any technology or combination of technologies could compensate." Nevertheless, it's possible that biochar could help mitigate those emissions, he says (Lehmann *et al.*, 2007).We must answer certain questions before recommending large-scale use of biochar for agriculture purposes (Jha *et al.*, 2010).

1. Before production of biochar, the question strike in the mind that the amount of carbon sequestered in the biochar biomass either taking into account of net carbon balance or not. The production in such a way

that the amount of CO_2 evolved for producing biochar must be considerably less than the amount of carbon sequestered in charcoal. There must be positive carbon balance for producing biochar biomass.

2. The research on the many complex issues related to biochar production systems is growing very quickly and will be needed to fully understand the implications for food systems, the environment and bioenergy production. Finally, biochar could play an important basis for rural economic development because its production can be scaled down for smaller communities closer to biomass sources.
3. Does biochar has any effect on the soil microbial community and how they behave under the presence of a non-degrading carbon source? The known concerns of decomposers present in the soils have driven their energy from the organic matter present in the soil, *i.e.* soil organic matter (SOM), particularly the soil heterotrophs. Thus, microbial dynamics must be taken in to account for the understanding, particularly in the presence of non-degrading carbon source and vice-versa.
4. The degradation of biochar is extremely slow; therefore the mechanism behind nutrients release/availability must be understood.
5. What will be the enzymatic activity under the influence of a non-degrading substrate?
6. What should be the optimum rate and time for the application of biochar?
7. The long-term application effect of biochar on crop yield and soil quality must be pointed out? As the fact indicated that biochar also apply as soil amendment for improving soil quality and soil-carbon sequestration has attracted global attention, hence an inadequate knowledge on the long-term application, production, quality, properties and feed stock use in biochar must be ensured before using it as soil amendment and also ensured that it may be beneficial in amelioration of problematic soils.
8. Biochar has very promising potential for the further development of sustainable agriculture production systems. Also, biochar production provides a great potential for worldwide climate change mitigation that goes beyond its uses in agricultural production alone.

REFERENCES

Akiyama K, Matsuzaki KI and Hayashi H (2005) Plant sesquiterpenes induce hyphal branching in arbuscular mycorrhizal fungi. *Nature* **435:** 824–827.

Brodowski S, John B, Flessa H and Amelung W (2006) Aggregate-occluded black carbon in soil. *European Journal of Soil Science* **57:** 539–546.

Cao X, Harris W (2010) Properties of dairy-manure-derived biochar pertinent to its potential use in remediation. *Bioresource Technology* **101(14):** 5222–5228.

Chan KY, Zwieten VL, Meszaros I, Downie A and Joseph S (2007) Agronomic values of green-waste biochar as a soil amendment. *Australian Journal of Soil Science Research* **45:** 629–634.

Chan KY, Zwieten VL, Meszaros IA, Downie C and Joseph S (2008) Using poultry litter biochars as soil amendments. *Australian Journal of Soil Science Research* **46:** 437–444.

Chidumayo EN (1994) Effects of wood carbonization on soil and initial development of seedlings in miombo woodland, Zambia. *Forest Ecological Management* **70:** 353–357.

Collins H (2008) Use of biochar from the pyrolysis of waste organic material as a soil amendment: Laboratory and greenhouse analyses. Quarterly progress report prepared for the Biochar Project, December, 2008.

Day D, Evans RJ, Lee JW and Reicosky D (2005) Economical CO_2, SOx, and NOx capture from fossil-fuel utilization with combined renewable hydrogen production and large-scale carbon sequestration. *Energy* **30:** 2558–2579.

DeLuca TH, MacKenzie MD, Gundale MJ and Holben WE (2006) Wildfire-produced charcoal directly influences nitrogen cycling in ponderosa pine forests. *Soil Science Society of America Journal* **70:** 448–453.

Demirbas A (2004) Combustion characteristics of different biomass fuel. *Progress in Energy and Combustion Science* **30:** 219–230.

Downie A, Crosky A and Munroe P (2009) Physical properties of biochar. (In) Biochar for Environmental Management: Science and Technology by Lehmann J and Joseph S (Eds). Earthscan Ltd. London. pp 13–32.

Fearnside PM, Graca P and Rodrigues FJA (2001) Burning of Amazonian rainforests: burning efficiency and charcoal formation in forest cleared for cattle pasture near Manaus, Brazil. *Forest Ecology and Management* **146:** 115–128.

Galinato SP, Yoder JK and Granatstein D (2011) The economic value of biochar in crop production and carbon sequestration. *Energy Policy* **39(10):** 6344–6350.

Glaser B, Haumaier L, Guggenberger G and Zech W (2001) The Terra Preta phenomenon: a model for sustainable agriculture in the humid tropics. *Nature wissenschaften* **88:** 37–41.

Glaser B, Lehmann J and Zech W (2002) Ameliorating physical and chemical properties of highly weathered soils in the tropics with charcoal- a review. *Biology and Fertility of Soils* **35(4):** 219–230.

Gundale MJ and De Luca TH (2006) Temperature and substrate influence the chemical properties of charcoal in the ponderosa pine/Douglas-fir ecosystem. *Forest Ecology and Management* **231:** 86–93.

Hunt J, Du Ponte M, Sato D and Kawabata A (2010) The basics of biochar: A natural soil amendment. *Soil and Crop Management* **30:** 1–6.

Iswaran V, Jauhri KS and Sen A (1980) Effect of charcoal, coal and peat on the yield of moong, soybean and pea. *Soil Biology and Biochemistry* **12:** 191–192.

Jha P, Biswas AK, Lakaria BL and Subba Rao A (2010) Biochar in agriculture – prospects and related implications. *Current Science* **99 (9):** 1218–1225.

Keech O, Carcaillet C and Nilsson MC (2005) Adsorption of allelopathic compounds by wood-derived charcoal: the role of wood porosity. *Plant Soil* **272:** 291–300.

Kishimoto S and Sugiura G (1985) Charcoal as a soil conditioner. *Int. Achieve Future* **5:** 12–23.

Krishnakumar S, Rajalakshmi AG, Balaganesh B, Manikandan P, Vinoth C and Rajendran V (2014) Impact of Biochar on Soil Health. *International Journal of Advanced Research* **2(4):** 933–950.

Laird DA, Fleming PD, Davis DD, Horton R, Wang B, Karlen DL (2010) Impact of biochar amendments on the quality of a typical Midwestern agricultural soil. Geoderma 158: 443-449.

Lehman J, Da Silva Jr JP, Steiner C, Nehls T, Zech W and Glaser B (2003) Nutrient availability and leaching in an archaeological Anthrosol and a Ferralsol of the Central Amazon basin: fertilizer, manure and charcoal amendments. Plant and Soil **249:** 343–357.

Lehmann J (2007) A handful of carbon. *Nature* **447:** 143–144.

Lehmann J, Gaunt J and Rondon M (2006a) Biochar sequestration in terrestrial ecosystems - A review. *Mitigation and Adaptation Strategies for Global Change* **11:** 403–427.

Lehmann J, Kern DC, German LA, McCann J, Martins GC and Moreira A (2003b) Soil fertility and production potential. In: Lehmann J, Kern DC, Glaser B, Woods WI (eds) Amazonian Dark Earths: origin, Properties, Management. Kluwer Academic Publishers, The Netherlands. pp. 105–124.

Liang B, Lehmann J, Solomon D, Sohi S, Thies JE, Skjemstad JO, Luizão FJ, Engelhard MH, Neves EG and Wirick S (2008) Stability of biomass-derived black carbon in soils. *Geochimicaet Cosmochimica Acta* **72:** 6078–6096.

Major J, Lehmann J, Rondon M and Goodale C (2010a) Fate of soil-applied black carbon: downward migration, leaching and soil respiration. *Global Change Biology* **16**: 1366–1379.

Major J, Rondon M, Molina D, Riha SJ and Lehman J (2010) Maize yield and nutrition during 4 years after biochar application to a Colombian savannah oxisol. *Plant Soil* **333:** 117–128.

Matsubara YI, Hasegawa N and Fukui H (2002) Incidence of *Fusarium* root rot in asparagus seedlings infected with arbuscular mycorrhizal fungus as affected by several soil amendments. *Journal of the Japanese Society for Horticultural Science* **71:** 370–374.

Mukherjee A and Zimmerman AR (2013) Organic carbon and nutrient release from a range of laboratory-produced biochars and biochar–soil mixtures. *Geoderma* **193-194:** 122–130.

Nguyen BT and Lehmann J (2009) Black carbon decomposition under varying water regimes. *Organic Geochemistry* **40:** 846–853.

Noguera D, Rondón M, Laossi KR, Hoyos V, Lavelle P, de Carvalho MHC and Barot S (2010) Contrasted effect of biochar and earthworms on rice growth and resource allocation in differentsoils. *Soil Biology & Biochemistry* **42:** 1017–1027.

Oguntunde P, Fosu M, Ajayi A and van de Giesen A (2004) Effects of charcoal production on maize yield, chemical properties and texture of soil. *Biology and Fertility of Soils* **39:** 4.

Prabha B, Pugalendhi S and Subramanian P (2015) Design and development of semi-indirect non-electric pyrolytic reactor for biochar production from farm waste. *Indian Journal of Agricultural Sciences* **85(4):** 585–591.

Rondon M, Ramirez JA and Lehmann J (2005) Charcoal additions reduce net emissions of greenhouse gases to the atmosphere", in Proceedings of the 3rd USDA Symposium on Greenhouse Gases and Carbon Sequestration, Baltimore, USA, March 21–24 2005, p. 208.

Spokas KA, Baker JM and Reicosky DC (2010) Ethylene: potential key for biochar amendment impacts. *Plant Soil* **333:** 443–452.

Steiner C, Glaser B, Teixeira WG, Lehman, J, Blum WEH and Zech W (2008) Nitrogen retention and plant uptake on a highly weathered central Amazonian Ferralsol amended with compost and charcoal. *Journal of Plant Nutrition and Soil Science* **171(6):** 893–899.

Steiner C, Teixeira WG, Lehmann J, Nehls T, de Macedo JLV, Blum WEH, Zech W (2007) Long term effects of manure, charcoal and mineral fertilization on crop production and fertility on a highly weathered Central Amazonian upland soil. *Plant and Soil* **291(1-2):** 275–290.

Tim Lenton (2009) http://climate change psychology. blogspot. com/2009/03/chris-goodall-johanneslehmanntim.htm (Retrived on15th Oct,2015).

Venterea RT and Baker JM (2008) Effects of soil physical no uniformity on chamber-based gas flux estimates. *Soil Science Society of America Journal* **72:** 1410-1417.

Verheijen FGA, Jeffery S, Bastos AC, Velde M van der and Diafas I (2010) Biochar application to soils: acritical scientific review of effects on soil properties, processes and functions, EurComm, Luxembourg, p 149.

Warnock DD, Lehmann J, Kuyper TW and Rillig MC (2007) Mycorrhizal responses to biochar in soil – concepts and mechanisms. *Plant Soil* **300:** 9–20.

Woolf D, Amonette JE, Street-Perrott FA, Lehmann J and Joseph S (2010) Sustainable biochar to mitigate global climate change. *Nature Communications 1*, Article number: 56 (online journal).

Yamato M, Okimori Y, Wibowo IF, Anshiori S and Ogawa M (2006) Effects of the application of charred bark of Acacia mangium on the yield of maize, cowpea and peanut, and soil chemical properties in South Sumatra, Indonesia. *Soil Science and Plant Nutrition* **52:** 489–458.

Zech W, Haumaier L and Hempfling R (1990) Ecological aspects of soil organic matter in tropical land use. In: McCarthy MR, B.P. (Ed.), Humic Substances in Soil and Crop Sciences: Selected Readings. American Society of Agronomy and Soil Science Society of America, Madison, pp. 187–202.

Zwieten LV, Kimber S, Morris S, Chan KY, Downie A, Rust J, Joseph S and Cowie A (2010) Effects of biochar from slow pyrolysis of papermill waste on agronomic performance and soil fertility. *Plant Soil* **327:** 235–246.

7

Enhancing Crop Competitiveness Through Sustainable Weed Management Practices

Kairovin Lakra[1], *S.K.Verma*[2*], *Avinash Chandra Maurya*[3], *S.B. Singh*[3], *Ram Swaroop Meena*[2] *and U.N. Shukla*[4]

[1] *Department of Agronomy, CSAUAT, Kanpur – 280 002, India*
[2]*Department of Agronomy, Institute of Agricultural Sciences, Banaras Hindu University, Varanasi – 221 005, India*
[3]*Department of Agronomy, KVK, Jakhdhar, Rudraprayag, GBPUA&T, Pantnagar, Uttrakhand.*
[4]*College of Agriculture, Agriculture University, Jodhpur – 342 304.*
**Corresponding author email: suniliari@gmail.com*

ABSTRACT

Weeds are the major threat in crop production and their management in modern agriculture is crucial to avoid yield losses and ensure food security. Intensive agricultural practices, changing climate, and natural disasters affect weed dynamics and that requires a change in weed management protocols. The existing manual control options are no longer viable because of labour shortages; chemical control options are limited by ecodegradation, health hazards, and development of herbicide resistance in weeds. We are, therefore, reviewing some potential alternative weed management strategies for modern agriculture that are viable, feasible, and efficient. Enhancing crop competitiveness through appropriate planning of crop and cropping system, preventive methods, cultural and mechanical methods, developing competitive cultivars, allelopathy, biological control and weed seeds distraction at harvest of crop will be the central thesis in sustainable weed management. Improvement in tillage regimes has long been identified as an impressive weed control measure. Harvest weed seed control and seed predation have been shown as potential tools for reducing weed emergence and seed bank reserves. Development in the field of allelopathy for weed management has led to new techniques for weed control. The remarkable role of biotechnological advancements in developing herbicide resistant crops, and harnessing the allelopathic potential of crops is also worth mentioning in a modern weed management program. Thermal weed management has also been observed as a useful technique, especially under conservation agriculture systems. Last, sustainable weed management has been elaborated with sufficient details. The role of bioherbicides as an integral part of sustainable weed management has been highlighted in a realistic manner. All these strategies are viable for today's agriculture; however, site-specific selection and the use of right combinations will be the key to success. No single strategy is perfect, and therefore an integrated approach may provide better results. The adoption of such methods may improve the efficiency of cropping systems under sustainable agriculture.

Key words: Allelopathy, Bioherbicides, Herbicide resistant crop, Objective of sustainable weed management, Weed and its effect, Weed management methods

1. INTRODUCTION

In many agricultural systems around the world, competition from weeds is one of the major factors reducing crop yield and farmers' income. In developed countries, despite the availability of high tech solutions (e.g. selective herbicides and genetically-modified herbicide resistant crops), the share of crop yield loss to weeds does not seem to reduce significantly over time (Cousens and Mortimer, 1995). In developing countries, herbicides are rarely accessible at a reasonable cost, hence farmers often need to rely on alternative methods for weed management. Worldwide limited success in weed control is probably the result of an over simplification in tackling the problem. Too much emphasis has been given to the development of weed control tactics (especially synthetic herbicides) as the solution for any weed problems, while the importance of sustainable weed management strategy has long been neglected. Sustainable weed management is based on knowledge of the biological and ecological characteristics of weeds to understand how their presence can be modulated by the different weed management practices. Based on this knowledge, the farmer must first build up a global weed management strategy within her/his crop and crop sequence, and then choose the best method for weed control during crop growing cycles. Besides this, it must be remembered that weed management is always strictly embedded in crop management itself. As such, the interactions between weed management practices must be duly taken into account.

The word weed means any wild plant that grows at an unwanted place especially in fields or in gardens where it interferes with the growth of cultivated plants. A number of definitions of weeds have been proposed, but none has come to universal satisfaction. Weeds may be defined as plants with little economic value and possessing the potential to colonize disturbed habitats or those modified by human activities. Aldrich (1984) defined weeds as *those plants that originated in a natural environment and in response to human (imposed) or natural conditions interfere with the crops and human activities*. Weeds have also been defined to be *objectionable plants interfering with the activities and the welfare of man* (WSSA, 1994). Cousens and Mortimer (1995) described weeds as those vascular plants that are natural impediments to human activities or health or are otherwise known to cause unacceptable changes to natural plant communities. Mohler (2001) defined weeds as *plants that are especially succe-*

ssful in colonizing disturbed (but potentially productive) sites and maintaining their abundance under repeated disturbed conditions. A weed is any plant not intentionally sown or propagated by the grower that requires management to prevent it from interfering with crop or livestock production (Schonbeck, 2011). Most of the definitions on weeds have one similarity that they are unwelcome guests in cultivated areas and need to be controlled or managed. Ecologically, weeds represent pioneers of secondary succession, whereas from an economic perspective they are the plants whose virtues have not been assessed fully.

2. WEED AND ITS EFFECTS

Weeds are nature's way of covering soil that has become exposed by fire, flood, landslide, windstorms, clear-cutting, clean tillage, herbicides, overgrazing, or other disturbance. Bare soil is hungry and at risk. The soil life, so vital to soil fertility, goes *hungry* because the normal influx of nourishing organic compounds from living plant roots has been cut off for the time being. The exposed soil surface is *at risk* of erosion by rain or wind, especially if root systems have also been removed or disrupted. Pioneer plants 'weeds' are those species that can rapidly cover bare soil and begin performing a number of vital ecological functions:

1. Protect the soil from erosion.
2. Replenish organic matter, feed and restore soil life.
3. Absorb, conserve, and recycle soluble nutrients that would otherwise leach away.
4. Absorb carbon dioxide from the atmosphere.
5. Restore biodiversity.
6. Provide habitat for insects and animals.

At the same time, weeds cause a number of harms in agro-ecosystems. Due to their interference with crops (both competitive and allelopathic), they reduce crop yields leading to huge losses on a global scale. About 240 weed species are reported to be allelopathic and interfere with the growth and development of crop plants (Qasem and Foy, 2001). In addition, they deteriorate crop quality, clog water ways, cause health problems in humans, and look unsightly in amenity areas, such as garden, parks, pathways, and pavements, etc. Further, the costs of weed eradication are also enormous (Sharma, 2014). Manual weed control is labour intensive and therefore limits the production area (Dubey, 2014 and Bahadur *et al.*, 2015). In many rural Indian communities, it has becomes increasingly difficult to hire labour for weeding and other farming activities, due to a swindling labour force as consequence of outmigration of the male population. As a result, farm operations are often delayed and labour cost increases (Singh *et al.*, 2012). Besides being detrimental to crop yields and unappealing, they also cause fire hazards (Zimdahl, 1999). Weeds may also harbour insects and pathogens, adding more complications to their control.

In light of these characteristics of weeds and their hazards, it becomes imperative to control them. A number of management practices are available. In the earlier times, since no synthetic chemicals were known, crop rotations, polyculture, and other management practices were tried that were low input but sustainable. With the discovery of synthetic herbicides in the early 1930s, there was a shift in the weed management practices towards high input, and target oriented ones. DNOC (4,6-dinitro-o-cresol) was the first synthetic herbicide that was patented, and later a number of synthetic growth regulators like 2,4-D (2,4-dichlorophenoxyacetic acid) and MCPA (2-methyl-4-chlorophenoxyacetic acid) completely replaced other herbicides due to their selective nature and thus became the basis of the agrochemical industry. These have enhanced crop production by minimizing competition between crops and weeds, and reduced interference risk from weeds (Das, 2008 and Sharma, 2014). As a result, agricultural practices started relying on these chemicals heavily. However, during 1980s the chemical weed control started becoming unpopular worldwide due to rising costs and various environmental and health hazards associated with their use (Singh *et al.*, 2003). Public concern over safety has put tremendous pressure to reassess toxicological and environmental impacts of synthetic chemicals and more stringent protocols are being developed (Dayan *et al.*, 1999). Fast-developing herbicide resistant ecotypes of weeds is also posing serious threat to agricultural production, and so far 392 weed biotypes resistant towards herbicides have been identified (Kewat, 2014). Even the cross resistance is fast developing in weeds, where they have become resistant to those chemicals that are tried for the first time or belong to chemically disparate classes (Kewat, 2014). Kupatt *et al.* (1993) opined that resistance or cross-resistance among weeds is a serious threat to herbicide industry and weed management practices.

From the beginning of agriculture until the introduction of herbicides, weed management in agriculture depends largely on crop rotation, tillage, seed cleaning, growing competetive cultivars and change in sowing methods. These practices some time reduce weeds population below economic injury level, but some time it will filed. Due to all these problems, efforts are being made to find out alternative strategies for weed management, which is low cost and eco-friendly. One key sustainable strategy for dealing with weeds is to minimize open niches for weeds in crops and cropping systems, while maintaining satisfactory crop yields. In annual cropping systems and less competitive perennials like asparagus and cut flowers, some open niches are unavoidable, whereas others can be eliminated through cover cropping, tighter crop rotations, closer row spacing, and intercropping. Well-managed pasture, forage, orchard, permaculture, and agroforestry systems are generally more closed, and usually require less intensive weed control efforts (Schonbeck, 2011). Sustainable weed management is needed in the wake of a huge decline in crop outputs due to weed pressure. Diversity in weed management tools ensures sustainable weed control and reduces chances of herbicide resistance development in weeds (Jabran *et al.*, 2015).

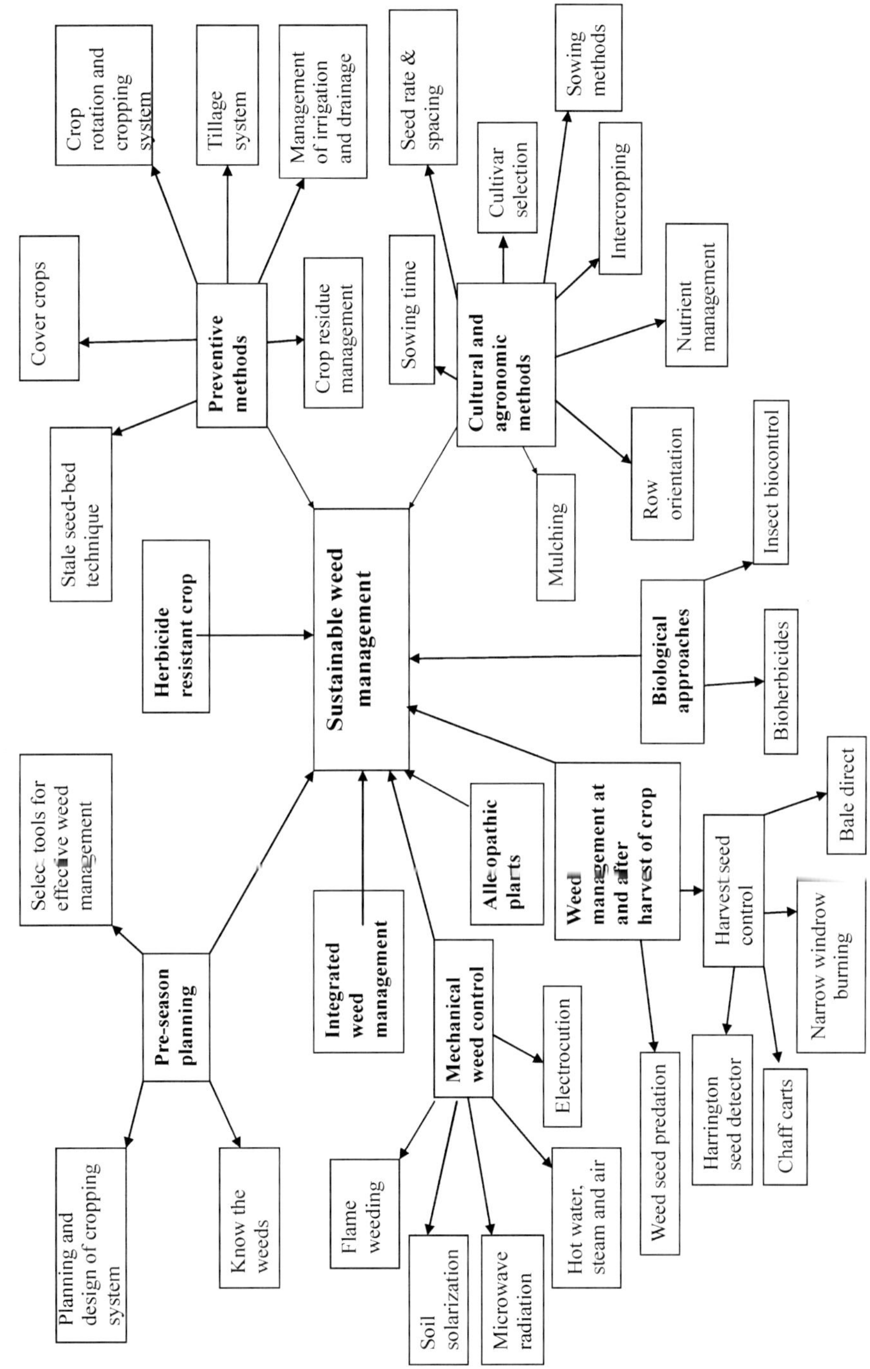

Fig. 1. Sustainable Weed Managment

A long term effective weed management strategy is based on the practical application of the ecological concept of 'maximum diversification of disturbance', which means diversifying crops and cultural practices in a given agro-ecosystem as much as possible. This results in a continuous disruption of weed ecological niches (Liebman and Davis, 2000) and hence in a minimized risk of weed flora evolution towards the presence of a limited number of highly competitive species (Verma *et al.*, 2009). Besides this, a highly diversified cropping system also reduces risk of the development of herbicide resistant weed populations. In practice, weed management strategies should involved indirect (pre-season planning and preventive) methods with direct (cultural/curative, mechanical, biological and chemical) methods. The first category includes any method used before a crop is sown, while the second includes any methods applied during a crop growing cycle. Methods in both categories can influence either weed density (*i.e.* the number of individuals per unit area) and/or weed development (biomass production and soil cover). However, while indirect methods aim mainly to reduce the numbers of plants emerging in a crop, direct methods also aim to increase crop competitive ability against weeds.

3. OBJECTIVES OF SUSTAINABLE WEED MANAGEMENT

There are several basic objectives of the sustainable weed management. The main objectives are as follows:

1. To make best use of the available resources for weed management.
2. To develop cultivation methods that manage weeds and improve soil health and to determine the impact of weed management systems.
3. To minimize use of non-renewable resources like herbicides and to use of renewable energy and recycled mineral resources.
4. To protect the health and safety of farm workers and animals, local communities and society from the application of chemicals.
5. To protect and enhance the environment and natural resources.
6. To protect the economic viability of farming operations.
7. Provide technical knowledge to the farmer to enable continued production and contribute to the well-being of the community.
8. To produce sufficient high-quality and safe food.
9. To build on available weed control technology, knowledge and skills in ways that suit local conditions and capacity.

4. SUSTAINABLE WEED MANAGEMENT PRACTICES

In the following sections, we have discussed various ways for practical implementation of sustainable weed management practices.

4.1 Pre-season planning

4.1.1 *Know the farm weeds*

Important step is to correctly identify the weed that is giving you headaches and causing your blood pressure to rise. Broadleaf weeds, grassy weeds and sedges are three different classes of weeds that are selectively controlled in different ways. Broadleaf weeds and grassy weeds are easy to tell apart, but weeds such as purple and yellow nutsedge can be a real challenge to control. "Know the weeds" is listed first because it informs most of the succeeding steps. At the same time, gaining a thorough knowledge of the farm's weed flora is an ongoing process over many seasons, and is important for fine-tuning the farm's weed management system. The more you know about the weeds in your fields, the more precisely you can design effective strategies for weed management.

Not all herbicides kill all weeds. It is for this reason that weed identification is a critical step in controlling certain weeds. This is not to say that you must identify every weed in your farm. Herbicides kill groups of weeds in many cases; however, you may notice a few weeds that are not killed. These are the weeds you must identify, because there is likely a control recommendation out there that requires a more targeted approach.

No-till planting of row crops into mowed, rolled, roll-crimped, undercut, or winterkilled cover crops, eliminates the bare soil period at the cover crop. Whereas, continuous no-till is not currently feasible in crop production, tillage can be reduced, thereby minimizing soil degradation and flushes of weed germination. The no-till cover crop strategy works best when weed populations are moderate, and the primary weeds are annuals, especially small-seeded annual broadleaf weeds, which are readily blocked by the mulch.

4.1.2 *Planning of cropping systems*

Weed prevention begins at the planning stage of any cropping system. Plan the crop rotation and cropping system to keep the soil fully occupied by desired living vegetation, or mulch as much of the season as possible. Schedule planting of horizontally orientation or cover crop as soon as possible after the preceding crop is finished. An idle soil is the weed devil's playground. For example, growing continuous corn each summer with winter fallow leaves the entire field available for weeds from harvest in early fall until crop emergence late the following spring. Between-row spaces remain open for weed growth until crop canopy closure which may take two months or more for corn. This is why continuous corn is economically feasible only for conventional producers who use synthetic herbicides and many of them now strive to save soil, money, and chemicals by planting a winter rye cover crop after corn harvest.

4.1.3 *Design the cropping system and select tools for effective weed management*

Once a crop rotation has been planned that minimizes opportunities for weed growth, the next step is to design the system to facilitate weed control throughout the season. Develop control strategies to address anticipated weeds in major crop, and select tools for pre-plant, between-row, and within-row weed removal. Plan bed layout, row spacing, and plant spacing to facilitate precision cultivation. Choose irrigation methods and other cultural practices that are compatible with planned weed control operations. Because row spacing vary as per the crop nature in a diversified rotation, matching cultivation tools and row spacing can be challenging. Use row spacing that are multiples of one another and are compatible with equipment dimensions to facilitate mechanized cultivation (Josh Volk, 2008).

4.2 Preventive methods

Prevention is the most essential aspect of weed management. Once a noxious weed infestation becomes established, any increase in size and density creates increasingly more expensive management efforts. Awareness of weed seed sources and plant identification is the must. Weed prevention embodies all measures to deny the entry and establishment of new weeds in an area not infested with it yet. This can be achieved by use of weed free crop seeds, seed certification, weed laws, and by quarantine laws. Preventive methods include stale seedbed technique, crop rotation, cover crops, tillage systems, seed bed preparation, soil solarization, management of drainage and irrigation systems, and of crop residues on the soil surface.

4.2.1 *Stale seedbed technique*

Stale seed bed technique, a preventive method with the specific aim of reducing weed emergence in the next crop cycle. Cultivation for seed bed preparation has two contrasting effects on weeds: (i) it eliminates the emerged vegetation resulting after primary tillage; and (ii) it stimulates weed seed germination and consequent seedling emergence, due to soil mixing and reallocation of seeds towards shallower soil layers. This technique involves the soil preparation of a seedbed to promote germination of weeds, a number of days or weeks before the actual sowing or planting of the crop (Johnson *et al.*, 1995). After seedbed preparation, the field is irrigated and left unsown to allow weeds to germinate and which are killed either by a non-selective herbicide or by carrying out shallow tillage prior to the sowing (Singh *et al.*, 2012). Stale seedbed can also be implemented by submergence of rice field after 7 and 14 days of weed emergence (Sindhu *et al.*, 2010). This technique reduces weeds emergence (Rao *et al.*, 2007), weed density (Kumar *et al.*, 2015), delaying early crop-weed competition and also reduces weed seeds bank in the surface soil (Bahadur *et al.*, 2015). The success of stale seedbed depends on several factors like method of seedbed preparation, method of killing emerged weeds, weed species, duration of the stale seedbed and environmental condition (Singh, 2014). Depending on location and year, stale seedbed preparations followed by weed control prior to planting, reduced the amount of weeds during crop growth by 43–83 per cent in lettuce (Riemens *et al.*, 2007) and 53 per cent lower weed density in dry DSR (Singh *et al.*, 2009) after stale seedbed than without this practice. Bed orientation can make a difference in the suppression of weeds. Weeds often emerge on the sides of the beds, because they are shaded and, therefore, soil temperatures cannot rise to detrimental levels. Beds orientated north-south had reduced weed emergence compared to east-west orientation (McGovern *et al.*, 2004).

4.2.2 Crop rotation/cropping system

Crop rotations can help in controlling weeds, supplying soil nutrients, improving soil tilth, and reducing soil erosion. The positive effect of long term rotations on crop yields has been recognized and exploited for centuries. Although in the last few decades, its benefits in terms of yield seem to have been ignored by many farmers, it is now evident that crop rotation increases yield and that the practice is essential in sustainable agricultural systems (Filizadeh *et al.*, 2007). It is an essential practice in sustainable agricultural systems, because its effects on soil health and other benefits including reduction on weed competition (Marenco *et al.*, 1999), weed growth and reproduction, which may greatly reduce weed density. Other benefit of crop rotation may be associated with a smaller chance of selecting troublesome weeds, because crop rotation sequence also determines herbicide use and crop rotation and herbicide can interact to affect weed species (Marshall *et al.*, 2003). Crop rotation alone lowers weed infestation in crop fields, while it enhances the effectiveness of weed control when combined with other methods (Garrison *et al.*, 2014 and Baker and Mohler, 2014). Crop rotation becomes more effective when no weed seeds from the neighboring land invade the field under rotation (Gonzalez-Diaz *et al.*, 2012). Crop rotation has been shown to affect the composition, density, and extent of weed communities (Filizadeh *et al.*, 2007). Koocheki *et al.* (2009) reported that seed density in the rotations of wheat-sugarbeet was lower than under continuous wheat. In rotation of sugar beet-winter wheat, the seed density in soil was lower than in other rotations and this rotation caused 28 per cent reduction in the weed seed bank. Rotation of maize-winter wheat caused 12 per cent reduction in the seed bank compared with continuous wheat. Therefore, broadleaf crops in rotation with narrow-leaf crops and also crops with different agronomic practices can reduce the weed seed bank density. According to (Teasdale and Mohlet, 2000), growing of wheat, maize and soybeans in rotation tends to decrease the weed seed bank and abundance of broadleaf weeds.

Diversified crop rotations tend to reduce the development of a few primary weed species by offering different sowing and harvest times, different life cycles and different possibilities for weed control (Feizabady, 2013). For example, giant foxtail (*Setaria faberi* Herrm.) seed numbers increased in continuous maize compared to rotations of maize/soybean or maize/soybean/wheat. On the other hand, continuous cropping has led to increased weed pressure, weeding frequency, weed-related crop losses and low soil fertility, especially in farming systems where external inputs are not used (Akobundu and Ekeleme, 2002; and Kanmegne and Degrande, 2002).

Manipulation of cropping systems for the purpose of improving integrated weed management requires a good understanding of weed dynamics and influences of crop and soil-related factors on weed life cycles (Davis and Liebman, 2003). Weed flora have changed over the past century, with either increasing or decreasing species abundance depending on the management (Marshall *et al.*, 2003). Weed seed banks may reflect the status of weed population in the present and the past, and could be regarded as an indicator of the impact of soil and crop management (Buhler *et al.*, 2001). The shift from rice-wheat to rice-potato, rice-potato-wheat sequence or any other sequence reduced the population (Verma and Singh, 2008) and soil seed bank of *Phalaris minor* (Singh and Singh, 2006) and other weeds. Singh *et al.* (2012) studied that when rice-wheat cropping system is changed, there is reduction in weed density and weed dry matter production. Rice-wheat-greengram sequence recorded lowest population of all the three groups of weeds followed by rice-wheat, rice-chickpea and rice-pea sequence. But the success of crop rotation for weed suppression appears to be based on the use of crop sequences that create varying patterns of resource competition, allelopathic interference, soil disturbance and mechanical damage to provide an unstable and frequently inhospitable environment that prevents the proliferation of a particular weed species (Chamanabad *et al.,* 2009).

4.2.3 Cover crops

Cover crops are grown with the aim of maintaining the sustainability of an agro-ecosystem. It can prevent the development of weed population, control the soil disease, soil enrichment through nitrogen fixation in soil, improve soil structure, preventing absorption of nitrogen, increase the soil organic matter and decrease the soil, for these reasons they are cultivated (Kruidhof *et al.*, 2008 and Yeganehpoor *et al.*, 2015). Cover crops with allelopathic potential can suppress weeds. Several of the important cover crops include canola, rape seed, cereal rye, crimson clover, wheat, red clover, brown mustard, oats, cowpea, fodder radish, annual ryegrass, mustards, buckwheat, hairy vetch, and black mustard. The observations from farmers' fields and the results of experiments indicated that the release of allelochemicals from allelopathic cover crops and their physical effects were responsible for weed suppression in conservation organic farm fields (Mirsky *et al.*, 2010). Further, cover crops also possess several additional benefits other than weed management. For example, the results of a recent study indicated that along with suppressing weeds, the cover crops also improved soil moisture retention, soil fertility, and crop productivity (Altieri *et al.*, 2011). Mixtures of cover crops have been found more effective in suppressing weeds compared to a single cover crop. Haramoto and Gallandt (2004) explored the role of *Brassica* cover crops including white mustard and rape seed for weed suppression in agricultural systems. Bernstein *et al.* (2014) tested the efficacy of a rye cover crop to suppress weeds for planting soybean under a no-till system and concluded that soybean could be successfully sown in a standing rye cover crop in a no-till soil. Planting soybean in a standing rye crop resulted in long-lasting and effective weed control with no damage caused to the soybean crop. Moreover, cover crops can also reduce the weed seed bank in conservation till systems. For example, the cover crops hairy vetch and oat effectively reduced seed banks (30–70%) of weeds, including *Datura stramonium* L., *Digitaria sanguinalis* (L.) Scop., *Amaranthus retroflexus* L., and *E. indica* in the upper soil layer (Duby *et al.*, 2012).

Table 1. Suppression of weeds using cover crop and inter-crops

Cover crop	Main crop	Weed suppressed	Reference
Wheat	Cotton	*Eleusine indica, Amranthus* sp, *Ipomoea lacunosa*	Norsworthy *et al.* (2005)
Rye	Cotton	*E. indica, A. palmeri, I. lacunosa*	Norsworthy *et al.* (2011)
Rye	Soybean	*C. album, Abutilin theophrasti*	Bernstein *et al.* (2014)
Annual rye grass, rye, oat common vetch, radish	Common bean, tomato	*Brachiaria plantaginea* (Link) Hitchc., *Ipomoea grandifolia* (Dammer) O'Donell, *Bidens pilosa* L., *Euphorbia heterophylla* L.	Altieri *et al.* (2011)
Hairy vetch, oat	Maize	*D. sanguinalis, E. indica, A. retroflexus, Datura stramonium* L.	Dube *et al.* (2012)
Sorghum sudan-grass [*Sorghum bicolour* (L.) Moench *Sorgum sudanense* (Piper) Staph.]	Broccoli	Broad leaved weeds	Finney *et al.* (2009)
Bristle oat, hairy vetch	Cotton	*A. palmeri* L., *P. oleracea* L., *Helianthus annuus* L.	Moran and Greenberg, (2008)
Rye, hairy vetch, barley triticale, Austrian winter pea	Organically grown maize-soybean	*C. album* L., *Amaranthus hybridus* L., *Thlaspi arvense* L., *Taraxacum officinale* L., *Stellaria media* L., *Elymus repens* L. *Panicum crusgalli* L., *Setaria glauca* L.	Moran and Greenberg, (2008), Silva (2014)
White mustard	Olive groves (*Olea europaea* L.)	*Amaranthus blitoides* L., *C. album*	Alcantara *et al.* (2011)

Cover crop	Main crop	Weed suppressed	Reference
Hairy vetch, subterranean clover, oat/ hairy vetch	Tomato	*A. retroflexus* L.	Campiglia *et al.* (2010)
Sorghum	Cotton	*Cyperus rotundus*	Iqbal *et al.* (2007)
Soybean	Cotton	*Cyperus rotundus*	Iqbal *et al.* (2007)
Sesame	Cotton	*Cyperus rotundus*	Iqbal *et al.* (2007)
Barley	Soybean	Southern crab grass, barnyard grass	Farooq *et al.* (2011)

Table 2. Impact of cover crop residue and live cover crop on inhibition of weeds at various life cycle stages

Weed life cycle stage	Cover crop residue	Live cover crop
Germination	Moderate	High
Emergence/estabilishment	Moderate	High
Growth	Low	High
Seed production	Low	Moderate
Seed Survival	None	Moderate
Perennial structure survival	None	Low-moderate

Graglia *et al.* (2006)

When cover crops are used as dead mulch (*i.e.* they are left to decompose on soil surface), weed suppression seems mostly to be the result of the physical effects of mulch, rather than to nutrient or allelochemical-mediated effects (Teasdale and Mohler, 2000). In particular, weed suppression seems directly related to the mulch area index (mulch area divided by soil unit area), which influences light extinction through the mulch and consequently weed seed germination. Small-seeded weed species appear to be more sensitive than large-seeded species to mulch, physical effects as well as to allelochemicals (Liebman and Davis, 2000). Timely sowing of cover crops is very important to enhance biomass production and hence to increase their weed suppression potential. Cover crops can also interact with other biota; for example, they promote the establishment of vesicular-arbuscular mychorrhizae, which in turn may shift weed flora composition by favouring mycorrhizal plant species at the detriment of non-mycorrhizal species (Jordan *et al.*, 2000).

4.2.4 *Tillage systems*

Tillage influences weed life cycle processes by directly destroying seedlings, redistributing seeds vertically in the soil profile, and altering soil properties that influence seed persistence, dormancy, germination, and seedling survival

(Peigne, 2007). Therefore, shifts in weed community population dynamics frequently occur when any type of conservation tillage is adopted (Cardina *et al.*, 1991), including zero tillage (ZT). Understanding tillage effects on weed community dynamics can be challenging because the effects are variable and depend on interactions with other management tactics, environmental conditions, and weed biology (Derksen *et al.*, 2002). Diverse crop rotations indirectly impact weed population through tillage effects, because tillage timing and frequency varies with crop species (Smith, 2006). Changes in tillage systems may indirectly impact weed populations via altering processes that degrade or destroy weed seeds. Weed seeds remaining on or near the soil surface are subject to destruction via desiccation and herbivore or pathogen attack (Murphy *et al.*, 2006). Soil fungal and bacterial microorganisms have been shown to be more abundant in ZT *vs.* intensively tilled systems (Helgason *et al.*, 2009), but weed seed response to surface colonization by microorganisms is highly dependent on species (Chee-Sanford *et al.*, 2006). Therefore, weed seed decay may be enhanced by ZT, but the effect will probably vary greatly according to weed species (Carr *et al.*, 2013).

The effect of primary tillage on weeds is mainly related to the type of implement used and to tillage depth. These factors considerably influence weed seed and their propagules distribution over the soil profile and therefore they directly affect the number of weeds that can emerge in a field (Han *et al.*, 2013). Conservation tillage systems are frequently associated with weed infestations and shifts in weed communities, sometimes with the benefit of more difficult to control species (Legere and Samson, 2004). Conservation tillage (low disturbance) leaves more weed seeds on the surface, whereas high disturbance systems bury weeds. Weed seeds left on the surface are generally more susceptible to decay and ultimately reduce weeds seed banks (Chauhan *et al.*, 2006), it allowed early sowing and thus the competitive advantage remains in favour of crop not for weeds (Mishra *et al.*, 2010 and Sharma, 2014), lower emergence in conservation tillage might be due to association with higher soil strength (Dev *et al.*, 2013).

The lower degree of soil disturbance under chisel ploughing, as compared with moldboard ploughing, generally results in an increase in weed abundance in arable cropping systems (Gruber and Claupein, 2009), particularly those of perennials and annual monocotyledons (Torrensen *et al.*, 2003). For dicotyledonous species, the impact of tillage systems depends on the species (Moonen and Barberi, 2004). Mouldboard ploughing is very effective in reducing weed density and hence it is an important preventive method where farmers are forced (or are willing) to use partially suppressive direct weed control methods (mechanical weeding), and reduces the labour needed for subsequent hand-weeding (Sans *et al.*, 2011). Tillage systems may influence weed seed bank size and composition to a much greater extent than crop rotation (Mirsky *et al.*, 2010).

If the weeds have become too large, an intensive and aggressive adjustment of the implements is necessary to control the weeds, and by doing this one increases the risk of damaging the crop severely (Carter and Ivany, 2006). Stopping tillage practices has a positive impact on weed populations, because it can influence the weed seed viability and distribution, and it has a strong impact on weed emergence by burying weeds in the soil (Vasileiadis *et al.*, 2006). Continuous zero tillage reduced the total weed seed bank over continuous conventional tillage (Mishra and Singh, 2008) in soybean-linseed cropping system.

4.2.5 *Management of drainage and irrigation systems*

Careful choice and maintenance of drainage and irrigation systems are important preventive measures to reduce on-field weed infestation. Periodical clearance of weed vegetation established along ditches prevents it from invading the field. Where it is economically feasible, substitution of ditches with subterranean drains eliminates a potential source of weed infestation. Use of localized (drip) irrigation systems favour crop development to the detriment of weeds (Peter *et al.*, 2012). In contrast, broadcast irrigation systems often favour weeds because most of them have lower water use efficiency (dry biomass production per unit water used for evapotranspiration) than the crop. Weed dry weight was significantly influenced by weed management practices and the amount of water available (Towa and Xiangping, 2014).

Optimum time and number of irrigation reduces the density and weight of weeds (Das and Yaduraju, 2007 and Verma, 2014). Singh and Singh (2004) reported that pre-sowing irrigation reduced the dry weight of *C. album* and *C. murale* by 21 and 25 per cent, respectively, and subsequently grain yield was 12 per cent higher over post sowing irrigation. Wheat irrigated at CRI+tillering+ flowering stage reduced the dry weight of *Phalaris minor* over crop irrigated at RI+tillering+flowering+ dough, CRI+tillering, CRI+flowering and at CRI stage, respectively (Das and Yaduraju, 2007). Irrigation at 0.4 IW: CPE in Isabgul (Parmar *et al.*, 2010), 1.25 IW: CPE in wheat (Nadeem *et al.*, 2010) and 0.6 IW: CPE ratio in fenugreek (Mehta *et al.*, 2010) resulted lower weed population and higher yield over 0.8 and 1 IW: CPE. Dropping method caused sensible decrease of the amount of dry weight of weeds in comparison with surface water method (Ghanbari, 2013).

4.2.6 *Crop residue management*

High and sustainable crop production is linked to improved soil physical, chemical and biological properties, which they are a primary function of soil organic matter. Crop residues are a source of organic C for soil microorganisms, and also contribute to plant nutrients. Crop residue retention on the soil surface, substantially reduces runoff and soil erosion, and can decrease weed emergence (Dahal *et al.*, 2014), soil evaporation and land preparation costs. Residue retention can also, over the long term, improve soil structure (Singh and Malhi, 2006) and water-holding capacity, as well as improve long-term nutrient cycling.

These desirable outcomes of residue retention, some of which take several years to become evident, have been reported extensively for other mainly temperate environments (Feizabady, 2013).

Crop residues acting as mulches can influence weed seedling emergence and weed biomass. Globally there is mounting evidence that retention of crop residues from one season to the next suppresses the germination and development of weeds in minimum tillage systems, thus enhancing system productivity (Mashingaidze *et al.*, 2009). Residue retention has significantly influenced weed emergence (Sharma, 2014 and Bahadur *et al.*, 2015), although several interacting factors may determine the extent of this influence including residue nature, height, type and quantity, prevailing weed flora, soil type and weather conditions (Khankhane *et al.*, 2009). Residue mulching as a practical method for early season weed control in minimum tillage systems for smallholder farmers (Chauhan and Abugho, 2013). This is because surface application of residue 5 to 7 t/ha significantly suppressed weed growth and development as compared to incorporation and no-residue retention (Brar and Walia, 2010). In rice-wheat system, majority of the farmers (65%) were burning rice straw in north-western Indian plains (Chhokar *et al.*, 2007). Although seeds of many weed species can be devitalized by stubble burning, this technique is always to discourage because of its negative effect on soil organic matter content. For sustainability of rice-wheat system, instead of burning, residue should be either retained on surface or incorporated. Because residue retention on soil surface significantly contribute to the suppression of weeds (Liebman and Mohler, 2001 and Sharma *et al.*, 2008). (Chauhan and Abugho, 2013) reported the use of residue at high rates (6 t/ha) can help suppress seedling emergence (50%) and growth of many weeds. However, there is a need to integrate other weed management strategies with residue retention to achieve season-long weed control.

4.3 Cultural/agronomic practices

Cultural practices alter patterns of crop growth and can modify dynamics of weed-crop competition, and hence need to be investigated to evolve sustainable weed management.

4.3.1 *Crop sowing time*

Among the factors influencing weed dynamics and weed-crop interference, sowing date is of immense importance (Matloob *et al.*, 2014). Soil preparation at different sowing time inflicts a seasonal variation in temperature and moisture regimes, the two principle determinants of species survival and distribution, there by governing periodicity of weed seed emergence (Berzsenyi, 2000). Composition of weed communities are known to differ in mid and late season planted crops (Milberg *et al.*, 2001). The timing of soil disturbance will govern which weed seeds in what state of dormancy are available for germination; thereby, regulating germination and subsequent weed growth (Matloob *et al.*, 2014). Seed availability as well as their dormancy status is influenced by environmental attributes that may contrast within the same growing season. There is a relation between the timing of weed emergence and the pressure exerted to the crop through competition and resulting loss in crop yield (Khaliq and Matloob, 2011). Yield losses are usually higher when weeds emerge earlier or at the same time as the crop (Verma and Singh, 2008). Infield trials of direct-seeded irrigated rice, 95% of the weed-free rice yield was obtained when weeds were controlled up to 32 days after sowing (DAS) in the wet season and 83 DAS in the dry season (Johnson *et al.*, 2004). Early sown wheat (Last week of October) reduces *P. minor* infestation compared to late sown. In early sown wheat temperature is not optimum for *P. minor* germination. However, it is important not to deviate wheat seeding much from optimum planting time otherwise yields will be reduced (Chhokar *et al.*, 2012). Spandl *et al.* (1998) observed that, compared to autumn-sown wheat, control of *Setaria viridis* in the spring-sown cereal was favoured because the weed emerged in a single flush instead of several flushes, thus being more vulnerable to direct weed control methods (herbicides or cultivation). Nevertheless, infield crops, a yield penalty, often proportional to the duration of delayed period, are also observed when the planting is delayed beyond the optimum time (Matloob *et al.*, 2014). The tradeoff between weed suppression and crop yield reduction should be considered while using planting time as a tool for weed management.

4.3.2 *Seeding rate, row spacing and plant population*

Seed rate has also been suggested to be a prominent cultural strategy which could influence the ability of a crop to compete with weeds for resources and ultimately affect weed management (Arce *et al.*, 2009). The increased seed rate resulted in a higher crop plant population providing less space for weeds to grow and offering much higher competition for light, nutrient and other growth factors (Abdolraheem and Saeedipour, 2015). These factors collectively increased competitive ability of crop plant with weeds resulted in lower weed density (Daniya *et al.*, 2015). In this regard, weed dry matter and weed cover decreased with an increase in seed rate from 3 to 12 kg ha^{-1} in sesame (Imoloame *et al.* 2007). Olsen *et al.*, 2012 reported that increasing crop density reduces biomass and seed production in wild oat. Weed density and weed dry weight were significantly affected by seed rate; these two variables decreased with the increasing seed rate in the soybean (Place *et al.*, 2009 and Bell *et al.*, 2015) in wheat (Babaei and Saeedipour, 2015) resulted in higher yield and economic returns. Olsen *et al.* (2012) indicate that increased crop density in cereals can play an important role in increasing the crop competitive advantage over weeds. Kristensen *et al.* (2008) indicate that increased crop density was of negative effects on weed biomass and positive effects on crop biomass and yield.

A higher seeding rate may keep the weed flora under check through a smothering effect (Mahajan *et al.*, 2010). Mohler (1996) revealed that a higher seeding rate may provide a competitive advantage to crop over weeds because crop plants will absorb limited resources at a faster rate. However, an increased seeding rate may not always increase the weed competitiveness of a crop, and greater intra-crop competition may arise. This may leads to negative effects on crop production, especially under stressful environmental conditions (Krikland *et al.*, 2000). Therefore, an optimal seed rate, along with some weed control, is frequently practiced.

Crops can be favoured in competition against weeds by use of narrow rows and/or higher population. Use of narrow rows and/or higher population hastens the rapidity of closure of the canopy and enhances canopy radiation interception, thereby increasing crop growth rates, yield (Andrade *et al.*, 2002), suppressing weed growth and competitiveness (Mashingaidze, 2004); whereas, wide rows reduces early-season crop tolerance to weeds requiring earlier weed management programs than in narrower rows (Stevan *et al.*, 2003). Therefore, the use of narrow rows and/or higher population could be used by farmers as means of weed control through achieving full ground cover in the earlier season, thereby reducing the impact of weeds (Tracy *et al.*, 2004). Reduction in weed biomass with narrow rows is most likely a result of quicker and complete canopy cover with the narrow spacing, thereby depriving the weeds of photosynthetically active radiation (Fanadzo *et al.*, 2010). One theory for reduced weed growth in narrow rows is quicker row closure, which reduces light penetration to the weeds emerging below the crop canopy (Alford *et al.*, 2004). Weed growth suppression by narrow rows is mainly due to increased shading of the inter-row rather than the in-row (Mashingaidze, 2004). The higher density and closer spacing smother weeds due to better early canopy coverage (Singh and Singh, 2006).

4.3.3 *Cultivar selection*

Field crops vary widely in their ability to compete with weeds. Even within one species, different cultivars may have very different competitive abilities. In general, the competitive ability of a crop is related to its ability to access resources including light, water and nutrients. It may be discussed as the ability of a crop to tolerate weed pressure and maintain yield, or the ability of a crop to suppress weed growth and seed production. It is not always clear which plant traits make for a competitive crop; in most cases, it is probably a combination of traits that come into play. Generally speaking, crops with vigorous growth that reduce the quality and quantity of light beneath the crop canopy are the most competitive (Buhler, 2002). Specific characteristics that tend to influence a crop's competitive ability include: faster seedling emergence, canopy establishment, early fast growth, leaf characteristics (leaf area index, flag leaf length and angle), canopy structure (ability to intercept light) and height have better competitive ability against weeds (Mahmood *et al.*, 2013; Mahajan and Chauhan, 2013;Verma *et al.*, 2015 and Bhadur *et al.*, 2015). The weed suppressive cultivars can reduce weed infestation without incurring any extra cost, easy and environment friendly weed control, and would help to improve the efficacy of inputs and the method of weed control (Fragasso *et al.*, 2013and Jabran *et al.*, 2015).

Crop varieties can vary considerably in their ability to compete with weeds. Although specific varieties are not developed with competitiveness in mind, certain varieties do seem to compete better with weeds than others. Cultivars may also perform differently in different regions and growing conditions, so that the most competitive cultivars in one case may be less competitive in another case (Blackshaw *et al.*, 2002). There is some evidence that certain varieties, often older ones, perform better under high-stress environments such as drought and low soil fertility (Mason and Spaner, 2006). It is also important to remember that the most competitive cultivars are not always the highest yielding cultivars. All of these factors may influence the choice of crop cultivar for herbicide use reduction. While taller cereal varieties are more competitive with weeds than shorter or semi-dwarf ones (Watson *et al.*, 2006). Hybrid canola varieties tend to have more vigorous early-season growth than open-pollinated ones, resulting in better competition with weeds. Field pea cultivars that have long vines and rapid canopy development compete better with weeds than other cultivars (Blackshaw *et al.*, 2002; Nazarko *et al.*, 2005 and Mason and Spaner, 2006)

4.3.4 *Sowing/planting methods*

Over several years, the planting pattern/geometry of the crop has been modified to increase yield, because these methods significantly influenced weed dynamics (Dev *et al.*, 2013). Line sowing/drilling has replaced broadcasting to a large extent. Paired-row and criss-cross sowing are employed in certain cropping situations. Recently, a missing-row sowing method in which one drill row is left unsown after every four drill-rows sown with wheat in flat beds has been found to be superior to conventional sowing of wheat in continuously-drilled rows (Das and Yaduraju, 2011). The former method can be an alternative to FIRBS; it alters wheat plant density, spatial uniformity and competitiveness, providing gaps/spaces between the continuously-drilled rows of wheat (Das and Yaduraju 2012 and Jat, *et al.*, 2013) over conventional tillage and strip till drill system, in maize (Chopra and Angiras, 2008) over conventional tillage and flat bed system and in lentil (Manjunath *et al.*, 2010) over flat sowing. This is because of avoidance of wetting of whole cropped soil surface in bed sowing and the weed did not find congenial moisture conditions at the surface to germinate (Sharma, 2014). In zero till seeding by Happy Seeder machine with stubble mulching, undisturbed inter row space, where seeds lying at lower depths did not germinate (Bhullar *et al.*, 2006) and it saves time and energy. BBF method of sowing provides favourable environment for the growth and development of crop and reducing weed population over flat bed and ridge furrow methods. Bidirectional sowing in wheat gives fewer weeds compared to unidirectional sowing, although seed rate is same (Singh *et al.*, 2012). Transplanting under puddle condition had given detrimental impact on weed growth and resulted lowest producer of weed dry weight over direct sowing with zero till drill under unpuddled wet seed bed, direct drum seeding of pre-germinated seeds under puddle conditions, unpuddled transplanting, SRI (Hassan *et al.*, 2010), whereas, drum seeding + green manure significantly reduced weed density in direct seeded rice over drum seeding alone and broad casting (Sangeetha *et al.*, 2009).

4.3.5 *Intercropping*

Compatible crops are grown together in order to harvest higher net yield and economic benefits. Further, growing crops in mixtures improves resource (land, water, nutrients, and light) use efficiency. In addition to these benefits, intercropping can be used to suppress weeds for environment friendly and economical weed control (Yadollahi *et al.*, 2014). In particular, crops with allelopathic potential when intercropped with other crop plants help to reduce weed intensity, and hence improve crop productivity. For instance, intercropping maize and cowpea on alternate ridges helped to reduce weed [*Echinochloa colona* (L.) Link., *Portulaca oleracea* L., *Chorchorus olitorius* L., and *Dactyloctenium aegyptium* (L.) Willd.] intensity by 50 per cent as well as improves land use efficiency (Saudy, 2015). In another study, the relay-intercropping of legumes with wheat was evaluated for weed suppression in comparison with the sole wheat crop (Amosse *et al.*, 2013). The intercrops in the experiment included white clover, black medic, alfalfa, and red clover. The intercrops not only helped to suppress weeds compared with the sole wheat crop, but also reduced weed density in the following crop, while red clover was the most effective intercrop for suppressing weeds in wheat. Fernandez-Aparicio *et al.* (2010) reported that intercropping in berseem with legumes (broad bean and pea) reduced the intensity of *Orobanche crenata* Forssk. Kandhro *et al.* (2014) evaluated intercropping of two allelopathic crops (sorghum and sunflower) for weed management in cotton. Both intercrops suppressed weeds in cotton by 60 to 62 per cent, which resulted in a 17 to 22 per cent increase in seed cotton yield. Sorghum and sunflower were also harvested for grains, which resulted in improved crop productivity, land utilization, and economic benefits. Barley was intercropped with peas (main crop) and compared with the sole pea crop in terms of weed suppression. *Chenopodium album* L. and *Sinapis arvensis* L. were the two dominant weed species in the experimental sites. Pea-barley intercropping reduced weed intensity and weed biomass compared with the sole pea or fallow plots (Corre-Hellou *et al.*, 2011).

4.3.6 *Living mulch*

In modern agriculture, chemicals have become the most frequently used weed control strategy. However, environmental and economic costs, as well as increased weed resistance to herbicides have led to a desire for less herbicide use on farms (Yousefi and Rahimi, 2014). One environmentally sound way to decrease herbicide use is living mulch as a sustainable cultural weed control method. Such systems consist of a species pre-sown or intercropped with a main crop in order to serve as a soil cover, which contributes towards weed control

(Pouryousef *et al*., 2015). Living mulches are cover crops that are planted between the rows of a main crop and are maintained as a living ground cover throughout the growing season of the main crop, suppressing weed establishment and growth, and thereby reducing the number of weeds (Wang *et al*., 2006). It consists of one or more low growing ground cover species for example, low growing legumes such as greengram, blackgram, cowpea, *Sesbania*, white dutch clover; dwarf perennial ryegrass, and creeping (cucurbits) plants maintained between crop rows or beds. The goal of it is to reduce soil erosion (Arentoft *et al*., 2013), increase soil organic matter and nitrogen, improve water filtration, lower weed pressure (Hartwig and Ammon, 2002), and reduce weed seed bank (Gibson *et al*., 2011); while having minimal impact on crop yield (Talebbeigi and Ghadiri, 2012.). Some time living mulch reduces crop yield (Hiltbrunner *et al*., 2007), this yield reduction is very likely caused by the competition for light, water and nutrients between the living mulch and main crop (Thorsted *et al*., 2006). Plant density has been shown to be a main factor which changes the outcome of competition in plant community (Yousefi *et al*., 2012). In some cases, increased plant density leads to quicker canopy closure, increased crop interference and greater weed suppression, resulting in increased yields (Kolb *et al*., 2012 and Jamshidi *et al*., 2013).

Living mulches suppress weeds by competing for the use of growth resources, and changing environmental factors that affect weed germination and establishment, and can ultimately result in reduced herbicide application (Jamshidi *et al*., 2013). They may also suppress weeds through allelopathy, whereby alkaloids are released from both the roots and leaves of living plants; for example, it has been found that weed growth may be suppressed by the secretion of allelochemicals from living rice plants (Olofsdotter *et al*., 1999). But its efficacy depends most on soil coverage (> 50%), with light interception being the most important effect (Steinmaus *et al*., 2008). Maximum efficiency of living mulch is achieved by a rapid occupation of the open space between the main crop rows, preventing weed seed germination and reducing weed seedling growth and development. Weed seed germination may be inhibited by either complete light interception due to cover crop or allelochemical secretion. After weed seedling establishment, resource competition becomes the (cover crop's) main weed suppressing mechanism of living mulch (Hollander *et al*., 2007).

4.3.7 *Row orientation*

Manipulation of row orientation is an ideal method to incorporate into an sustainable weed management program because it does not cost growers anything to implement, and it is environmental friendly compared with chemical weed control tactics (Mohler, 2001). During early growth stages, there is interference between crop and weed plants because of reflected light. The reflection of red photons by the stem of one plant lowers the red photon ratio of light experienced by the stems of neighbouring plants. This modifies the light environment in the plant stem tissue, which results in an increased stem elongation rate. As plants age, the crop canopy closes, and mutual shading further increases the competition for photosynthetic light (Alam *et al.*, 2015). Shaded leaves lower in the canopy have access to low levels of photosynthetically active radiation and a low-red to far-red photon ratio. Crops can be manipulated to increase shading of weeds by the crop canopy, to suppress weed growth, and to maximize crop yield. One possible way to reduce light interception by weeds and to increase light interception by the crop canopy is to manipulate the crop row orientation (Drews *et al.*, 2009). Reducing the space between crop rows or orientating crop rows at a near right angle to the sunlight direction increases the shading of weeds between the rows. The growth of poison ryegrass (*Lolium temulentum* L.), little seed canary grass (*Phalaris minor* Retz.), wild oat (*Avena fatua* L.), and common vetch (*Vicia sativa* L.) in wheat crops and black nightshade (*Solanum nigrum* L.) in vineyards were influenced by crop row orientation (Angiras and Sharma 1996; Sharma and Angiras, 1996 and Shrestha and Fidelibus, 2005). Borger *et al.* (2010) reported that wheat and barley crop oriented east–west, weed biomass was reduced by 51 and 37 per cent, and grain yield increased by 24 and 26 per cent compared with crops oriented north–south in Australia. This reduction in weed biomass and increase in crop yield likely resulted from the increased photosynthetically active radiation interception by crops oriented at east–west direction.

4.3.8 *Fertilizer application*

Adequate availability of essential nutrients needed for all crops, but plants differ in the way they respond to nutrient availability mainly because of differences in their root structures or growth stage. The role of nutrient management through fertilizer application in crop production is substantial and very clear. Weeds take up significant amounts of nutrients, just like crops. But the comparative effects on weed growth, population, distribution, and proliferation are generally ignored. Fertilizer application strategy to crop play important role in reducing weed population in crop as weeds are always strong competitor of crops for available resources (Blackshaw *et al.*, 2005 and Mtambanengwe *et al.*, 2015). It is important to develop fertilization strategies for crop production that enhance the competitive ability of crop, minimize weed competition, and reduce the risk of nonpoint source pollution (Cathcart and Swanton, 2003). Significant research in this area has shown that there exists a strong relationship between nutrient management and weed behaviour and management. Weed growth, development, distribution, dynamics, persistence, emergence, and competitiveness were highly affected by fertilizer application (Bajwa *et al.*, 2014). Proper crop nutrient management can play a pivotal role in weed management. Abouziena *et al.* (2008) reported that increasing amount of nitrogen can cause the enhancement of plant performance, but weed may have a negative effect on performance. Nasrollahzadeh *et al.* (2015) observed that biomass reduces harvesting index and final corn performance under low nitrogen condition in contrast to high nitrogen and this is a short time after greening weed and its intervention with corn. Different weed species show a variable response to nutrient management. The dynamics of Persian darnel (*Lolium persicum* Boiss. & Hohen. ex Boiss.), wild oat (*Avena fatua* L.), and spineless Russian thistle (*Salsola collina* Benth.) were not affected by N fertilization, whereas redroot pigweed (*Amaranthus retroflexus* L.) was significantly affected (Blackshaw and Brandt, 2008). The possible interactions might be due to the effect of fertilizer on weed crop competition (Evans *et al.*, 2003). Nutrient availability may alter the weed crop competition duration. In a study, the application of N fertilizer changed the emergence pattern, density, and competitive ability of different weeds (Sweeney *et al.*, 2008). Increased supply of nutrients over a period of time may reduce weed density but increase total weed biomass (Mohammaddoust-e-Chamanadad *et al.*, 2006). Variable weed responses to fertility suggest that weeds can be controlled (suppress the emergence and persistence) through regulating fertilizer management. Varying fertilizer doses, application timings, and methods can modify weed crop competition (Blackshaw *et al.*, 2004; Sweeney *et al.*, 2008 and Arif *et al.*, 2013). The nature of fertilizers may affect weed biology and ecology. Yin *et al.* (2005) reported that the abundance of weeds was highly variable in cropped field because of variation in the nutrients application source. Toler *et al.* (2004) observed that normally weeds respond positively to the starter fertilizer dose and grow well. It is suggested that a specific amount of fertilizer can provide better crop growth, but an over or under application may facilitate the competing weeds, resulting in yield losses (Major *et al.*, 2005). Shifting the N application from the spring season to the fall season reduced the density and biomass of four noxious weeds, including wild oat, green foxtail (*Setaria viridis* (L.), wild mustard (*Sinapis arvensis* L.), and common lambsquarters (Blackshaw *et al.*, 2004). Therefore, proper consideration must be given to fertilizer type, dose, and application timing when devising weed management strategies. Above all, the role of nutrient placement in weed management is crucial. Most of the weed seeds are present near the soil surface and fertilizer application in that zone may promote their emergence and subsequent growth as well (Guza *et al.*, 2008). Blackshaw *et al.* (2004) reported up to 68 per cent weed reduction in cases where N was injected rather than broadcast. Surface banding of N and P reduced weed pressure because of less availability to weeds as compared with broadcasting (Blackshaw, 2005). Significant reductions in the shoot biomass of wild oat and green foxtail were observed when N fertilizer was applied through banding and injection rather than broadcasting (Blackshaw *et al.*, 2004). Subsurface application of fertilizer in dry direct seeded rice significantly reduces weed biomass reported a by Chauhan and Abugho (2013).

The timing of N fertilizer application in early planted crops, such as sugar beet and corn, may especially influence the germination, emergence, and competitiveness of weeds that might otherwise remain dormant early in the growing season (Sweeney *et al.*, 2008). Pre-sowing N fertilization can increase crop competitive ability against weeds in crops having high growth rates at early stages, but this effect is modulated by the type of weeds prevailing in a field. For example, in sunflower grown in Mediterranean conditions, a pre-sowing application of synthetic N fertilizer increased the suppression of late-emerging weeds like *Chenopodium album*, *Solanum nigrum* and *Xanthium strumarium* compared to split application (Paolini *et al*., 1998). In contrast, the same technique resulted in a competitive advantage for early-emerging weeds like *Sinapis arvensis*. Similarly, anticipation or delay of top-dressing N application in sugar beet increased crop competitive ability with dominance of late- or early-emerging weeds, respectively (Paolini *et al*., 1999). Modulation of crop-weed competitive interactions through crop fertilization is unlikely to be feasible when organic fertilizers or amendments (e.g. manure) are used, because of the slow release of nutrients from these sources. However, application of fertilizers (either synthetic or organic) along with, or in close proximity to the crop row, can improve weed management because it increases the relative chances of the crop to capture nutrients (especially N) to the detriment of weeds (Rasmussen, 2000).

4.3.9 *Biofertilizer*

Azolla has multiple uses, such as biological herbicide, animal feed, and as a water purifier (Watanabe and Hove, 1996). Weed growth is suppressed when Azolla forms a thick, virtually light-proof mat. There are probably two mechanisms for this suppression, the most effective being the light-starvation of young weed seedlings by the blockage of sunlight (Lumpkin and Plucknett, 1980). The other mechanism is the physical resistance to weed seedling emergence created by a heavy, interlocking Azolla mat, which does not affect the growth of rice (Pons, 1987). Bangun and Syam (1988) showed that an Azolla cover could significantly reduce weed infestation without harming the rice yield. Biswas *et al*. (2005) reported the full area coverage by the *Azolla pinnata* mat and the highest biomass production with superphosphate and cow manure treated plots were able to completely inhibit two weed species (*Scirpus juncoides* Roxb and *Monochoria vaginalis* Burm, and significantly suppressed the density of *Cyperus serotinus* Rottb, *Echinochloa oryzicola* Vasing, and *Eclipta prostrate* L. and fresh weight of these weeds significantly reduced by 13, 29, 34, and 9 per cent, respectively over control. Combined use of *Azolla* and loach was successful in weed suppression (*M. vaginalis*) and increase in rice yield (Cheng *et al*., 2015). (Boyetchko, 1996) reported mycorrhizal fungi are beneficial symbionts that can impart a competitive advantage to their plant hosts, particularly if mycorrhizal dependency is exhibited in weeds as opposed to crops. It may be possible to exploit various soil microbes by directly or indirectly reducing weed competition and tipping the competitive advantage in favour of the crop.

4.4 Mechanical weed control

Mechanical control consists of methods that kill or suppress weeds through physical disruption. The success of mechanical weed control method depends on the life cycle of target weed species. Most mechanical weed control methods, such as hoeing, tillage, harrowing, torsion weeding, finger weeding and brush weeding, are used at very early weed growth stages (Singh, 2014 and Kewat, 2014). Hoeing can be effective on older weeds, and remains selective, many mechanical control methods become difficult after the cotyledon stage and their selectivity decreases with increasing crop and weed age. Thus, if the weeds have become too large, an intensive and aggressive adjustment of the implements is necessary to control the weeds, and by doing this one increases the risk of damaging the crop severely (Carter and Ivany, 2006). Stopping tillage practices has a positive impact on weed populations, because it can influence the weed seed viability and distribution, and it has a strong impact on weed emergence by burying weeds in the soil (Vasileiadis *et al.*, 2006). Continuous zero tillage reduced the total weed seed bank over continuous conventional tillage (Mishra and Singh, 2008) in soybean-linseed cropping system. Conservation tillage (low disturbance) leaves more weed seeds on the surface, whereas high disturbance systems bury weeds. Weed seeds left on the surface are generally more susceptible to decay and ultimately reduce weeds seed banks (Chauhan *et al.*, 2006), it allows early sowing and thus the competitive advantage remains in favour of crop not for weeds (Mishra *et al.*, 2010 and Sharma, 2014), lower emergence in conservation tillage might be due to associated with higher soil strength (Dev *et al.*, 2013).

4.4.1 *Thermal weed management*

Plant tissues are susceptible to high temperatures, when most of the physiological functions are disrupted because of membrane rupture, protein denaturation, and enzyme inactivation. This led to the development of weed management strategies involving high temperature. Most of the plants die after exposure temperatures between 45 and 55ºC (Zimdahl, 2013). To control weeds, heat may be used in different ways, including direct flaming, water steam, solarization and microwave technology (Ascard *et al.*, 2007 and Singh, 2014), which provide rapid weed control without leaving chemical residues in the soil and water, selective towards the weeds, they do not disturb the soil as in case of cultivation methods (Zimdhal, 2007), but its effectiveness depends on the temperature, exposure time and energy input (Ascard *et al.*, 2007). Thermal weed control methods kills above ground plant parts, they may regenerate and repeated treatments may be required.

4.4.1.1 *Flame weeding*

Flaming is a unique technique to kill weeds through the use of direct heat in the form of fire. Flame weeding is most prevalent in European countries (Bond and Grundy, 2001). The temperature of about 55ºC is used to kill the weeds by destroying the cell wall structure. Fuel and temperature requirements depend on weed growth stage and biomass. However, for effective weed control, frequent flaming is often needed (Ascard, 1994). Commonly, propane is used as fuel, but relatively renewable alternatives like hydrogen, are also under consideration (Andersen, 1997). Weeds that have thin leaves like common lambsquarters, nettle, and chickweed are readily burnt through flaming, whereas shepherd's purse, barnyard grass, and annual bluegrass (*Poa annua* L.) could not be burnt in a single operation (Ascard, 1995). Flaming has shown good results after weed emergence, but before crop emergence in potato, sugarbeet, carrot, and chilli (Melander, 1998). Rask *et al.* (2012) studied the effect of flaming on grasses and shared some positive results regarding the control of grasses. Knezevic and Ulloa (2007) evaluated broadcast flaming in different agronomic crops to manage barnyard grass, green foxtail, velvetleaf, and redroot pigweed and concluded that flaming offers best control for broad-leaved weeds, whereas the grasses are less susceptible. Weeds of corn were significantly controlled through integration of tillage and flaming (Knezevic *et al.*, 2011). Flaming has provided effective weed control in different ecosystems and has led to system stability. In many regions, fire is not a threat, but a tool to reduce competition and to improve nutrient cycling (Kyser and DiTomaso, 2002). A quick response and prompt results are also the distinct features of flame weeding. With advancement in this subject, logistic models have been developed to estimate the efficiency of flame weeding and species response to flaming (Ascard, 1995).

4.4.1.2 *Soil solarization*

The impact of pesticides use on the environment is now well documented, and a more wide spread adoption of integrated weed management strategies and tactics is recommended in sustainable agriculture systems. Solarization is a useful nonchemical technique for controlling weeds, nematodes, and several soil-borne diseases (Stapleton, 2000). This hydrothermal process occurs in moist soil which is covered by a transparent plastic film for 4–6 weeks and exposed to sunlight during the warm summer months (Singh, 2014). Soil solarization increases soil temperatures by 8°12°C over the corresponding non-mulched soil (Hosman and Meti, 1993). Solarization has been shown to be most effective in regions that are cloudless and have hot weather. This technique has also been applied in regions with humid climates (McGovern *et al.*, 2004 and Chase, 2007), except when a prolonged period of rain occurred (Wang *et al.*, 2006). Solarization was found to be a cost-effective and low-risk management practice for small farmers and has the potential to increase crop yield (Culman *et al.*, 2006) by decreasing the weeds about 36 per cent per square meter.

Soil propagules, as weed seeds, offer a wide range of tolerance to high temperatures, conditioning the long term success of weed control by solarization. To increase the solarization effect as much as possible, the soil surface must be smooth and must contain enough water to favour heat transfer down the profile and to make reproductive structure of pests, diseases and weeds more sensitive to heat damage. For this reason, prior to solarization the soil is usually irrigated and a plastic mulch film is laid down onto the soil to further increase soil heating and to avoid heat dissipation to the atmosphere.

Soil solarization will decrease weed seed germination more at the top layer than deeper layers because temperature increases more slowly at deeper depths. Thus only heat sensitive weeds may be killed at deeper soil layers, while dormant and more heat tolerant weed seeds escape the soil solarization effect. Soil solarization effectively controls broomrapes (*Orobanche* sp.) and many other weeds, but not *Cuscuta* species, bindweed, or purple nutsedge. Efficacy of soil solarization for weed control in the field is increased by providing irrigation at least 2–3 week prior to solarization, letting the weeds grow, and incorporating them in soil before establishing the solarization treatment. Benlloglu *et al.* (2005) reported solarization in combination with other weed management practices effectively control annual weeds, such as annual bluegrass (*Poa annua*), common purslane (*Portulaca oleracea*), redroot pigweed (*Amaranthus retroflexus*), and barnyard grass (*Echinochloa crusgalli*), but not horseweed (*Conyza canadensis*). Solarization is also an important tool to manage nutsedge, which is often hard to control with regular mulches because it grows as a rhizome (not requiring sunlight) until it encounters light, then pierces mulches with its sharp growing point and thereafter expands its leaves above the plastic film (Johnson *et al.*, 2007).

The success of soil solarization is affected due to intensification and the length of exposure to sunlight. Significant reduction in weed emergence was observed over the following 12 months after one month's solarization (Singh, 2014). The times of the year corresponding to these conditions occur during the warm summer months. Soil solarization has been attempted at other times of the year, but with limited success. Also, climate with cool air temperatures are less likely to provide adequate weather conditions conducive to the success of soil solarization (Peachey *et al.*, 2001).

4.4.1.3 *Microwaves and radiations*

Use of microwave energy to kill weeds has gained popularity in the recent past. It is based on the high energy of microwaves, which can kill weeds very efficiently. This method is highly targeted and there is no fear of non-targeted damage (Rask and Kristofferson, 2007). Microwaves were successfully used in Denmark for the control of little mallow (*Malva parviflora* L.), hairy fleabane (*Conyza bonariensis* L.) and gooseberry gourd (*Cucumis myriocarpus*) (Brodie *et al.*, 2007). This technology is effective against many weeds, but the energy required is very high, which increases cost of production. However, its efficiency and energy budget may be decreased by flux configuration and through induction of thermal runaway in weed plants, making it comparable with other weed-control tools in terms of cost (Brodie *et al.*, 2011). Similarly, laser radiation may be used effectively to kill weeds (Rask and Kristofferson, 2007). In United States, laser beams were used to kill water hyacinth plants. Lasers transfer high energy to plant tissues and raise the water temperature at the cellular level, resulting in cell death. Mathiassen *et al.* (2006) studied the biological efficacy of laser treatment against common chickweed and scentless chamomile (*Tripleurospermum inodorum* L.) under varying levels of exposure time and observed that weeds were significantly suppressed. Use of ultraviolet radiation for weed management has also been tested (Day *et al.*, 1993 and Andreasen *et al.*, 1999) and it was observed that the ultraviolet energy acts severely on plant tissues. It kills weeds just like flaming; however, limited development of this technology is due to possible health hazards. Further research is needed in this particular aspect to develop economically viable options on a sustainable basis.

Depending on the dose, microwave treatment can kill or stimulate the germination of gorse seeds (Moore *et al.*, 2009), and other species (Brodie *et al.*, 2009). Brodie *et al.* (2009) found that both annual and perennial ryegrass germination in the top 10cm of soil was reduced to less than 2.5 per cent of that in untreated areas when the soil was wet and the irradiation time was 8 minutes. No seeds within 5cm of the soil surface in wet soil germinated. In dry sand, they found that unless the seed was within 2cm of the soil surface there was no significant effect on germination. Germination was also not affected until irradiation was carried out for 12 mins in dry soil, irradiation at this rate reduced germination to 2 per cent of those in untreated areas. High temperatures can kill seeds and microwaves can heat soil or seeds containing water. Lower doses of microwaves may affect protein integrity which may increase or decrease germination.

4.4.1.4 *Hot water, steam, and hot air*

Heat can also be used to kill weeds through hot water application. Hot water treatment for weed control has been trialed in many countries with a great deal of success (Rask and Kristofferson, 2007 and De-Cauwer *et al.*, 2015). In the 1990s, a commercial tool, Aqua Heat, was developed in the United States to apply hot water for weed control (Berling, 1992). Hot water application proved effective against most of the annual and a large number of perennial weeds. The effects were even comparable against a glyphosate application. Similar kinds of equipment were successfully used against weeds in New Zealand, where hot water remained in contact with weeds for a longer period of time (Rask and Kristofferson, 2007). Hot water equipment for weed control is also available in Denmark and the Netherlands. Hot water treatment is safe and has no side effects like flame weeding or radiation methods. Its effectiveness is greater under dense weed population because of increased penetration ability (Hansson and Ascard, 2002). Due to greater success rate, this technique is being considered in precision weed management strategies in European countries. Use of steam instead of hot water has been observed as a more effective, quick, and sustainable method, especially in cases where weed control is on relatively hard surfaces (Rask and Kristofferson, 2007). Engineering efforts are needed in this area to improve the efficiency of availability of equipment and to introduce new equipment for weed management in crop production regimes.

4.4.1.5 *Electrocution*

The practice of weed control via electric shock is called electrocution. Although it is a less researched domain, evidence supports the fact that weeds can be killed by spark discharge or electrical contact (Diprose and Benson, 1984 and Parish, 1990). The strength of electric shock, contact or exposure duration, weed species, morphological features, and growth stage significantly affect the success of electrocution. The severity of damage is aggravated in cases of dry soil conditions (Diprose and Benson, 1984). However, because of higher costs involved, energy crises, and hazards to operators, its application in agriculture is limited. In the future, this particular method may have practical implications, especially in sustainable agriculture.

4.5 Allelopathic plant

The term allelopathy, however, was first coined in 1937 by the Austrian Professor Hans Molisch from two Greek words: allelon 'of each other' and pathos 'to suffer' and means the "injurious effect of one organism upon the other". Today, the term is generally accepted to cover both inhibitory and stimulatory effects of one plant on another plant (Rizvi and Rizvi, 1992). In 1996, the International Allelopathy Society defined allelopathy as follows: "The science that studies any process involving secondary metabolites produced by plants, micro-organisms, viruses, and fungi that influence growth and development of agricultural and biological systems" (Xuan *et al.*, 2005). Nowadays, allelopathy has a significant role in research involving sustainable agriculture, like biological weed and pest control (Sodaeizadeh and Hossein, 2012). The current trend is to find a biological solution to minimize the perceived hazardous impacts from herbicides and insecticides in agriculture production. In this regards, the harmful impact of allelopathy can be exploited for pest and weed control (Sodaeizadeh *et al.*, 2009). The chemicals responsible for the phenomenon of allelopathy are generally referred to as allelochemicals or phytotoxins (Kholi *et al.*, 2004). Allelochemicals are usually classified as secondary metabolites and are produced as offshoots in the primary metabolic pathways of plants (Kruse *et al.*, 2000). Many such natural compounds have the potential to induce a wide array of biological effects and can provide great benefits to agriculture and weed management (Macias *et al.*, 2006 and Sodaeizadeh and Hossein, 2012).

Sustainable weed management is needed in the wake of a huge decline in crop outputs due to weed pressure. Diversity in weed management tools ensures sustainable weed control and reduces chances of herbicide resistance development in weeds. Allelopathy as a tool, can be importantly used to combat the challenges of environmental pollution and herbicide resistance development (Jabran *et al.*, 2015). Suppressing weeds by harnessing the allelopathic phenomenon is included among the important innovative weed control methods (Jabran and Farooq, 2013 and Zeng, 2014). Plant hormones and defense mechanisms are manipulated to control weeds in different agro-ecosystems (Pickett *et al.*, 2014). Research has shown that allelopathic practices may meet all these requirements; hence in future, allelopathy may provide a basis to sustainable agriculture (organic, alternative, regenerative, biodynamic, low input, or resource conserving agriculture). To achieve the goals of sustainable agriculture, current research involves, plant breeding, soil fertility, tillage, crop protection, and cropping systems. Allelopathy, being an important phenomenon in agriculture, is also important in sustainable agriculture (Bruinsma, 2003 and Farooq *et al.*, 2011). Thus for sustainability, future weed control practices must minimize the use of herbicides and use allelopathic strategies and other practices for weed management (Farooq *et al.*, 2011).

Oat, rye, sorghum, rice, sunflower, rape seed, barley and wheat have been documented as important allelopathic crops (Yenish *et al.*, 1995; Smeda and Weller, 1996; Kong, 2005; Wu, 2005; Yongqing, 2008; Xu *et al.*, 2010; Farooq *et al.*, 2011; and Narwal and Haouala, 2013). These crops express their allelopathic potential by releasing allelochemicals which not only suppress weeds, but also promote underground microbial activities. Weed suppression in summer crops followed this order: pearl millet>maize>sorghum>cluster bean> cowpea and the order in winter crops was: Egyptian clover>oat>lentil>wheat has been reported by (Narwal and Haouala (2013). Root exudates of barley inhibited the germination and growth of *S. media, C. bursapastoris, A. hybridus* and *C. album* through phytotoxins, but had no effect on wheat (Narwal and Haouala, 2013). Smeda and Weller (1996) reported that rye residues mowed or desiccated with glyphosate herbicide controlled more than 89 per cent weeds and did not exert any harmful effect on tomato crop. Hoffman *et al.* (1996) applied rye residues on soil surface in greenhouse experiments to determine effect on early growth of barnyard grass and velvet bean, using capillary mat irrigation to maintain a constant soil moisture and fertilizer supply. The presence of rye residues delayed emergence and decreased height of barnyard grass. Rye residues were surface applied in maize reduces the emergence and growth of many weeds (Yenish *et al.,* 1995). Yongqing (2008) found that wheat up to 60 days age has significant inhibitory effects on *Asphodelus tenuifolius* weed. It was attributed to release of untriacontane compound present in root exudates of wheat plants. Mahadeveppa and Kulkarni (1996) reviewed their work on suppression effect of *C. sericea* on parthenium and concluded that the weed is replaced by *C. sericea* mainly through allelopathic effects. Different lines of *Beta vulgaris, Cucumis sativus, H. vulgare, Lupinus* sp., *O. sativa, Pisum sativum*, and *T. aestivum*, were reported to inhibit different weed species, and differences among these lines were detected biologically and at the molecular level (Kong, 2005 and Wu, 2005).

Higher weed smothering efficiency of *Capsicum annuum, G. hirsutum*, and *Phaseolus vulgaris* was found attributed to canopy coverage offered by *Arachis hypogea* and/or *P. vulgaris* as evident for higher light interception of intercrops (Shesshadri and Prabhakarasetty, 2001). Cheema *et al.* (1997) tested the allelopathic potential of water extract of sorghum and sunflower against weeds in the field of wheat crop and found spray of 100 per cent water extracts of sorghum and sunflower applied at 30 DAS significantly reduced total weed density up to 48 and 32 per cent, respectively and weed dry weight up to 51 per cent. The inhibition varied between weed species. In another work, Cheema *et al.* (2000) showed that spray of 'sorgaab' on cotton crop suppressed weed density by 13–54 per cent and weed dry weight by 87 per cent. Additional work indicated that spraying of sorgaab on wheat, maize, alfalfa, soybean, and raya (*B. juncea*) at different time of sowing significantly reduced total density and dry biomass of weeds in these crops (Khaliq *et al.*, 1999; Cheema and Khaliq 2000; Ahmad *et al.*, 2000 and Bahatti *et al.*, 2000). Plots amended with sunflower or sorghum residue recorded least total weed density and biomass in faba bean and barley fields and resulted in similar crop yield or even better than was noticed with recommended herbicides dose (Alsaadawi *et al.*, 2007, Alsaadawi *et al.*, 2011 and Alsaadawi and Al-Temimi, 2011).

Momilactone A and B are growth inhibitors exuded from rice roots, which act as potent allelochemicals for barnyard grass with a concentration of 0.21–1.5 and 0.66–3.8 μmol L^{-1}, respectively (Kato-Noguchi and Ino, 2005 and Kato-Noguchi *et al.*, 2008). Khan *et al.* (2007) evaluated rice bran compost for weed control in organically grown spinach at 10, 20, and 30 per cent of soil in the greenhouse and 2 kg m^{-2} in the field. Rice bran compost significantly reduced weeds density and dry weight in both greenhouse and field experiments. Javaid *et al.* (2011) found that wheat cultivars were inhibitory to the invasive weed Parthenium. Barley cultivars, most were found allelopathic in nature and inhibited the germination and growth of weeds like barnyard grass, sterile oat (*A. sterilis* sp. Sterilis L.), black grass (*Alopecurus myosuroides*), wild mustard (*Sinapis arvensis*), and hood canary grass (*Phalaris paradoxa*) plants (Dhima *et al.*, 2008; Ashrafi *et al.*, 2008 and Vasilakoglou *et al.*, 2009). Brown mustard and white mustard, grown as green manure, suppressed weeds in cowpea and improved yields by about 415 kg ha^{-1} compared with controls (Norsworthy *et al.*, 2005). Sunflower crop residues and the allelopathic water extracts were effective in controlling lentil weeds and increasing grain yield. Sunflower allelopathic potential has a tremendous scope for practical utilization for the sake of weed management in field crops without reliance on herbicides and in some instances less use of herbicides (Jabran *et al.*, 2008 and 2010 and Farooq *et al.*, 2011). For example, sunflower allelopathic water extracts synergistically worked with lower herbicide rates to offer weed control and increase in maize grain yield, statistically comparable with the full dose of herbicide (Khan *et al.*, 2011). Similarly, the application of sunflower residues for weed management in the rice crop not only offered attractive weed control potential but also delivered a significant increase in rice grain yield over the unweeded control (Rehman *et al.*, 2007).

4.6 Biological approaches of weed management

Biological control, which is the use of biotic agents to suppress pests, is an underutilized pest management strategy that can play a significant role in developing low-input sustainable agriculture (Hoy, 1990). It can provide additional means for controlling pests, such as weeds, while reducing the reliance on chemical pesticides, maintaining economic viability and environmental quality, and addressing the consumer's need for safe and nutritious food. During recent history, the primary method of weed control has been with chemical herbicides, particularly because they are relatively cheap and effective, and their application is not labour intensive (Singh and Singh 2006; Verma and Singh, 2008 and Singh *et al.*, 2009). Unfortunately, the reliance on chemicals has promoted the development of monoculture cropping rather than a systems approach to weed management and, although weed eradication may not be an achievable goal, millions of dollars have been invested to develop chemicals to control or eradicate noxious weeds (Burnside, 1993). Biocontrol should be perceived as a complementary tool for weed management where it would be integrated with other weed control methods. However, for biological control to be successful as a weed management tool, it will be necessary to develop an environment that will sustain biological control agents. This strategy is an applied technology that enhances or mimics nature and can be utilized to maintain pest populations at non-damaging levels, but not necessarily eradicate them. The two main types of strategies for controlling weeds through biological control are the classical (inoculative), and inundative (augmentative) approaches (Verma *et al.*, 2015). The classical approach involves the release of the biotic agent, (natural enemy), followed by its establishment and dissemination. The agent reduces the weed population below the socio-economic or ecological threshold and provides long-term control, with no requirement for reintroducing the agent. The inundative approach, often referred to as the bioherbicide or mycoherbicide approach when fungi are used, involves the periodic application of a pathogen to control a weed. In this latter case, a host-specific pathogen is mass-produced artificially and applied at high concentrations to a target weed. The biocontrol agent is not expected to provide control beyond one season after application (Tiwari *et al.*, 2013 and Kumar, 2014).

4.6.1 *How are biological controls agents identified and introduced?*

Step 1: Identifying target weeds

To be considered a good candidate for biological control, a weed should be: (i) non-native, present in numbers and densities greater than in its native range and numerous enough to cause environmental or economic damage, (ii) the weed should also be present over a broad geographic range have few or no redeeming or beneficial qualities have taxonomic characteristics sufficiently distinct from those of economically important and native plant species and (iii) for classical biocontrol, the weed should occur in relatively undisturbed areas to allow for the establishment of biological control agents. Cultivation, mowing and other disturbances can have a destructive effect on many arthropod biocontrol agents. Inundative biocontrol agents, such as bacteria and fungi are less sensitive to these types of disturbances so may be used in cropland.

Step 2: Identifying control agents and assessing level of specialization

Scientists observe weeds in their areas of origin and collect the insects and other organisms attacking the plants and affecting their survival. These organisms are subjected to a multi-level screening process to assess their host range and their effect on the weed. These screening efforts do two things: they ensure the safety of any valuable crop, forage or native plant species that the agents may encounter when released, and they assess the efficacy of the agent. Assessing the host range (how specific the potential biocontrol agent is to a particular plant) is probably the most important step in this process. Over very long periods, some plants and herbivores have evolved to form very close associations. Plants have developed a number of defenses, such as toxic chemicals, that plant-eating organisms (herbivores) must overcome. Some herbivores have evolved the ability to bypass only certain host plants defenses such that they cannot feed or develop on anything else. To find out how specialized a particular agent is, scientists collect and expose them to a wide assortment of plants. These plants include crop and forage species as well as species native to the intended release area, especially if the species are close relatives of the weed. Screening potential biological control agents ensures that only those with a very narrow host range (i.e., those that represent no threat to crop, forage or native species) are released.

Step 3: Controlled release

All biological control agents must be approved under the plant protection act and are reviewed by an advisory panel of Agriculture and Agri-Food Canada before they can be released. Following federal approval, the agents are released on their target weeds at selected experimental sites, which are closely monitored. Data from these sites help to assess both the agents potential for survival under field conditions and their potential to cause damage to the target weed.

Step 4: Full release and identifying optimal release sites

If the agents survive our harsh climate and damage or suppress the weeds in the controlled tests, they can be released on other weed infestations. The agents are released on as many varied sites as agent numbers, time and resources will permit. Note that instructions for the proper method and timing of agent releases should be followed closely, and these instructions will vary with the species of agent. The agents are then carefully monitored. Release on a large variety of sites allows scientists to find the habitats best suited to each species.

Many types of biological control agents spend at least part of their lives underground and can be very sensitive to the soil types and conditions present in the various areas. For example, the black dot leafy spurge beetle, *Aphthona nigriscutis*, spends its entire larval stage underground feeding on leafy spurge roots. Information gathered from a number of release sites, indicates that these beetles prefer dry, sandy soils and will not do nearly as well in soils with higher moisture levels.

Step 5: Monitoring release sites (classical)

Release sites should be monitored periodically to assess the size of the biological control agent population and the effect of the agent on the weed. A sweep net is particularly useful for sampling insect biological control agents that feed on the foliage of the weed. Monitoring release sites are very important to determine if agents have established on a site. Some insect biocontrol agents may need two to five years before their populations increase to a sufficient size to have a visible impact on weed numbers. When the classical control agent is establishing on a weed infestation and increasing its numbers, the site should not be mowed or disturbed. Herbicide may be used along the boundaries of the weed infestation to help contain the weed, while the biological agent is increasing in population and spread.

Step 6: Redistribution (classical)

If a biological control agent is released on a site with favourable conditions, its population can grow quickly. A large number of insects can suppress the weed in this area and may allow for the establishment of an 'insectary'. This is an area where the insect population has grown to the point where the insects can be collected in very large numbers. Insects collected from these sites can be distributed to other weed-infested areas. Once established, the insects can suppress the weeds in these areas as well. An important point to remember is that it can take years before a biocontrol agents population reaches optimum. The agents should not be collected for redistribution before this time.

Step 7: Maintaining control agent populations (classical)

What happens to the control agents as their food supply dwindles over time? Since the weeds are never completely eradicated, a small population of insects stays on the few remaining plants. Insects that cannot be supported by the weed population at this lower level either move to new stands or die. If weed numbers increase for some reason, the number of insects will increase as well. Weed numbers fluctuate around a new, much lower equilibrium point after control is implemented.

a. Insect biocontrol

Several insects have been tried to control parthenium weed in the different countries, of which the leaf-feeding beetle (*Zygogramma bicolorata*) and the stem galling moth (*Epiblema strenuana*), both imported from Mexico, have shown good potential to control this weed (Kaur *et al.*, 2014). The beetle, *Z. bicolorata*, an effective leaf eater, was imported from Mexico for the management of parthenium in Australia in 1980, and in Indian Institute of Horticulture Research (Jayanth, 1987). Both the adults and larvae of this insect feed on leaves and flower (Kumar, 2014). The research effort in the use of fish to control excessive aquatic weed growth in irrigation canal has steadily gained ground in recent years (Center *et al.*, 1997). The list of successful insects given in Table 3.

Table 3. Successful insects using as a bioagent

Name of insect bioagent	Kind of bioagent	Target weed
Bactra verutana	Shoot boring moth	*Cyperus rotundus*
Hydrellia pakistanae	Shoot fly	*Hydrilla verticillata*
Zygogramma bicolorata	Leaf eating beetle	*Parthenium hysterophorus*
Conotrachels sp.	Stem galling insect	*Parthenium hysterophorus*
Gastrophysa viridula	Beetle	*Rumex* spp.
Microlarinus lareynii	Pod weevil	*Tribulus terrestris*
M. lypriformis	Pod weevil	*Tribulus terrestris*
Bucculatrix parthenica	Leaf mining moth	*Parthenium hysterophorus*
Conotrachelus albocinereus	Stem galling weevil	*Parthenium hysterophorus*
Epiblema strenuana	Stem galling moth	*Parthenium hysterophorus*
Listronotus setosipennis	Stem boring weevil	*Parthenium hysterophorus*
Platphalonidia mystica	Stem boring moth	*Parthenium hysterophorus*
Smicronyx lutulentus	Seed feeding weevil	*Parthenium hysterophorus*
Stobaera concinna	Sap feeder	*Parthenium hysterophorus*
Bactra minima	Shoot boring moth	*Cyperus rotundus*
Teleonemia scrupulosa	Bug	*Lantana camara*
Cecidochares connexa	Stem galling moth	*Chromolanea odorata*
Procecidochares utilis	Stem galling moth	*Ageratina adenophora*
Aspidomorpha miliaris	Beetle	*Ipomoea carnea*
Nupserha lenita	Beetle	*Parthenium hysterophorus*
Epiblema strenuana	stem boring moth	*Parthenium hysterophorus*
Neochetina bruchi	Beetle	*Eichhornia crassipes*
N. eichhorniae	Beetle	*Eichhornia crassipes*
Cyrtobagous salviniae	Beetle	*Salvinia molesta*
Agasicles hygrophila	Flea beetle	*Alternanthra philoxeroides*
Dactylopius tomentosus	Bug	*Opuntia* sp.
Cactoblastis cactorum	Bug	*Opuntia* sp.

Agfact (2003), Kaur *et al.* (2014) and Kumar (2014)

b. Bioherbicide

The concept of mycoherbicide was introduced by Daniel *et al.* (1973), who demonstrated that an endemic pathogen might be rendered completely destructive to its weedy host by applying a massive dose of inoculum at a particularly susceptible growth stage. To achieve success, the pathogen must be culturable in artificial media; the inoculum must be capable of abundant production using conventional methods such as liquid fermentation; the final product must be genetically stable and specific to the target weed; storage (shelf-life), handling, and methods of application must be compatible with current agricultural practices; and the pathogen must be efficacious under sufficient different environment conditions to allow a feasible application window (Daniel *et al.*, 1973 and Gnanavel, 2015).

The potential for successful use of bio-herbicides in managing herbicides-resistant biotypes was demonstrated where growth of an imazaquin-resistant common cocklebur biotype originating soybean field was suppressed with the mycoherbicides, *Alternaria helianthi* (Abbas and Barrentine, 1995). The fungus *Colletotrichum gleosporioides* attack cuscutta (Zhang, 1985) and has been used to control Cuscuta selectively in soybean (Li, 1987). *Fusarium oxysporum* was found to be the best resulting in killing of inoculated water hyacinth in about 15 days (Pathak and Kannan, 2011). Saxena and Kumar (2010) worked on the mycoherbicidal potential of *Alternaria alternata* in northern India to control parthenium weed and reported 50 per cent damage of plants *in vitro* detached leaf and whole plant bioassay at 96 hours after treatment at a concentration of 1 × 106 spores mL^{-1}. *Sclerotium rolfsii* (teleomorph: *Athelia rolfsii*) incites a severe collar rot disease on parthenium (Pandey *et al.*, 1998 and Shukla and Pandey, 2006). The list of different bio-herbicides available for controlling weeds is given in Tabel 4.

Table 4. Bio-herbicides available for controlling weeds

Microorganism	Target weed	Ecosystem	Commercial product
Bipolaris sorghicola	*Sorghum halepense*		Biopolaris
Colletotrichum gleosp-orioides aeschynomene	*Aeschynomene viriginica*	Rice, soybean	Collego
Colletotrichum gloeos-porioides f.sp. *malva*	*Malva pusilla*	Wheat, horticultural crops	Biomal, Mallet
Colletotrichum gloeosp-orioides f.sp. *cuscutae*	*Cuscuta* sp.	Soybean	Lubao
Colletotrichum truncatum	*Sesbania exaltata*	Soybean, Cotton, rice	Coltru
Colletotrichum coccodes	*Abutilon theophrasti*	Maize, soybean	Velgo
Phytopthora palmivora	*Morrenia odorata*	Citrus groves	De Vine
Alternaria cassiae	*Cassia obtusifolia*	Soybean	CASST
Alternaria destruens	*Dodders*	Cranberry	Smolder
Puccinia canaliculata	*Cyperus esculentus*	Rice, horticultural crops	Dr.Biosedge

Microorganism	Target weed	Ecosystem	Commercial product
Cercospora rodmanii	*Eichhornia crassipes*	Water ways, impoundments	ABG 5003
Chondrostereum purpureum	*Prunes serotina*	Forest	Biochon
Cylindrobasidium leave	*Acacia* sp.	Forest, rangelands	Stumpout
Nectria ditissima	Red alder	Forest	PFC-Alderkill
Sclerotinia sclerotiorum and *S. minor*	*Taraxacum officinale*	Turf	Formulation development
Rhizoctonia solani	*Euphorbia esula-virgata*	Rangelands	Formulation development
Fusarium solani f.sp. *cucurbitae*	*Cucurbita texana*	Cotton, soybean	Formulation development
Fusarium oxysporum f.sp. *erythroxyli*	*Erythroxylum coca*	Illicit narcotic crops	Formulation development
Trichoderma virens	Several	Row and horticultural crops	Formulation development
Xanthomonas campestris pv. *poae*	*Poa annua*	Turf, athletic fields	Camperico
Streptomyces hygroscopicus	General vegetation	Row and horticultural crops	Biolophos
Pseudomonas syringae pv. tagetis	Composite weeds	Maize, soybean	Field evaluation
Pseudomonas syringae pv. *phaseolicola*	*Pueraria lobata*	Non-crop lands, pastures	Field testing
Pseudomonas syringae strain 3366	*Epilobium angustifolium*	Cranberry	Formulation development
Ralstonia solanacearum	*Solanum nigrum*	Non-crop lands, pastures	Formulation development
Pseudomonas fluorescens D7	*Bromus tectorum*	Cereal grain crops	Formulation development
Tobacco mild green mosaic virus U2	Tropical soda apple	Non crop lands, pastures	Formulation development
Colletotrichum acutatum	*Hakea gummosis* and *H. sericea*	native vegetation	Hakatak

(Kaur *et al.*, 2014 and Gnanavel, 2015)

4.7 Herbicide resistant crops

Herbicide resistance is the inherited ability of the plant to survive and reproduce following exposure to a dose of herbicide that would normally be lethal to the wild type (Kewat, 2014). Crops made resistant to herbicides by biotechnology and are being widely adopted in various parts of the world. Those containing transgenes that impart resistance to post-emergence, non-selective herbicides, such as glyphosate and glufosinate will have the major impact. These products allow the farmer to more effectively use, reduced or no-tillage cultural practices, eliminate use of some of the more environmentally suspect herbicides and use fewer herbicides to manage nearly the entire spectrum of weed species. In some cases, non-selective herbicides used with herbicide resistant crops reduce plant pathogen problems because of the chemicals' toxicity to certain microbes. Herbicide tolerant crops can be produced by either insertion of a 'foreign' gene (transgene) from another organism into a crop, or by regenerating herbicide tolerant mutants from existing crop germplasm. Biotech crops reached 160 m ha, on 8 per cent growth, from 2010 and 94 fold increase in hectares from 1.7 m ha in 1996 to 160 m ha in 2011, makes biotech crops the fastest adopted crop technology in the history of modern agriculture. From the genesis of commercialization in 1996 to 2011, herbicide tolerance has consistently been the dominant trait. In 2011, herbicide tolerance deployed in soybean, maize, canola, cotton, sugar beet and alfalfa, occupied 59 per cent or 93.9 m ha of the global biotech area of 160 m ha. Over the past few years, several herbicide resistant crops (HRCs), both transgenic and non-transgenic, have become available in many countries for commercial cultivation (Chinnusamy *et al.*, 2014).

Non-selective herbicides such as glyphosate and glufosinate aid in broadening the spectrum of weeds controlled, which is particularly important in no-till systems, and those weedy fields. Genetically modified herbicide tolerant maize and spring oil seed rape cultivars tolerant to glufosinate ammonium, were used which gives post-emergence broad spectrum control of annual grasses and broad-leaved weeds (Firbank, 2003). In general, glyphosate is the most widely used herbicide in the world and literature about its use and characteristics is extensive (Woodburn, 2000). Grichar *et al.* (2004) who had found that single application of glyphosate as early or late post-emergence effectively controlled the broad spectrum of weeds. According to AICRPWC (2011) trials, total weed density was significantly lowered with post-emergence application of glyphosate in transgenic cotton and corn hybrids when compared to hand weeding plots in transgenic cotton and national and state checks in transgenic maize. Keeling *et al.* (1998) also observed that, weed control is often excellent (95%) with the application glyphosate as post emergence in cotton. Main *et al.* (2007) reported that Roundup Ready Flex cotton could provide producers with acceptable weed control without compromising cotton yield.

4.8 Weed management at and after harvest of crop

Most of the annual weeds produce a majority of seeds after completing their vegetative growth. Many seeds are retained in the soil seed bank, which creates problems after emergence in standing crops. Weed populations can be decreased by removing their seeds at maturity (Walsh *et al.*, 2013). This strategy eliminates potential seeds from the system that can be deposited in soil or may germinate in coming seasons.

4.8.1 *Harvest weeds seed control*

The harvesting time of grain crops is very important as weed seeds are retained by plants and easy to remove and discard (Walsh and Powles, 2007). Weed seeds can be collected and destroyed during or soon after harvesting of a crop. Harvest weed seed control has been developed in Australia and has shown promising results. This strategy can be implemented with the Harrington seed destructor (HSD), chaff carts, narrow windrow burning, and bale direct (Walsh *et al.* 2013). Each technology is based on the principle of weed-seed collection during grain crop harvest and seed destruction to avoid the replenishment of the seed bank.

4.8.2 *Harrington seed destructor*

Weed seeds present in annual grain crops remained a major concern for growers. A progressive grain producer, Ray Harrington, from Australia tested a cage mill for seed destruction in 2005 (Walsh *et al.*, 2013). Cage mills are normally used to crush stone materials. Further research enabled scientists to destroy up to 90 per cent of seeds of annual ryegrass (*Lolium rigidum* Gaudin) contained by wheat chaff during harvesting (Walsh *et al.*, 2012). HSD is a modified cage mill having a chaff-and-straw transfer system along with power source. According to Walsh *et al.* (2012), field studies showed that HSD offers an impressive destruction rate of above 95 per cent for annual ryegrass, ripgut brome (*Bromus diandrus* Roth), wild radish (*Raphanus raphsnistrum* L.), and wild oat seeds. With such impressive results, HSD is a pragmatic option for the destruction of weed seeds.

4.8.3 *Chaff carts*

Weed seeds that remain intact during conventional harvesting operations that are then added to the crop residues become randomly distributed to the whole field. To avoid this problem, chaff carts were introduced. Chaff carts are simply a cart on a trailer that is attached to a harvester that collects the chaff and places weed seeds in a specified bin (Walsh *et al.*, 2013). Chaff carts effectively collect large amount of seeds of obnoxious weeds like annual ryegrass, wild oat, and wild radish (Shirtliffe and Entz, 2005 and Walsh and Powles, 2007). Collected weed seeds and chaff are then dumped in piles to be burned.

Table 5. Efficacy of harvested weed seed control systems (HWSC) in targeting weed seeds during cereal crop harvest

HWSC system	Weed seed control (%)	Weed species	Reference
Chaff collection	60–80	Annual ryegrass	Gill (1996)
	56–63	Annual ryegrass	Matthews *et al.* (1996)
	73–86	Annual ryegrass	Walsh and Powles (2007)
	95	Wild radish	Walsh and Powles (2007)
	74	Wild oat	Shirtliffe and Entz (2005)
Bale direct	95	Annual ryegrass	Walsh and Powles (2007)
Narrow windrow burning	99	Annual ryegrass	Walsh and Newman (2007)
	99	Wild radish	Walsh and Newman (2007)
Harrington Seed Destructor	95	Annual ryegrass	Walsh *et al.* (2012)
	93	Wild radish	Walsh *et al.* (2012)
	99	Wild oat	Walsh *et al.* (2012)
	99	Brome grass	Walsh *et al.* (2012)

4.8.4 *Narrow windrow burning*

This technique for controlling harvested weed seed is where chaff and residues containing weed seeds are concentrated in a narrow windrow during harvesting (Walsh *et al.*, 2013). The harvester-mounted chute makes about a 60-cm windrow that is burned later, while keeping ecoprotection in mind (Walsh and Newman, 2007). The limited burning of the windrow avoids pollution hazards as the whole field is not burned. Windrow burning is promising as it offers maximum weed control after harvesting of wheat, canola (*Brassica napus* L.), and garden lupin (*Lupinus polyphyllus* Lindl.).

4.8.5 *Bale direct*

This is another sophisticated method of harvest weed seed control, in which chaff and residues from the harvester are converted into bales by a mechanized baler attached to the harvester (Walsh *et al.*, 2013). Despite its efficient and clean function, the adoption rate is lower, which may be due to the marketing issue with bales. However, it offers remarkable weed seed destruction. Hence, weed seed control during crop harvest is an encouraging prospect in the field of weed management. It helps to reduce weed seed bank and to minimize the chances of weed infestations in subsequent seasons. The development of technology in this sector may enable the farming community to manage weeds efficiently.

4.8.6 *Weed seed predation*

Seed predation through granivorous insects and small mammals is a useful tactic for weed control. Seed predation may be pre- (seeds still attached with plant) or post- (seeds dispersed after maturity) dispersal. Insects, birds, and small mammals are the major post dispersal weed seed predators (Heggenstaller *et al.*, 2006 and Menalled *et al.*, 2007). Most *Coleoptera* and *Hymenoptera* insects are involved in weed seed feeding. Several species of carabid beetles and field cricket have been observed as potential predators of redroot pigweed, velvetleaf, large crabgrass (*Digitaria sanguinalis* L.) and giant foxtail seeds (Carmona *et al.*, 1999). Mice consumed up to 20 per cent of seeds of barnyard grass (*Echinochloa crusgalli* L.) and common lambsquarters in no-till soybean and corn (Carmona *et al.*, 1999). Feeding habits and preferences of the predators significantly affect weed seed destruction and concomitant weed emergence. White *et al.* (2007) observed feeding choice of three beetle species (*Amara aenea*, *Anisodactylus sanctaecrucis*, and *Harpalus pensylvanicus*) and the field cricket (*Gryllus pennsylvanicus*). The beetles consumed more seeds of redroot pigweed than of giant foxtail. Experimental and modeling studies have clearly shown that predation significantly affects weed seed demographics (Mauchline *et al.*, 2005). Meiss *et al.* (2010) reported that the vegetative cover influences weed-seed predators by altering the habitat quality. Effective predation can be achieved by delaying tillage or increasing interval between land preparation and seeding (Chauhan, 2012). Seed predation is a potential nonchemical weed management option. It can be used alone or in combinations with other cultural management practices.

4.9 Integrated weed management

One of the definitions of integrated weed management (IWM) implies methods of controlling weed that require no herbicide or rational use of herbicides. IWM includes more than one method of control *viz.,* seed purity, crop varieties, spacing and methods of planting, cultivations, soil solarization, intercropping, crop rotation, water management, manure application, biological control and herbicides. According to FAO, "the integrated campaign against pests is a method whereby all economically, ecologically and toxicologically justifiable methods are employed to keep the harmful organisms below the threshold level of economic damage, keeping in the foreground the conscious employment of natural limiting factors. Integrating fish culture and dual culture of *Azolla* in transplanted rice is observed to compliment weed control in transplanted rice (Kathiresan *et al.*, 2001). Off-season ploughing and mulching the inter row space enhanced the weed control in combination with herbicide in cotton (Vijayabaskaran and Kathiresan, 1993).

5. BENEFITS OF SUSTAINABLE WEED MANAGEMENT

The benefits are reviewed in the context of the environment, society and economics, (1) improved soil and water conservation, (2) mitigation of global warming, (3) enhanced biodiversity, (4) reduction of persistent pollution, (5) increased food nutrient density, (6) reduced toxic load in adults and children who eat organic, (7) better conditions for farm workers, (8) competitive yields, (9) price premiums, (10) direct-to-consumer marketing channels, (11) lower input costs, (12) higher per farm income, (13) improved resilience or lower volatility, (14) energy savings and (15) income from carbon markets.

6. CONCLUSIONS AND FUTURE NEED

As we know the sustainable farming has the ability to save the natural resources for the future and develop the farm in the little expense, a transition to sustainable weed control is required for environmental, social and economic reasons. Fortunately, sustainable farming is a robust business model, delivering superior economics over conventional farming on a wide variety of metrics such as crop yields, gross and net income per acre, cost of inputs, per farm income and more.

Farmers have several preventive and cultural methods in their arsenal that they can put together to build up a good weed management strategy. The convenience of using one method instead of another depends on local attitudes and constraints, such as availability of money and labour, access to technical means (*e.g.* seeds, fertilizers, herbicides), environmental, social and economic features that may limit the range of feasible agronomic choices (*e.g.* length of the growing season, rainfall and temperature patterns, soil mineralization rate, farm and market structure, cultural heritage, existence of advisory services, etc.). However, highest diversification of the cropping system (*i.e.* crop sequence and associated cultural practices) based on agro-ecological principles is the key to effective long-term weed management in any situations. In this respect, the systematic inclusion of preventive, cultural and other possible methods as per the growing condition for weed management must always be pursued. This obviously implies that farmers must be educated to acquire a higher level of knowledge and technical skills. Simple solutions, such as monocropping and reliance on herbicides as the only direct weed-control method may be successful in the short term, but are never rewarding in the long term.

Weed management practices such as tillage and application of organic amendments can change the incidence and severity of weed density, and the weed community competition in long term management, unless other effective alternative treatments are found. Long and diverse crop rotations with careful weed control in low-input and organic systems are fundamental in sustainable and ecological crop production systems. Cropping system and crop rotation strategies are fundamentally significant considerations in development of sustainable and environmentally safe strategies for weed control.

The non-judicious use of chemical herbicides is causing-environmental damage, health hazards, herbicide resistance in weeds, and non target actions. Thus, a set of alternative weed management tools is needed under the prevailing conditions. The use of possible weed management strategies discussed above in an integrated manner can help on a sustainable basis. All these methods focus on environmental protection, practical viability, compatibility for integrated programs, and ecological stability. The right choice of one or more of these strategies according to geographic, agricultural, and socioeconomic conditions may offer an impressive weed control. None of them has the potential to comprehensively replace chemical weed management; however, an integrated approach may lead to success. The diversified nature of these strategies may be very useful against invasive and resistant weeds. Further research is needed to optimize these tools for improvement in efficiency and practical suitability. In the long run, a single weed control measure may not remain effective and, thus, integrated weed management on the basis of advanced nonconventional strategies will be a pragmatic option in modern intensive sustainable agriculture.

REFERENCES

Abbas HK and Barrentine WL (1995) *Alternaria helianthi* and imazaquin for control of imazaquin susceptible and resistant cocklebur biotypes. *Weed Sci* 43: 425–428.

Abdolraheem S and Saeedipour (2015) Influence of seeding rate and reduced doses of super gallant herbicide on weed control, yield and component yield of mungbean. *Research J Env Sci* **9:** 241–248.

Abouziena HF, Hafez OM, El-Metwally IM, Sharma SD and Singh (2008) Comparison of weed suppression and mandarin fruit yield and quality obtained with organic mulches, synthetic mulches, cultivation, and glyphosate. *Hort Sci* **43:** 795–799.

Agfact (2003) Bob trounce, weeds agronomist, former noxious weeds advisory officer, Grafton. 2nd edition, June, Pp. 1–4.

Ahmad S, Cheema ZA and Ahmad R (2000) Evaluation of sorgaab as natural weed inhibitor in maize. *J Anim Plant Sci* **10:** 141–146.

AICRPWC (2011) Annual Report All India co-ordinated research programme on weed control. In: DWSRC, Department of Agronomy, Tamil Nadu Agricultural University, Coimbatore.

Akobundu IO and Ekeleme F (2002) Weed seedbank characteristics of arable fields under different fallow management systems in the humid tropical zone of southern Nigeria. *Agrofor Syst* **54:** 161–170.

Alam MA,Tipu MMH, Chowdhury MMI, Rubel MH and Razzak MA (2015) Effect of weeding regime and row direction on growth and yield of rice in Bangladesh. Research and Reviews: *J Crop Sci Tech* **4(2):** 1–11.

Alcantara C, Pujadas A and Saavedra M (2011) Management of *Sinapis alba* subsp. mairei winter cover crop residues for summer weed control in southern Spain. *Crop Prot* **30:** 1239–1244.

Aldrich RJ (1984) Weed crop ecology. Principles in weed management. Breton Publisher, MA.

Alford CM, Miller SD and Cecil JT (2004) Using narrow rows to increase crop competition with weeds. Proc. 4th Int. Crop Sci. Congress. Brisbane, Australia 26 Sept. to 1 Oct.

Alsaadawi IS, Al-Ekeelie MH and Al-Hamzawi MK (2007) Differential allelopathic potential of grain sorghum genotypes to weeds. *Allelopathy J.* **19:** 153–160.

Alsaadawi IS and Al-Temimi AA (2011) Use of sunflower residues in combination with sub recommended dose of herbicides for weeds control in barley field. *Herbilogia* **12:** 83–93.

Alsaadawi IS, Khaliq A, Al-Temimi AA and Matloob A (2011) Integration of sunflower residues with a pre-plant herbicide enhances weed suppression in broad bean field. *Planta Danninah* **29:** 849–859.

Altieri MA, Lana MA, Bittencourt HV, Kieling AS, Comin JJ and Lovato PE (2011) Enhancing crop productivity via weed suppression in organic no-till cropping systems in Santa Catarina, *Brazil J Sustain Agric* **35:** 855–869.

Amosse C, Jeuffroy MH and David C (2013) Relay intercropping of legume cover crops in organic winter wheat: Effects on performance and resource availability. *Field Crops Res* **145:** 78–87.

Andersen J (1997) Experimental trials and modelling of hydrogen and propane burners for use in selective flaming. *Biol Agric Hort* **14:** 207–219.

Andrade FH, Calvinho P, Cirilo A and Barbieri P (2002) Yield responses to narrow rows depend on increased radiation interception. *Agron J* **94:** 975–980.

Andreasen C, Hansen L and Streibig JC (1999) The effect of ultraviolet radiation on the fresh weight of some weeds and crops. *Weed Tech* **13:** 554–560.

Angiras N and Sharma V (1996) Influence of row orientation, row spacing and weed control methods on physiological performance of irrigated wheat. *Indian J Agron* **41:** 41–47.

Arce GD, Pedersen P and Hartzler RG (2009) Soybean seeding rate effects on weed management. *Weed Tech* **23:** 17–22.

Arentoft BW, Ali A, Streibig JC and Andreasen C (2013) A new method to evaluate the weed-suppressing effect of mulches: a comparison between spruce bark and cocoa husk mulches. *Weed Res* **53:** 169–175.

Arif M, Ali K, Haq MS and Khan Z (2013) Biochar, FYM and nitrogen increases weed Infestation in wheat. Pak. J. *Weed Sci* Res **19(4):** 411–418.

Ascard J (1995) Effects of flame weeding in weed species at different developmental states. *Weed Res* **35:** 397–411.

Ascard J, Hatcher PE, Melander B and Upadhyay MK (2007) Thermal weed control. (Eds M. K. Upadhaya & R.E. Blackshaw). Non-chemical weed management. Principles, Concepts and technology. CABI, London, UK.

Ascard J. 1994. Dose-response models for flame weeding in relation to plant size and density. *Weed Res.* **34:** 377–385.

Ashrafi ZY, Sadeghi S, Mashhadi HR and Alizade HM (2008) Study of allelopathic effects of barley on inhibition of germination and growth of seedling green foxtail. *J SAT Agric Res* **6:** 1–6.

Babaei M and Saeedipour S (2015) Effect of seed rate and post emergence herbicide application on weed infestation and subsequent crop performance of wheat. *Walia J* **31(3):** 158–162.

Bahadur S, Verma SK, Prasad SK, Madane AJ, Maurya SP,Gaurav, Verma VK and Sihag SK (2015) Eco-friendly weed management for sustainable crop production-A review. *J Crop and Weed* **11(1):** 181–189.

Bahatti MQL, Cheema ZA and Mahmood T (2000) Efficacy of sorgaab as natural weed inhibitor in raya. *Pak J Biol Sci* **3:** 1128–1130.

Bajwa AA, Ehsanullah, Anjum SA, Nafees W, Tanveer M and Saeed HS (2014) Impact of fertilizer use on weed management in conservation agriculture-a review. *Pak J Agric Res* **27:** 69–78.

Baker BP and Mohler CL (2014) Weed management by upstate New York organic farmers: strategies, techniques and research priorities. Renew. Agric Food Syst http://dx.doi.org/10.1017/S1742170514000192.

Bangun P and Syam M (1988) Weed management in rice crops. PADI-2, pp. 579–599.

Bell HD, Jason KN, Robert CS and Michael P (2015) Effect of row spacing, seeding rate, and herbicide program in glufosinate-resistant soybean on palmer amaranth management. *Weed Tech* **29(3):** 390–404.

Benlloglu S, Boz O, Yildiz A, Kaskavalci G and Benlioglu K (2005) Alternative soil solarization treatments for the control of soil-borne diseases and weeds of strawberry in the Western Anatolia of Turkey. *J Phytopathol* **153:** 423–430.

Berling J (1992) Getting weeds in hot water. A new hot water weed sprayer and soy-based oil help cut herbicide use. *Farm Indus News* **26:** 44–49.

Bernstein ER, Stoltenberg DE, Posner JL and Hedtcke JL (2014) Weed community dynamics and suppression in tilled and no-tillage transitional organic winter rye-soybean systems. *Weed Sci* **62:** 125–137.

Berzsenyi Z (2000) Methods of weed control. (Ed. Hunyadi K, Breres I and Kazinczi G-weed species, weed control, weed biology), Mezogazda Kiado, Budapest pp. 334–378.

Bhullar MS, Kamboj A and Singh GP (2006). Weed management in spring planted sugarcane-based intercropping system. *Indian J Agron* **51:** 183–185.

Biswas M, Parveen S, Shimozawa H and Nakagoshi N (2005) Effects of Azolla species on weed emergence in a rice paddy ecosystem. *Weed Bio Manag* **5:** 176–183.

Blackshaw RE (2005) Nitrogen fertilizer, manure and compost effects on weed growth and competition with spring weed. *Agron J* **97:** 1612–1621.

Blackshaw RE and Brandt RN (2008) Nitrogen fertilizer rate effects on weed competitiveness is species dependent. *Weed Sci* **56:** 743–747.

Blackshaw RE, Molnar LJ and Janzen HH (2004) Nitrogen fertilizer timing and application method affect weed growth and competition with spring wheat. *Weed Sci* **52:** 614–622.

Blackshaw RE, Molnar LJ and Larney FJ (2005) Fertilizer, manure and compost effects on weed growth and competition in western Canada. *Crop Prot* **24:** 971–980.

Blackshaw RE, O-Donovan JT, Harker KN and Li X (2002) Beyond herbicides: New approaches to managing weeds. ICESA.

Bond W and Grundy AC (2001) Non-chemical weed management in organic farming systems. *Weed Res* **41:** 383–405.

Borger CPD, Hashem A and Pathan S (2010) Manipulating crop row orientation to suppress weeds and increase crop yield. *Weed Sci* **58:** 174–178.

Boyetchko SM (1996) Impact of soil microorganisms on weed biology and ecology. *Phytoprotection* **77 (1):** 41–56.

Brar AS and Walia US (2010) Rice residue management and planting techniques on herbicidal efficiency in relation to wheat productivity. In National Symposium on Integrated Weed Management in the Era of Climate Change, held at NAAS, New Delhi on 21–22 August, pp. 7.

Brodie G, Hamilton S and Woodworth J (2007) An assessment of microwave soil pasteurization for killing seeds and weeds. *Plant Prot* **22:** 143–149.

Brodie G, Harris G, Pasma L, Travers A, Leyson D, Lancaster C and Woodworth J (2009) Microwave soil heating for controlling ryegrass seed germination. *Transactions of the ASABE* **52:** 295–302.

Brodie G, Ryan C and Lancaster C (2011) Microwave technologies as part of an integrated weed management strategy: a review. *Int J Agron* doi:10.1155/2012/ 636905.

Bruinsma J (2003) World agriculture: towards 2015/2030 An FAO Perspective. Ed. Earthscan, London.

Buhler DD (2002) Challenges and opportunities for integrated weed management. *Weed Sci* **50:** 273–280.

Buhler DD, Kohler KA and Thompson RL (2001) Weed seed bank dynamics during a five-year crop rotation. *Weed Tech* **15:** 170–176.

Burnside OC (1993) Weed science - the step child. *Weed Tech* **7:** 515–518.

Campiglia E, Mancinelli R, Radicetti E and Caporali F (2010) Effect of cover crops and mulches on weed control and nitrogen fertilization in tomato. *Crop Prot* **29:** 354–363.

Cardina J, Regnier E and Harrison K (1991) Long-term tillage effects on seed banks in three Ohio soils. *Weed Sci* **39:** 186–194.

Carmona DM, Menalled FD and Landis DA (1999) Northern field cricket, Gryllus pennsylvanicus Burmeister (Orthoptera : Gryllidae): laboratory weed seed predation and within field activity density. *J Econ Entomol* **92:** 825–829.

Carr PM, Gramig GG and Liebig MA (2013) Impacts of organic zero tillage systems on crops, weeds, and soil quality. *Sustainability* **5:** 3172–3201.

Carter MR and Ivany J A (2006) Weed seed bank composition under three long term tillage regimes on a fine sandy loam in Atlantic Canada. *Soil and Tillage Res* **90:** 29–38.

Cathcart RJ and Swanton CJ (2003) Nitrogen management will influence threshold values of green foxtail in corn. *Weed Sci* **51:** 975–986.

Center TD, Frank JH and Dray FA (1997) Strangers in paradise. In: biological control. (Eds.- Simberloff D, Schmitz DC and Brown TC). Island Press, Washington DC., pp. 245–266.

Chamanabad HRM, Ghorbanil A, Asghari A, Tulikov AM and Zargarzadeh F (2009) Long-term effects of crop rotation and fertilizers on weed community in spring barley. *Turk J Agric* **33:** 315–323.

Chase CA (2007) Soilborne plant pathogens and pest management with soil solarization. <http://www.imok.ufl.edu/LIV/groups/cultural/pests/solar.htm>. 12 Jan.

Chauhan BS (2012) Weed ecology and weed management strategies for dry-seeded rice in Asia. *Weed Tech* **26:** 1–13.

Chauhan BS and Abugho SB (2013) Fertilizer placement affects weed growth and grain yield in dry-seeded rice systems. *Am J Plant Sci* **4:** 1260–1264.

Chauhan BS, Gill GS and Preston C (2006) Influence of tillage systems on vertical distribution, seedling recruitment and persistence of rigid ryegrass. *Weed Sci* **54:** 669–676.

Cheema ZA, Asim M and Khaliq A (2000) Sorghum allelopathy for weed control in cotton. *Int J Agric Biol* **2:** 37–41.

Cheema ZA and Khaliq A (2000) Use of sorghum allelopathic properties to control weeds in irrigated wheat and semiarid region of Punjab. *Agric Ecosys Env* **79:** 105–112.

Cheema ZA, Luqman M and Khaliq A (1997) Use of allelopathic extracts of sorghum and sunflower herbage for weed control in wheat. *J Plant Anim Sci* **7:** 91–93.

Chee-Sanford JC, Williams MM, Davis AS and Sims GK (2006) Do soil microorganisms influence seed-bank dynamics. *Weed Sci* **54:** 575–587.

Cheng W, Miwa T, Kautsar V, Sasaki Y, Satoru S, Keitaro T and Yasuda H (2015) Combined use of Azolla and loach suppressed paddy weeds and increased organic rice yield: second season results. *J Wetlands Env Manag* **3(1):** 1–13.

Chhokar RS, Sharma RK, Jat GR, Pundir AK and Gathala MK (2007) Effect of tillage and herbicides on weeds and productivity of wheat under rice-wheat growing system. *Crop Prot* **26:** 1689–1696.

Chhokar RS, Sharma RK and Sharma I (2012) Weed management strategies in wheat-A review. *J. Wheat Res* **4(2):** 1–21.

Chinnusamy C, Nithya C and Ravishankar D (2014) Herbicide C.–tolerant GM crops in India: challenges and strategies. *Ind J Weed Sci* **46(1):** 86–90.

Chopra P and Angiras MN (2008) Influence of tillage and weed control methods on weeds, yield and yield attributes of maize. *Indian J Weed Sci* **40:** 47–50.

Corre-Hellou G, Dibet A, Hauggaard-Nielsen H, Crozat Y, Gooding M, Ambus P, Dahlmann C., Fragstein P, Pristeri A and Monti M (2011) The competitive ability of pea-barley intercrops against weeds and the interactions with crop productivity and soil N availability. *Field Crops Res* **122:** 264–272.

Cousens R, Mortimer M (1995) Dynamics of weed populations. Cambridge University Press, Cambridge, UK. 332 pp.

Culman SW, Duxbury JM, Lauren JG and Thies JE (2006) Microbial community response to soil solarization in Nepal's rice-wheat cropping system. *Soil Biol Biochem* **38:** 3359–3371.

Dahal, Saugat and Tika BK (2014) Conservation agriculture based practices affect the weed dynamics in spring maize. *World J Agric Res* **2(6A):** 25–33.

Daniel JT, Templeton GM, Smith RJ and Fox WT (1973) Biological control of northern jointvetch in rice with an endemic fungal disease. *Weed Sci* **21(4):** 303–307.

Daniya E, Salihu AD, Williams BN, Chiroma K and Bashir AB (2015) Effect of seed rate and nitrogen fertilizer on weed species composition, density and diversity in two sesame varieties. *Archi Agron and Soil Sci* **61:** 553–567.

Das TK (2008) Weed science: basics and application. Jain Brothers Pub, New Delhi, First edition.

Das TK and Yaduraju NT (2007) Effect of irrigation and nitrogen levels on grassy weed competition in wheat and comparative eco-physiology of Phalaris minor Retz. and Avena sterilis ssp. leudoviciana in wheat. Indian J *Weed Sci* **39:** 178–184.

Das TK and Yaduraju NT (2011) Effects of missing-row sowing supplemented with row spacing and nitrogen on weed competition and growth and yield of wheat. *Crop Pasture Sci* **62(1):** 48–57.

Das TK and Yaduraju NT (2012) The effects of combining modified sowing methods with herbicide mixtures on weed interference in wheat crops. *Int J Pest Manag* **58(4):** 310–319.

Davis AS and Liebman M (2003) Cropping system effects on giant foxtail (Setaria faberi) demography: green manure and tillage timing. *Weed Sci* **51:** 919–929.

Day TA, Martin G and Vogelmann TC (1993) Penetration of UV-B radiation in foliage: evidence that the epidermis behaves as a non-uniform filter. *Plant Cell Env* **16:** 735–741.

Dayan F, Romagni J, Tellez M, Rimando A and Duke S (1999) Managing weeds with natural products. Pestic Outlook (October): 185–188.

De-cauwer B, Bogaert S, Claerhout S, Bulcke R and Reheul D (2015) Efficacy and reduced fuel use for hot water weed control on pavements. *Weed Res* **55:** 195–205.

Derksen DA, Anderson RL, Blackshaw RE and Maxwell B (2002) Weed dynamics and management strategies for cropping systems in the northern Great Plains *Agron J* **94:** 174–185.

Dev D, Vivek, Singh SP and Kumar R (2013) Weed management studies in wheat with herbicides under different establishment methods. *Indian J Agron* **58(2):** 215–219.

Dhima K, Vasilakoglou I, Lithourgidis A, Mecolari E, Keco R, Agolli XH and Eleftherohorinos I (2008) Phytotoxicity of 10 winter barley varieties and their competitive ability against common poppy and ivy-leaved speedwell. *Exp Agric* **44:** 385–397.

Diprose MF and Benson FA (1984) Electrical methods of killing plants. *J Agric Eng Res* **29:** 197–209.

Drews S, Neuhoff D and Kopke U (2009) Weed suppression ability of three winter wheat varieties at different row spacing under organic farming conditions. *Weed Res* **49**: 526–533.

Dube E, Chiduza C, Muchaonyerwa P, Fanadzo M and Mthoko T (2012) Winter cover crops and fertiliser effects on the weed seed bank in a low-input maize-based conservation agriculture system. *South Afr J Plant Soil* **29:** 195–197.

Dubey RP (2014) Integrated weed management-an approach. In Training Manual Advance Training in Weed Management, held at DWSR, Jabalpur, India on 14–23 January, pp. 19–21.

Evans SP, Knezevic SZ, Shapiro C and Lindquist JL (2003) Nitrogen level affects critical period for weed control in corn. *Weed Sci* **51:** 408–417.

Fanadzo M, Chiduza C and Mnkeni PNS (2010) Effect of inter-row spacing and plant population on weed dynamics and maize yield at Zanyokwe irrigation scheme, Eastern Cape, South Africa. *African J Agric Res* **5(7):** 518–523.

Farooq M, Flower KC, Jabran K, Wahid A and Siddique KHM. 2011. Crop yield and weed management in rainfed conservation agriculture. *Soil Till Res.* **117:**172–183.

Farooq M, Jabran K, Cheema ZA, Wahid A and Siddique KHM (2011) Role of allelopathy in agricultural pest management. *Pest Manag Sci* **67:** 494–506.

Feizabady AZ (2013) Effects of crop rotation and residue management on bread wheat. *African J Plant Sci* **7(5):** 176–184.

Fernandez-Aparicio M, Emeran AA and Rubiales D (2010) Inter-cropping with berseem clover reduces infection by Orobanche crenata in legumes. *Crop Prot* **29:** 867–871.

Filizadeh Y, Rezazadeh A and Younessi Z (2007) Effects of crop rotation and tillage depth on weed competition and yield of rice in the paddy fields of northern Iran *J Agri Sci Tech* **9:** 99–105.

Finney DM, Creamer NG, Schultheis JR, Wagger MG and Brownie C (2009) Sorghum sudangrass as a summer cover and hay crop for organic fall cabbage production. *Renew. Agric. Food Syst.* **24:** 225–233.

Firbank LG (2003) An introduction to the farm-scale evaluations of genetically modified herbicide- tolerant crops. *J Applied Eco* **40:** 2–16.

Garrison AJ, Miller AD, Ryan MR, Roxburgh SH and Shea K (2014) Stacked crop rotations exploit weed-weed competition for sustainable weed management. *Weed Sci* **62:** 166–176.

Ghanbari M (2013) Comparing irrigation methods and weed control on yield of garlic cultivars. *World of Sci J* **3:** 70–78.

Gibson KD, McMillan J, Hallett SG, Jordan T and Weller SC (2011) Effect of living mulch on weed seed banks in tomato. *Weed Tech* **25(2):** 245–251.

Gill GS (1996) Managment of herbicide resistant ryegrass in Western Australia—research and its adoption. Pages 542–545 in R.C.H. Shepherd, ed. 11th Australian Weeds conference. Melbourne, Victoria, Australia: Weed Science Society of Victoria.

Gnanavel I (2015) Eco-friendly weed control options for sustainable agriculture. *Sci International* **3(2):** 37–47.

Gonz-alez-Diaz L, Van Den Berg F, Van Den Bosch F and Gonzalez-Andújar JL (2012) Controlling annual weeds in cereals by deploying crop rotation at the landscape scale: *Avena sterilis* as an example. *Ecol Appl* **22:** 982–992.

Graglia E, Melander B and Jensen RK (2006) Mechenical and cultural stratigies to control Cirsium arvense in organic arable cropping system. *Weed Res* **46:** 304–312.

Grichar WJ, Besler BA, Brewer KD and Minton BW (2004) Using soil-applied herbicides in combination with glyphosate in a glyphosate-resistant cotton herbicide program. *Crop Prot* **23:** 1007–1010.

Grubers, Claupein W (2009) Effect of tillage intensity on weed infestation in organic farming. *Soil and Tillage Res* **105:** 104–111.

Guza AE, Renner KA, Laboski C and Davis AS (2008) Effect of early spring fertilizer nitrogen on weed emergence and growth. *Weed Sci* **56:** 714–721.

Han H, Ning T and Li Z (2013) Effects of tillage and weed management on the vertical distribution of microclimate and grain yield in a winter wheat field. *Plant Soil Env* **59 (5):** 201–207.

Hansson D and Ascard J (2002) Influence of developmental stage and time of assessment on hot water weed control. *Weed Res* **42:** 307–316.

Haramoto ER and Gallandt ER (2004) Brassica cover cropping for weed management: a review. *Renew. Agric. Food Syst* **19:** 187–198.

Hartwig NL and Ammon HU (2002) Cover crops and living mulches. *Weed Sci* **50:** 688–699.

Hassan D, Upasani RR, Thakur R, Singh MK and Puran AN (2010) Productivity of rice as affected by planting method and weed management. Indian J *Weed Sci* **42:** 67–69.

Heggenstaller AH, Menalled FD, Liebman M and Westerman PR (2006) Seasonal patterns in post-dispersal seed predation of Abutilon theophrasti and Setaria faberi in three-cropping systems. *J Appl Ecol* **43:** 999–1010.

Helgason BL, Walley FL and Germinda JJ (2009) Fungal and bacterial abundance in long term no-till and intensive-till soils of the Northern Great Plains. *Soil Sci Soc Am J* **73:** 120–127.

Hiltbrunner J, Streit B and Liedgens M (2007) Are seeding densities an opportunity to increase grain yield of winter wheat in a living mulch of white clover. *Field Crops Res* **102:** 163–171.

Hoffman MJ, Weston LA, Snyder JC and Reigner EE (1996) Separating the effects of sorghum and rye root and shoot residues on weed development. *Weed Sci* **44:** 402–407.

Hollander NG, Bastiaans L and Kropff MJ (2007) Clover as a cover crop for weed suppression in an intercropping design. I. Characteristics of several clover species. *Eur J Agron* **26(2):** 92–103.

Hosmani MM and Meti SS (1993) Non-chemical means of weed management in crop production. Proceedings of the conference on integrated weed management for sustainable agriculture. Department of Agronomy, CCS Haryana Agricultural Universtiy, Hisar, Haryana, India, **1:** 299–305.

Hoy MA (1990) Commentary: The importance of biological control in U.S. Agriculture. *J Sust Agric* **1:** 59–79.

Imoloame EO, Gworgwor NA and Joshua SD (2007) Sesame weed infestation, yield and yield components as influenced by sowing method and seed rate in a Sudan Savanna agro-ecology of Nigeria. *African J Agric Res* **2:** 528–533.

Iqbal J, Cheema ZA and Mushtaq MN (2007) Allelopathic crop water extracts reduce the herbicide dose for weed control in cotton. *Int J Agric Biol* **11:** 360–366

Jabran K, Cheema ZA, Farooq M, Basra SMA, Hussain M and Rehman H (2008) Tank mixing of allelopathic crop water extracts with pendimethalin helps in the management of weeds in canola field. *Int J Agric Biol* **10:** 293–296.

Jabran K, Cheema ZA, Farooq M and Hussain M (2010) Lower doses of pendimethalin mixed with allelopathic crop water extracts for weed management in canola. *Int J Agric Biol* **12:** 335–340.

Jabran K and Farooq M (2013) Implications of potential allelopathic crops in agricultural systems, allelopathy. Springer Berlin Heidelberg, pp. 349–385.

Jabran K, Mahajan G, Sardana V and Chauhan BS (2015) Allelopathy for weed control in agricultural systems. *Crop Prote* **72:** 57–65.

Jamshidi K, Yousefi AR and Oveisi M (2013) Effect of cowpea intercropping on weed biomass and maize yield. *New Zeal. J Crop Hort* **41(4):** 180–188.

Jat RK, Banga RS and Yadav A (2013) Resource conservation techniques and pendimethalin for control of weeds in durum wheat cultivars. Indian J *Weed Sci* **45:** 93–98.

Javaid A, Jabeen K, Samad S and Javaid A (2011) Management of parthenium weed by extracts and residue of wheat. *Afr J Biotech* **10:** 14399–14403.

Jayanth KP (1987) Introduction and establishment of *Zygogramma bicolorata* on *Parthenium hysterophorus* at Bangalore, India. *Current Sci* **40:** 568–569.

Johnson DE, Wopereis MCS, Mbodj D, Diallo S, Powers S and Haefele SM (2004) Timing of weed management and yield losses due to weeds in irrigated rice in the Sahel. *Field Crops Res* **85:** 31–42.

Johnson WC, Benjamin G and Mullinix JR (1995) Weed management in peanut using stale seedbed techniques. *Weed Res* **43(2):** 293–297.

Johnson WC, Davis RF and Mullinix Jr BJ (2007) An integrated system of summer solarization and fallow tillage for *Cyperus esculentus* and nematode management in the southeastern coastal plain. *Crop Prot* **26:** 1660–1666.

Jordan NR, Zhang J and Huerd S (2000) Arbuscular-mychorrhizal fungi: potential roles in weed management. *Weed Res* **40:** 397–410.

Josh Volk (2008) Basic tools for mechanical cultivation. *Growing for Market* **17:** 7–10.

Kandhro MN, Tunio S, Rajpar I and Chachar Q (2014) Allelopathic impact of sorghum and sunflower intercropping on weed management and yield enhancement in cotton. *Sarhad J Agri Sci* **30:** 311–318.

Kanmegne and Degrande A (2002) From alley cropping to rotational fallow: farmers' involvement in the development of fallow management techniques in the humid forest zone of Cameroon. *Agrof Syst* **54:** 115–120.

Kathiresan RM, Ramah K and Sivakumar C (2001) Integration of azolla, fish and herbicides for rice weed management. Proceedings of the British Crop Protection Conference-Weeds, November 12–15, 2001, Brighton, UK., pp: 625–632.

Kato-Noguchi H and Ino T (2005) Concentration and release level of momilactone B in the seedlings of eight rice cultivars. *J Plant Physiol* **162:** 965–969.

Kato-Noguchi H, Ino T and Ota K (2008) Secretion of momilactone A from rice roots to the rhizosphere. *J Plant Physiol* **165:** 691- 696.

Kaur M, Aggarwal NK, Kumar V and Dhiman R (2014) Effects and Management of Parthenium hysterophorus: a weed of global significance. Hindawi Publishing Corporation, http://dx.doi.org/10.1155/2014/368647, pp.1–12.

Keeling JW, Dotray PA, Osborne TS and Asher BS (1998) Annual and perennial weed management strategies in Roundup Ready cotton with Roundup Ultra. P. 49. In: Proceedings of the Southern Weed Science Society, Champaign, IL, Vol. 51.

Kewat ML (2014) Improved weed management in Rabi crops. National Training on Advances in Weed Management, pp. 22–25.

Khaliq A, Cheema ZA, Mukhtar MA and Ahmad SM (1999) Evaluation of sorghum water extract for weed control in soybean. *Int J Agric Biol* **1:** 23–26.

Khaliq A and Matloob A (2011) Weed crop competition period in three fine rice cultivars under direct seeded rice culture. Pak J *Weed Sci* Res **17:** 229–243.

Khan MAI, Ueno K, Horimoto S, Komai F, Tanaka K and Ono Y (2007) Evaluation of the use of rice bran compost for eco-friendly weed control in organic farming systems. *Am J Environ Sci* **3:** 234–239.

Khan MB, Ahmad M, Hussain M, Jabran K, Farooq S and Waqas-Ul-Haq M (2011) Allelopathic plant water extracts tank mixed with reduced doses of atrazine efficiently control weeds in maize. In: International science conference on prospects and challenges to sustainable agriculture, University of Azad Jammu and Kashmir, Faculty of Agriculture, Rawalako, 14–16 July.

Khankhane PJ, Barman KK and Varshney JG (2009) Effect of rice residue management practices on weed density, wheat yield and soil fertility in a swell-shrink soil. *Ind J Weed Sci* **41:** 41–45.

Kholi RK, Singh HP and Batish DR (2004) Allelopathy in agro-ecosystems. Food Products Press. New York, USA.

Knezevic S, Datta A, Stepanovic S, Bruening C, Neilson B and Gogos G (2011) Weed control with flaming and cultivation in corn. *Phytopathology* **101:** 81–92.

Knezevic S and Ulloa S (2007) Potential new tool for weed control in organically grown agronomic crops. *J Agric Sci Belg* **52:** 95–104.

Kolb LN, Gallandt ER and Mallory EB (2012) Impact of spring wheat planting density, row spacing, and mechanical weed control on yield, grain protein, and economic return in Maine. *Weed Sci* **60:** 244–253.

Kong C (2005) Allelopathy in China. In: Harper JDI, An M, Wu H, Kent JH (eds.), Proceedings of fourth world congress on allelopathy. Establishing the scientific base, 21–26 Aug, Charles Strut University, Wagga Wagga, NSW pp 314–317.

Koocheki A, Nassiri M, Alimoradi L and Ghorbani R (2009) Effect of cropping systems and crop rotations on weeds. *Agron Sust Dev* **29:** 401–408.

Krikland KJ, Holm FA and Stevenson FC (2000) Appropriate crop seeding rate when herbicide rate is reduced. *Weed Tech* **14:** 692–698.

Kristensen L, Olsen J and Weiner J (2008) Crop density, sowing pattern, and nitrogen fertilization effects on weed suppression and yield in spring wheat. *Weed Sci* **56:** 97–102.

Kruidhof HM, Bastiaans L and Kropff MJ (2008) Ecological weed management by cover cropping: effects on weed growth in autumn and weed establishment in spring. *Weed Res* **48:** 492–502.

Kruse M, Strandberg M and Strandberg B (2000) Ecological effects of allelopathic plants: a review. Silkeborg, Denmark, National Environmental Research Institute.

Kumar M, Ghorai AK, Majumdar B, Mitra S and Kundu DK (2015) Integration of Stale Seedbed with Herbicides for Weed Management in Jute and their Impact on Soil Microbes. *J Agri Search* **2(1):** 24–27.

Kumar S (2014) Biological control of terrestrial weeds. In training manual advance training in weed management, held at DWSR, Jabalpur, on 14–23 January, pp. 91–95.

Kupatt CC, Bassi AB and Allemann DV (1993) Future methods for controlling weeds, plant diseases, and insects. In: Pesticide Interactions in Crop Production, pp. 545–565. Altman, B., Ed., CRC Press, Boca Raton, FL.

Kyser GB and DiTomaso JM (2002) Instability in a grassland community after the control of yellow starthistle (*Centaurea solstitialis*) with prescribed burning. *Weed Sci* **50:** 648–657.

Legere A and Samson N (2004) Tillage and weed management effects on weeds in barley-red clover cropping systems. *Weed Sci* **52:** 881–885.

Li YH (1987) Parasitism and integrated control of dodder on soybean. Proceedings of the 4th International Symposium on Parasitic Flowering Plants, August 2–7, 1987, Marburg, Germany, pp: 497–500.

Liebman M and Davis AS (2000) Integration of soil, crop, and weed management in low-external-input farming systems. *Weed Res* **40:** 27–47.

Liebman M and Mohler CL (2001) Weeds and the soil environment. pp. 210–268. In : ecological management of agricultural weeds. M Liebman *et al.* (eds.). Cambridge University Press, New York.

Lumpkin TA and Plucknett DL (1980) Azolla: botany, physiology and use as a green manure. *Econ Bot* **34:** 111–153.

Macias FA, Chinchilla N, Varela RM and Molinillo JMG (2006) Bioactive steroids from rice. *Steroids* **71:** 603–608.

Mahadeveppa M and Kulkarni RS (1996) Management of Parthenium through allelopathic interactions. In: Narwal SS, Tauro P (eds) Allelopathy in pests management for sustainable agriculture. Scientific Publishers (India), Jodhpur, pp 67–76.

Mahajan G and Chauhan BS (2013) The role of cultivars in managing weeds in dry seeded rice production systems. *Crop Prot* **49:** 52–57.

Mahajan G, Gill MS and Singh K (2010) Optimizing seed rate to suppress weeds and to increase yield in aerobic direct seeded rice in northwestern Indo-gangetic plains. *Journal of New Seeds* **11:** 225–238.

Mahmood K, Khaliq A, Cheema ZA and Arshad M (2013) Allelopathic activity of Pakistani wheat genotypes against wild oat. *Pak J Agric Sci* **50:** 169–176.

Main CL, Michael AJ and Murdock EC (2007) Weed response and tolerance of enhanced glyphosate-resistant cotton to glyphosate. *J Cotton Sci* **11:** 104–109.

Major J, Steiner C, Ditommaso A, Falcao NP and Lehmann J (2005) Weed composition and cover after three years of soil fertility management in the central Brazilian Amazon: compost, fertilizer, manure and charcoal applications. *Weed Biol Manag* **5:** 69–76.

Manjunath, Kumar R, Kumar S and Thakral SK (2010) Effect of irrigation and weed management on lentil under different planting techniques. Ind J *Weed Sci* **42:** 56–59.

Marenco RA, Maria A and Santos B (1999) Crop rotation reduces weed competition and increases chlorophyll concentration and yield of rice. *Pesquisa Agropecuaria Brasileira* **34(10):** 1–9.

Marshall EJP, Brown VK, Boatman ND, Lutman PJW, Squire GR and Ward LK (2003) The role of weeds in supporting biological diversity within crop fields. *Weed Res* **43:** 77–89.

Mashingaidze AB (2004) Improving weed management and crop productivity in maize systems in Zimbabwe. PhD Thesis, Wageningen University, The Netherlands p. 196.

Mashingaidze N, Twomlow SJ and Hove L (2009) Crop and weed responses to residue retention and method of weeding in first two years of a hoe-based minimum tillage system in semi-arid Zimbabwe. *Journal of SAT Agric Res* **7:** 1–11.

Mason HE and Spaner D (2006) Competitive ability of wheat in conventional and organic management systems: A review of the literature. *Can J Plant Sci* **86:** 333–343.

Mathiassen SK, Bak T, Christensen S and Kudsk P (2006) The effect of laser treatment as a weed control method. *Biosyst Eng* **95:** 497–505.

Matloob A, Khaliq and Chauhan BS (2014) Weeds of rice in Asia: problems and opportunities. Adv. Agron. p. 130 http://dx.doi.org/10.1016/bs.agron.

Matloob A, Khaliq and Chauhan BS (2014) Weeds of rice in Asia: problems and opportunities. Adv. Agron. p. 130 http://dx.doi.org/10.1016/bs.agron.2014.10.003 (in press).

Matthews JM, Llewellyn R, Powles S and Reeves T (1996) Integrated weed management for the control of herbicide resistant annual ryegrass. Pages 417–420 in Proceedings of the 8th Australian Agronomy Conference, Toowoomba, Queensland, Australia, 30 January to 2 February, 1996. Toowoomba, Australia: Australian Society of Agronomy.

Mauchline AL, Watson SJ, Brown VK and Froud-Williams RJ (2005) Post-dispersal seed predation of non-target weeds in arable crops. *Weed Res* **45:** 157–164.

McGovern RJ, McSorley R and Wang KH (2004) Optimizing bed orientation and number of plastic layers for soil solarization in Florida. *Soil and Crop Science Society of Florida Proc* **63:** 92–95.

Mehta RS, Patel BS, Meena SS, Anwar MM and Maheria SP (2010) Weed dynamics and yield of fenugreek as influence by irrigation and weed management practices. In National Symposium on IWM in the Era of Climate Change, held at NAAS, New Delhi on 21–22 August, pp. 56.

Meiss H, Le Lagadec L, Munier-Jolain N, Waldhardt R and Petit S (2010) Weed seed predation increases with vegetation cover in perennial forage crops. *Agric Ecosyst Envi* **138:** 10–16.

Melander B (1998) Interaction between soil cultivation in darkness, flaming, and brush weeding when used for in-row weed control in vegetables. Biol Agric Hort **16:** 1–14.

Menalled F, Smith R, Dauer J and Fox T (2007) Impact of agricultural management on carabid communities and weed seed predation. *Agric Ecosyst Envi* **118:** 49–54.

Milberg P, Hallgren E and Palmer MW (2001) Timing of disturbance and vegetation development: how sowing date affects the weed flora in spring sown crops. *J Veg Sci* **12:** 93–98.

Mirsky SB,Gallandt ER, Mortensen DA, Curran WS and Shumway DL (2010) Reducing the germinable weed seed bank with soil disturbance and cover crops. *Weed Res* **50:** 341–352.

Mishra JS, Singh V P and Namrata J (2010) Long term effect of tillage and weed control on weed dynamics, soil properties and yield of wheat in rice wheat system. *Ind* J *Weed Sci* **42:** 9–13.

Mishra JS and Singh VP (2008) Effect of tillage sequence and weed control on weed seed bank dynamics in soybean-linseed cropping system in Vertisols. *Ind J Agri Sci* **78:** 981–983.

Mohammaddoust-e-Chamanadad HR, Tulikor AM and Baghestani MA (2006) Effect of long term fertilizer application and crop rotation on the infestation of fields weeds. *Pak J Weed Sci* Res **12:** 221–234.

Mohler CL (1996) Ecological bases for the cultural control of annual weeds. *J Prod Agric* **9:** 468–474.

Mohler CL (2001) Mechanical management of weeds. In: ecological management of agricultural weeds. (Eds. Liebman M, Mohler CL and Staver CP). Cambridge University Press, Cambridge , pp. 139–209.

Moonen AS and Barberi P (2004) Size and composition of the weed seedbank after 7 years of different cover-crop-maize management system. *Weed Res* **44:** 163–177.

Moore (2009) Gorse control in Western Australia (including microwave eradication of seed banks) microwave machines developed by Graham Brodie of University of Melbourne.

Moran P and Greenberg S (2008) Winter cover crops and vinegar for early-season weed control in sustainable cotton. *J Sustain Agric* **32:** 483–506.

Mtambanengwe F, Hatirarami N, Tonny T, Christopher C, Muneta GM and Mapfumo P (2015) Mulching and fertilization effects on weed dynamics under conservation agriculture-based maize cropping in Zimbabwe. *Environments* **2:** 399–414.

Murphy SD, Clements DR, Belaoussoff S, Kevan PG and Swanton CJ (2006) Promotion of weed species diversity and reduction of weed seedbank with conservation tillage and crop rotation. *Weed Sci* **54:** 69–77.

Nadeem MA, Tanveer A, Ali A, Ayub M and Tahir M (2010) Effect of weed control practices and irrigation levels on weeds and yield of wheat. *Ind J Agron* **52:** 60–63.

Narwal SS and Haouala R (2013) Role of allelopathy in weed management for sustainable agriculture. Allelopathy, Current Trends and Future Application Ed. Cheema, Farooq M and Wahid A pp 217–250.

Nasrollahzadeh S, Hajebrahimi A, Shafagh J, Kolvanagh and Kouh SS (2015) Effect of nitrogen rates and weed interference on yield and yield components of corn. *Int J Agron and Agric Res.* **7(1):** 1–6.

Nazarko OM, Van ARC and Entz MH (2005) Strategies and tactics for herbicide use reduction in field crops in Canada: A review. *Can J Plant Sci* **85:** 457–479.

Norsworthy JK, Brandenberger L, Burgos NR and Riley M (2005) Weed suppression in *Vigna unguiculata* with a spring seeded brassicaceae green manure. *Crop Prot* **24:** 441–447.

Norsworthy JK, McClelland M, Griffith G, Bangarwa SK and Still J (2011) Evaluation of cereal and Brassicaceae cover crops in conservation-tillage, enhanced, glyphosate-resistant cotton. *Weed Tech* **25:** 6–13.

Olofsdotter M (2001) Rice - a step toward use of allelopathy. *Agron J* **93:** 3–8.

Olofsdotter M, Navarez D, Rebulanan M and Streibig JC (1999) Weed suppressing rice cultivars - does allelopathy play a role. *Weed Res* **39:** 441–454.

Olsen JM, Griepentrog HW, Nielsen J and Weiner J (2012) How important are crop spatial pattern and density for weed suppression by spring wheat. *Weed Sci* **60:** 501–509.

Pandey AK, Mishra and Hasija SK (1998) Effect of inoculum on mycoherbicidal potential of *Sclerotium rolfsii* against Parthenium. *J Myco and Plant Patho* **28:** 284–287.

Paolini R, Principi M, Del Puglia S and Lazzeri L (1998) Competitive effects between sunflower and six broad-leaved weeds. In Proc. 6th EWRS Mediterranean Symposium, Montpellier, France, pp. 81–88.

Paolini R, Principi M, Froud-Williams RJ, Del Puglia S and Biancardi E (1999) Competition between sugarbeet and *Sinapis arvensis* and *Chenopodium album*, as affected by timing of nitrogen fertilization. *Weed Res* **39:** 425–440.

Parish S (1990) A review of non-chemical weed control techniques. Biol Agric Hort **7:** 117–137.

Parmar JR, Patel MV, Patel JJ, Parmar AB and Shah SN (2010) Influence of irrigation schedules based on IW/CPE ration and herbicidal weed control in Isabgul. *Ind J Weed Sci* **42:** 87–90.

Pathak, Kannan C (2011) Isolation and pathogenicity of some native fungal pathogens for the biological management of water hyacinth. *Ind J Weed Sci* **43:** 178–180.

Peachey RE, Pinkerton JN, Ivors KL, Miller ML and Moore LW (2001) Effect of soil solarization, cover crops, and mentha on field emergence and survival of buried annual bluegrass (*Poa annua*) seeds. *Weed Tech* **15:** 81–88.

Peigne J, Ball BC, Roger EJ and David C (2007) Is conservation tillage suitable for organic farming. *Soil Use Mgmt* **23:** 129–144.

Peter JD, DMonks W and Katherine MJ (2012) Effect of drip-applied herbicides on yellow nutsedge in plasticulture. *Weed Tech* **26(2):** 243–247.

Pickett JA, Aradottir GI, Birkett MA, Bruce TJ, Hooper AM, Midega CA, Jones HD, Matthes MC, Napier JA and Pittchar JO (2014) Delivering sustainable crop protection systems via the seed: exploiting natural constitutive and inducible defense pathways. *Philos. Trans. R. Soc. B Biol Sci* **369:** 201–202.

Place GT, Samuel CRH, Jim ED and Adam NS (2009) Seeding rate effects on weed control and yield for organic soybean production. *Weed Tech* 23(4): 497–502.

Pons TL (1987) Growth rates and competitiveness to rice of some annual weed species. In: Studies on Weeds and Rice Competition (ed. by Sorianegara I.). Biotrop, Bogor, Indonesia, 13–21.

Pouryousef M, Yousefi AR, Oveisi M and Asadi F (2015) Intercropping of fenugreek as living mulch at different densities for weed suppression in coriander. *Crop Prot* **69:** 60–64.

Qasem JR and Foy CL (2001) Weed allelopathy, its ecological impacts and future prospects: a review. *J Crop Prod.* **4:** 43–119.

Rao AN, Johnson DE, Sivaprasad B, Ladha JK and Mortimer AM (2007) Weed management in direct-seeded rice. *Adv Agron* **93:** 153–255.

Rask AM, Kristofferrsen P and Andreasen C (2012) Controlling grassy weeds on hard surface: effect of time intervals between flam treatments. *Weed Tech* **26:** 83–88.

Rask AM and Kristoffersen P (2007) A review of non-chemical weed control on hard surfaces. *Weed Res* **47:** 370–380.

Rasmussen K (2000) Can slurry injection improve the selectivity of weed harrowing in cereals? In Proc. of the 4th Workshop of the EWRS Working Group on Physical and Cultural Weed Control, Elspeet, *The Netherlands*, 33–34.

Rehman A, Cheema ZA, Khaliq A and Jabran K (2007) Combined effect of five allelopathic crop mulches and two irrigation levels on weed suppression in transplanted fine rice. In: International workshop on allelopathy: current trends and future applications, Weed Science-Allelopathy Lab, Department of Agronomy, University of Agriculture, Faisalabad, 18–21 March.

Riemens MM, Weide RY, Bleeker PO and Lotz LAP (2007) Effect of stale seedbed preparations and subsequent weed control in lettuce on weed densities. *Weed Res* **47:** 149–156.

Rizvi SJH and Rizvi V (1992) Allelopathy: basic and applied aspects. Champan & Hall. London SE1 8HN.

Sangeetha SP, Balakrishnan A, Priya RS and Maheswari J (2009) Influence of seeding methods and weed management practices on direct seeded rice. *Ind J Weed Sci* **41:** 210–212

Sans FX, Berner A, Armengot L and Mader P (2011) Tillage effects on weed communities in an organic winter wheat–sunflower–spelt cropping sequence. *Weed Res* **51:** 413–421.

Saudy HS (2015) Maize-cowpea intercropping as an ecological approach for nitrogen-use rationalization and weed suppression. *Archives of Agron Soil* Sci **61:** 1–14.

Saxena S and Kumar M (2010) Evaluation of *Alternaria alternata* ITCC4896 for use as mycoherbicide to control *Parthenium hysterophorus. J Plant Prote Res* **47:** 213–218.

Schonbeck M (2011) Principles of sustainable weed management in organic cropping systems. 3rd Edition.

Sharma AR (2014) Weed management in conservation agriculture systems-problems and prospects. In National Training on Advances in Weed Management, held at DWSR, Jabalpur on 14–23 January, pp. 1–9.

Sharma RK, Chhokar RS, Jat ML, Singh S, Mishra B and Gupta RK (2008) Direct drilling of wheat into rice residues: experiences in Haryana and western Uttar Pradesh. In: permanent beds and rice-residue management for rice-wheat systems in the Indo-Gangetic Plain, Humphreys, E. and Roth, C.H. (eds.). Proc. workshop held in Ludhiana, India, 7–9 Sept., 2006. ACIAR Proc No **127:** 147–158.

Sharma V and Angiras N (1996) Effect of row orientations, row spacing and weed-control methods on light interception, canopy temperature and productivity of wheat. *Ind J Agron* **41:** 390–396.

Shesshadri T and Prabhakarasetty TK (2001) Weed smothering efficiency as influenced by intercropping in chilli. In: 1st biennial conference of ecofriendly weed management options for sustainable agriculture, 23–24 May 2001 University of Agricultural Sciences, Bangalore, p 174.

Shirtliffe SJ and Entz MH (2005) Chaff collection reduces seed dispersal of wild oat by a combine harvester. *Weed Sci* **53:** 465–470.

Shrestha A and Fidelibus M (2005) Grapevine row orientation affects light environment, growth, and development of black nightshade (*Solanum nigrum*). *Weed Sci* 53: 802–812.

Shukla R and Pandey AK (2006) Maximization of production of oxalic acid from *Sclerotium rolfsii*, a mycoherbicidal agent against *Parthenium*. *An Plant Prote Sci* **14:** 202–205.

Silva EM (2014) Screening five fall-sown cover crops for use in organic no-till crop production in the upper mid-west. *Agroecol Sust Food Syst* **38:** 748–763.

Sindhu PV, Thomas CG and Abraham CT (2010) Seed bed manipulations for weed management in wet seeded rice. *Ind J Weed Sci* **42:** 173–179.

Singh A, Kaur R, Kang JS and Singh G (2012) Weed dynamics in rice-wheat cropping system. *Global J Biology, Agri and Health Sci* **1:** 7–16.

Singh AK, Parihar CM, Jat SL, Kumar A and Singh NK (2012) Weed management strategies against complex weed flora in maize-wheat cropping system. In Proceeding of 3rd International Agronomy Congress, Held at IARI, New Delhi, pp. 120–121.

Singh B and Malhi SS (2006) Response of soil physical properties to tillage and residue management on two soils in a cool temperate environment. *Soil Till Res* **85:** 143–153.

Singh HP, Daizy RB and Kohli RK (2003) Allelopathic interactions and allelochemicals: new possibilities for sustainable weed management. *Critical Revi Plant Sci* **22:** 239–311.

Singh MK and Singh A (2012) Effect of stale seedbed method and weed management on growth and yield of irrigated direct-seeded rice. *Ind J Weed Sci* **44(3):** 176–180.

Singh R (2014) Weed management in major kharif and rabi crops. In National Training on Advance in Weed Management, held at DWSR, Jabalpur, on 14–23 January, pp. 31–40.

Singh R and Singh B (2004) Effect of irrigation time and weed management practices on weeds and wheat yield. *Indian J Weed Sci* **36(1&2) :** 25–27.

Singh RP and Singh RK (2006) Ecological approaches in weed management. In National symposium on conservation Agriculture and Environment, held at BHU, Varanasi, India on 26–28 October, pp. 301–304.

Singh S (2014) Sustainable weed management in cotton. *Har J Agron* **30:** 1–14.

Singh Samar, Chhokar RS, Gopal R, Ladha JK, Gupta RK, Kumar V and Singh M (2009) Integrated weed management: A key to success for direct–seeded rice in the Indo-Genetic Plains, pp. 261–278. In: integrated crop and resource management in the rice–wheat system of South Asia (Eds. Ladha JK, Singh Y, Errenstein O and Hardy B), International Rice Research Institute, Los Banos, Philippines.

Singh VP (2014) Soil-solarization: a non-chemical technique of weed management. In National Training on Advance in Weed Management, held at DWSR, Jabalpur, on 14–23 January, pp. 56–63.

Smeda RJ and Weller SC (1996) Potential of rye for weed management in transplanted tomatoes. *Weed Sci* **44:** 596–602.

Smith RG (2006) Timing of tillage is an important filter on the assembly of weed communities. *Weed Sci* **54:** 705–712.

Sodaeizadeh H and Hossein Z (2012) Allelopathy an Environmentally Friendly Method for Weed Control. International Conference on Applied Life Sciences (ICALS 2012), Turkey, September 10–12, pp 387–392.

Sodaeizadeh H, Rafieiolhossaini M, Havlík J and Damme PV (2009) Allelopathic activity of different plant parts of *Peganum harmala* L. and identification of their growth inhibitors substances. *Plant Growth Regu* **59:** 227–236.

Stapleton JJ (2000) Soil solarization in various agricultural production systems. *Crop Prot.* **19:** 837–841.

Steinmaus S, Elmore CL and Smith RJ (2008) Mulched cover crops as an alternative to conventional weed management systems in vineyards. *Weed Res* **48:** 273–281.

Stevan Z. Knezevic, Evans P and Mainz M (2003) Row spacing influences the critical timing for weed removal in soybean. *Weed Tech* **17(4):** 666–673.

Sweeney AE, Renner KA, Laboski C and Davis A (2008) Effect of fertilizer nitrogen on weed emergence and growth. *Weed Sci* **56:** 714–721.

Talebbeigi RM and Ghadiri H (2012) Effects of cowpea living mulch on weed control and maize yield. *J Biol Env Sci* **6(17):** 189–193.

Teasdale JR and Mohler CL (2000) The quantitative relationship between weed emergence and the physical properties of mulches. *Weed Sci* **48:** 385–392.

Thorsted MD, Olesen JE and Weiner J (2006) Width of clover strips and wheat rows influence grain yield in winter wheat/white clover intercropping. *Field Crops Res* **95:** 280–290.

Tiwari A, Meena M, Zehra A and Upadhyay RS (2013) Efficacy of *Alternaria alternata* as bioherbicide against weed species. In International Conference on Global Scenario of Traditional System of Medicine, Ayurveda, Agriculture and Education, RGSC, Barkachha BHU, on 21–22 January, pp. 498–501.

Toler JE, Murdock EC and Camberato JJ (2004) Starter fertilizer effects on cotton development and weed interference. *J Cotton Sci* **8:** 33–41.

Torrensen KS, Skuterud R, Tandsaether HJ, Bredensen M and Hagemo MB (2003) Long-term experiments with reduced tillage in spring cereals. I. Effects on weed flora, weed seedbank and grain yield. *Crop Prot* **22:** 185–200.

Towa JJ and Xiangping G (2014) Effects of irrigation and weed control methods on growth of weed and rice. *Int J Agric and Biol Eng* **7(5) :** 22–33.

Tracy BF, Renne IJ, Gerrish J and Sanderson MA (2004) Effects of plant diversity on invasion of weed species in experimental pasture communities. *Basic Appl Ecol* **5:** 543–550.

Vasilakoglou I, Dhima K, Lithourgidis A and Eleftherohorinos I (2009) Allelopathic potential of 50 barley cultivars and the herbicidal effects of barley extract. *Allelopathy J* **24:** 309–320.

Vasileiadis VP, Froud- Williams RJ and Eleftherohorinos IG (2006) Vertical distribution, size and composition of the weed seedbank under various tillage and herbicide treatments in a sequence of industrial crops. *Weed Res* **47:** 222–230.

Verma S K (2014) Enhancing sustainability in wheat production though irrigation regimes and weed management practices in eastern Uttar Pradesh. *The Ecoscan*, Special Issue **6:** 115–119.

Verma SK, Bikramaditya, Siya Ram, Singh SB, Yadav AS, Kumar D, Meena H P, Roopkishore and Meena BP (2009) Crop diversification for higher productivity and profitability in Indian perspectives: A review. *Int J Trop Agric* **27(3&4):** 477–491.

Verma SK, Singh SB (2008) Enhancing of wheat production through appropriate agronomic management. *Indian Farming* **58:** 15–18.

Verma SK, Singh SB, Prasad SK, Meena RN and Meena RS (2015) Influence of irrigation regimes and weed management Practices on water use and nutrient uptake in. *Bangladesh J Bot* **44(3):** 437–442

Vijayabaskaran S and Kathiresan RM (1993) Integrated weed management in rice-cotton cropping system. Proceedings of the international symposium on integrated weed management for sustainable agriculture, November 18–20, Hisar, India, pp: 62.

Walsh M, Newman P and Powles SB (2013) Targeting weed seeds in-crop: a new weed control paradigm for global agriculture. *Weed Tech* **27:** 431–436.

Walsh MJ, Harrington RB and Powles SB (2012) Harrington seed destructor: a new nonchemical weed control tool for global grain crops. *Crop Sci* **52:** 1343–1347.

Walsh MJ and Newman P (2007) Burning narrow windrows for weed seed destruction. *Field Crops Res* **104:** 24–40.

Walsh MJ and Powles SB (2007) Management strategies for herbicide resistant weed popula-tions in Australian dryland crop production systems. *Weed Tech* **21:** 332–338.

Wang KH, McSorley R and Kokalis BN (2006) Effects of cover cropping, solarization, and soil fumigation on nematode communities. *Plant Soil* **286:** 229–243.

Watanabe I and Hove CV (1996) Phylogenetic, molecular and breeding aspects of Azolla *Anabaena symbiosis*. In: Pteridology in Perspective (ed. by Camus JM, Gibby M and Johns RJ). *Royal Botanic Gardens, Kew*, 611–619.

Watson PR, Derksen DA and Van-Acker RC (2006) The ability of 29 barley cultivars to compete and withstand competition. *Weed Sci* **54:** 783–792.

White SS, Renner KA, Menalled FD and Landis DA (2007) Feeding preferences of weed seed predators and effect on weed emergence. *Weed Sci* **55:** 606–612.

Woodburn AT (2000) Glyphosate: Production, pricing and use worldwide. *Pest Manag Sci* **56:** 309–319.

WSSA (1994) Herbicide Hand book. Weed Science Society of America, Champaign, IL. 7th ed. pp. 147–149,

Wu H (2005) Molecular approaches in improving wheat allelopathy. In: Harper JDI, An M, Wu H, Kent JH (eds.), Proceedings of fourth world congress on allelopathy. Establishing the scientific base, 21–26 Aug, Charles Strut University, Wagga Wagga, NSW pp 201–208.

Xu GF, Zhang FD, Li TL, Wu D and Zhang YH (2010) Induced effects of exogenous phenolic acids on allelopathy of a wild rice accession (*Oryza longistaminata* S37). Rice Sci **17:** 135–140.

Xuan TD, Tawata S, Khanh TD and Chung IM (2005) Decomposition of allelopathic plants in soil. *J Agron and Crop Sci* **191:** 162–171.

Yadollahi P, Borjibbad AR, Khaje M, Reza M, Asgharipour and Amiri A (2014) Effect of intercropping on weed control in sustainable agriculture. *Int J Agric and Crop Sci* **7(10):** 683–686.

Yeganehpoor F, Salmasi SZ, Abedi G, Samadiyan F and Beyginiya V (2015) Effects of cover crops and weed management on corn yield. *J Saudi Society of Agric Sci* **14(2):** 178–181.

Yenish JP, Worsham AD and Chilton WS (1995) Disappearance of DIBOA- glucoside, DIBOA and BOA from rye cover crop residues. *Weed Sci* **43:**18–20.

Yin L, Cai Z and Zhong W (2005) Changes in weed composition of winter wheat crops due to long-term fertilization. *Agric Ecosyst Env* **107:** 181–186.

Yongqing MA (2008) Allelopathic studies of common wheat. *Weed Biol Manag* **5:** 93–104.

Yousefi AR, Gonzalez-Andujar JL, Alizadeh H, Baghestani MA, Rahimian H and Karimmojeni H (2012) Interactions between reduced rate of imazethapyr and multiple weed species-soya bean interference in a semi-arid environment. *Weed Res* **52:** 242–251.

Yousefi AR and Rahimi MR (2014) Integration of soil-applied herbicides at the reduced rates with physical control for weed management in fennel (*Foeniculum vulgare* Mill.). *Crop Prot* **63:** 107–112.

Zeng RS (2014) Allelopathy-the solution is indirect. *J Chem Ecol* **40:** 515–516.

Zhang TY (1985) A forma specialis of *Colletotrichum gloeosporioides* on *Cuscuta* spp. *Acta Mycol Sin* **4:** 234–239.

Zimdahl RL (1999) Fundamentals of Weed Science, 2nd ed., Academic Press, San Diego, pp. 450.

Zimdahl RL (2007) Fundamental of weed science. Academic Press, Burlington, MA.

Zimdahl RL (2013) Fundamentals of weed science. Waltham MA: Academic Press, pp 295–344.

■■■■

8

Carbon Sequestration for Agriculture Production System and Climate Change

Ram Swaroop Meena and Ekta Kumari

Department of Agronomy, Institute of Agricultural Sciences, Banaras Hindu University, Varanasi – 221 005, India

ABSTRACT

To combat the devastating effect of carbon dioxide there is strong and emergent need to divert the way of produced CO_2 incurred in the atmosphere direct to soil absolutely for increasing the juvenescence of soil. The panacea is (1) to prevent carbon emissions produced by human activities from reaching the atmosphere by capturing and diverting them to secure storage, or, (2) to remove carbon from the atmosphere by various means and 'storing' it in the soil. Carbon sequestration refers to the removal of carbon dioxide from the atmosphere into a long-lived stable form that does not affect atmospheric chemistry. Currently, the only viable way to trap atmospheric carbon dioxide is via photosynthesis, where carbon dioxide is absorbed by plants and turned into carbon compounds for plant growth. Carbon is considered sequestered if it ends up in a stable form, such as wood or soil organic matter. Soil carbon sequestration is an important and immediate sink for removing atmospheric carbon dioxide and slowing global warming. Carbon sequestration is the capture and secure storage of carbon that would otherwise be emitted to or remain in the atmosphere. Therefore, in this chapter we would study in detail the potential cause and ultimate solution to combat the burning issue of the present era *i.e.*, Carbon Capture and Storage (CCS).

1. INTRODUCTION

Soils are the largest carbon reservoirs of the terrestrial carbon cycle which contained about three times more carbon than in the world's vegetation; hold double the amount of carbon that is present in the atmosphere and important sinks for carbon sequestration. Soils play a key role in the global carbon budget and greenhouse effect (Jha *et al.*, 2003). Soil carbon sequestration is win-win-win strategy because it offset fossil fuel emissions, improves water quality, enhances and sustains agronomic productivity. Agricultural soils are among the earth's largest terrestrial reservoirs of carbon (C) and hold potential for expanded C sequestration and crop due to photosynthetic ability are major sink for atmospheric CO_2. The soil organic carbon (SOC) pool in soils of India is 2.2 per cent of the world pool for 1 m depth and 2.6 per cent to 2 m depth. Worldwide, SOC in the top 1 meter of soil comprises about 3/4 of the earth's terrestrial carbon; nevertheless, there is tremendous potential to sequester additional carbon in soil. For example, many cropland soils of the United States have lost as much as 50 per cent of their original SOC due to the effects of land clearing and tillage. Such conventional farming practices "burn" SOC just as we burn fossil fuels today. However, in the case of SOC this historical decline can be reversed, which is not the case for fossil fuel reserves. Improved management of crop, grazing, and forest lands are estimated to potentially offset 30,000–60,000 million metric tons of the carbon released by fossil fuel combustion over the next 50 years (Lal, 2004).

Potential of soil C sequestration in India is estimated 6 to 7 Tg C y^{-1} for adoption of recommended management practices (RMPs) on agricultural soils, and 22 to 26 Tg C y^{-1} for secondary carbonates, while total potential of soil C sequestration is 39 to 49 (44 ± 5) Tg C y^{-1} (Lal 2004a). In cultivated land manures alone or in combination with fertilizers not only improve soil fertility, physical environment of soil, enhance biomass production but also help to build up of SOC content. Behind this, studies showed that SOC sequestration is a major challenge in soils of the tropics and sub-tropics, where climate is harsh and the rate of C (carbon) mineralization is high; because of high temperature and low humification efficiency (Ladha *et al.*, 2003; Pathak *et al.*, 2011).

2. PRESENT SCENARIO OF WORLD COMBATING CLIMATE CHANGE

Global surface temperatures have increased by 0.8ºC since the late nineteenth century, and 11 out of the 12 warmest years on record have occurred since 1995 (IPCC 2007). Earth's mean temperature is projected to increase by 1.5–5.8ºC during the twenty-first century. The rate of increase in global temperature has been 0.15ºC per decade since 1975. In addition to the sea-level rise of 15–23 cm during the twentieth century (IPCC 2007), there have been notable shifts in ecosystems and frequency, and intensity of occurrence of wild fires. These and other observed climate changes are reportedly caused by emission of greenhouse gases (GHGs) through anthropogenic activities including land-use change, deforestation, biomass burning, draining of wetlands, soil cultivation and fossil fuel combustion. Consequently, the concentration of atmospheric GHGs and their radiative forcing have progressively increased with increase in human population, especially since the onset of industrial revolution around 1850. The concentration of carbon dioxide (CO_2) has increased by 31 per cent from 280 ppmv in 1850 to 398 ppmv in 2015. Concentrations of methane (CH_4) and nitrous oxide (N_2O) have also increased steadily over the same period. Total radiative forcing, externally imposed perturbation in the radiative energy budget of the Earth's climate system, of all GHGs since 1850 is estimated at 2.43 W mK^2 (IPCC, 2007).

3. IMPORTANCE OF SOIL CARBON, SOIL ORGANIC CARBON AND SOIL ORGANIC MATTER

Soil carbon is used for growing crops and pastures are also very important for storing carbon. Soils can store large amounts of carbon, up to 50–300 tonnes per hectare, which is equivalent to 180–1100 tonnes of carbon dioxide. The above-ground component (or above-ground biomass) of pastures and crops store between 2 and 20 tonnes of carbon per hectare. The above-ground component of plantation forests can store 250 tonnes of carbon per hectare. Soil organic carbon, the major component of soil organic matter, is extremely important in all soil processes. Soil organic carbon refers to all the different carbon compounds found in soils that are, or were previously, a living organism (Prasad *et al.*, 2005).

4. SOURCE OF CARBON AND ENERGY FOR SOIL MICRO-ORGANISMS

- Cation exchange capacity which affects the retention (prevents the leaching of essential plant nutrients), release and availability of plant nutrients
- Major source of and a temporary sink for plant nutrients (such as N, P and S),
- Improvement in soil buffering capacity, against acidification and toxicities
- Formation and maintenance of desirable soil structure
- Improvement of water percolation into and retention by the soil
- Absorption of solar radiation which influences soil temperatures
- Ability to stimulate plant growth.

Formation of SOC from plant and animal residues is a complex phenomenon that includes the decomposition, humification, accumulation and distribution of the various organic substances within the soil profile. Soil organic carbon is made up of a continuum of soil organic compounds in terms of the ease of their decomposition and turnover time. Due to its complexity, for modelling purposes, SOC is conceptually divided into a number of pools based on their turnover rates (lability).

5. SOIL ORGANIC CARBON - THE BASIS OF SUSTAINABLE AGRICULTURE

Soil organic carbon (SOC) is the carbon associated with soil organic matter. Soil organic matter is the organic fraction of the soil that is made up of decomposed plant and animal materials as well as microbial organisms, but does not include fresh and un-decomposed plant materials, such as straw and litter, lying on the soil surface. Soil carbon can also be present in inorganic forms, *e.g.* lime or carbonates in some soils in the drier areas.

Table 1. Carbon pool size and changes due to human activities

Carbon pool size	
Vegetation	610Gt*
Atmosphere	750 Gt
Soil	1,580 Gt
Ocean	39,000 Gt
Carbon changes due to human activities	
Fossil fuel use	+ 5.5 Gt/year
Land Use	+ 1.6 Gt/year
Rate of carbon increase in the atmosphere	
+ 3.3 Gt/year	

Source: 'The carbon cycle, climate and the long-term effects of fossil fuel burning', J F Kastings, Consequences Vol 4, No 1, 1998.

6. CHARACTERISTICS OF HEALTHY SOIL

Healthy soils with high organic carbon content contribute to the following soil characteristics, which are important for soil fertility:

- Nutrient availability. Decomposition of soil organic matter releases nitrogen, phosphorus and a range of other nutrients for plant growth.
- Soil structure and soil physical properties. Soil organic carbon promotes a healthy soil structure by holding the soil particles together, thereby improving soil physical properties, such as water-holding capacity, water infiltration, gaseous exchange, root growth and ease of cultivation.
- Biological soil health. Soils contain microscopic plants and animals which live between, and feed on, the many soil particles. Soil organic matter plays an important role in the soil food web by controlling the number and types of soil inhabitants. These inhabitants serve important functions, such as cycling nutrients, making nutrients available for other organisms, assisting root growth, assisting plant nutrient uptake, creating burrows, and even suppressing crop diseases.
- Buffer for toxic and harmful substances. Soil organic matter can lessen the effect of harmful substances, for example toxins such as heavy metals, by acting as a buffer. Heavy metal toxins can bond very tightly

to soil particles, preventing their release into waterways where contamination affecting the food chain may occur. Soil organic matter also increases the adsorption of pesticides, thus reducing the amount of chemicals that may enter groundwater.

7. SOIL CARBON SEQUESTRATION

Plants sequester (remove) carbon from the atmosphere through growth. Carbon dioxide is converted into plant tissue through photosynthesis. After a plant dies, the plant material is decomposed primarily by soil microorganisms, and carbon is released back into the atmosphere through respiration or it is left behind as humus. So plants and the microorganisms in the soil provide the link between the carbon in the atmosphere and how it can be stored or fixed to biological matter in the soils - this process is called soil carbon sequestration (Singh *et al.*, 2008).

Carbon is found in all living organisms and is the major building block for life on Earth. Carbon exists in many forms, predominately as plant biomass, soil organic matter, and as the gas carbon dioxide (CO_2) in the atmosphere and dissolved in seawater. Carbon sequestration is the long term storage of carbon in oceans, soils, vegetation (especially forests), and geologic formations (Fig. 1). Although oceans store most of the Earth's carbon, soils contain approximately 75 per cent of the carbon pool on land — three times more than the amount stored in living plants and animals. Therefore, soils play a major role in maintaining a balanced global carbon cycle. Carbon dioxide (CO_2) capture and sequestration (CCS) is a set of technologies that can greatly reduce CO_2 emissions from new and existing coal- and gas-fired power plants and large industrial sources. CCS is a three-step process that includes:

- Capture of CO_2 from power plants or industrial processes
- Transport of the captured and compressed CO_2 (usually in pipelines)
- Underground injection and geologic sequestration (also referred to as storage) of the CO_2 into deep underground rock formations. These formations are often a mile or more beneath the surface and consist of porous rock that holds the CO_2.

7.1. Importance of carbon sequestration

Carbon dioxide capture and sequestration (CCS) could play an important role in reducing greenhouse gas emissions, while enabling low-carbon electricity generation from power plants (Singh *et al.*, 2011).

CCS technologies are currently available and can dramatically reduce (by 80–90%) CO_2 emissions from power plants that burn fossil fuels. Applied to a 500 MW coal-fired power plant, which emits roughly 3 million tons of CO_2 per year, the amount of GHG emissions avoided (with 90% reduction efficiency) would be equivalent to:

- Planting more than 62 million trees, and waiting at least 10 years for them to grow.
- Avoiding annual electricity-related emissions from more than 300,000 homes.

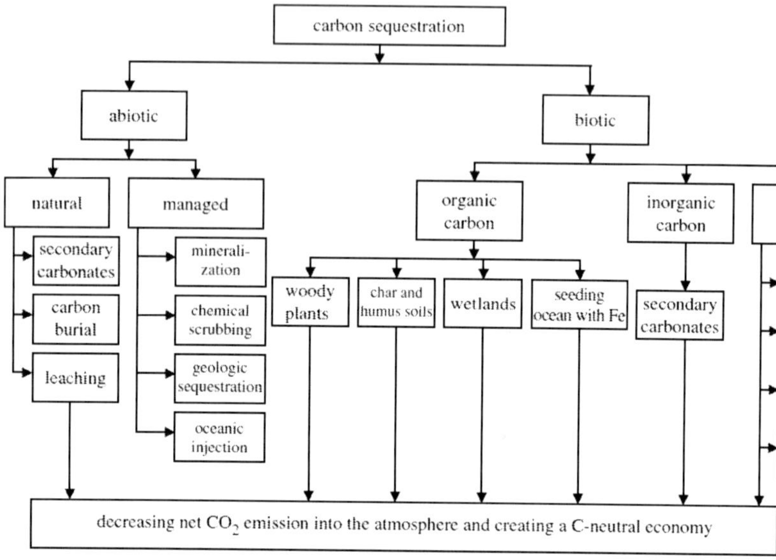

Fig. 1. Sequestration of C through various processes

7.2 Types of carbon sequestration

There are two major types of CO_2 sequestration: terrestrial and geologic (Fig. 2).

7.2.1. *Geologic sequestration*

Geologic sequestration is putting CO_2 into long-term storage in geologic zones deep underground. Geologic sequestration is the method of storage that is generally considered for carbon capture and storage (CCS) projects. CCS is the practice of capturing CO_2 at anthropogenic sources before it is released to the atmosphere and then transporting the CO_2 gas to a site where it can be put into long-term storage.

7.2.2. *Terrestrial carbon sequestration*

A carbon sink absorbs CO_2 from the atmosphere, and stores it as carbon; in the case of a growing forest, carbon storage is in the form of wood and other vegetation and soil carbon. Young fast-growing forests absorb carbon dioxide more rapidly than older forests. An old forest is characterized by slow-growing trees and carbon losses due to death and decay that may translate to a net loss of carbon over time. A carbon sink such as an old forest "may not be capturing any new carbon but can continue to hold large volumes of carbon as biomass over long periods of time." It is estimated that without the removal of CO_2 from the atmosphere via carbon sinks, the present concentration of CO_2 (398 parts per million) in 2015. Carbon sinks can help offset environmental damage of energy intensive activities. Because forests are sequestering carbon, forest managers are sometimes paid by corporations for maintaining carbon sinks, creating a "credit" on emissions for that corporation.

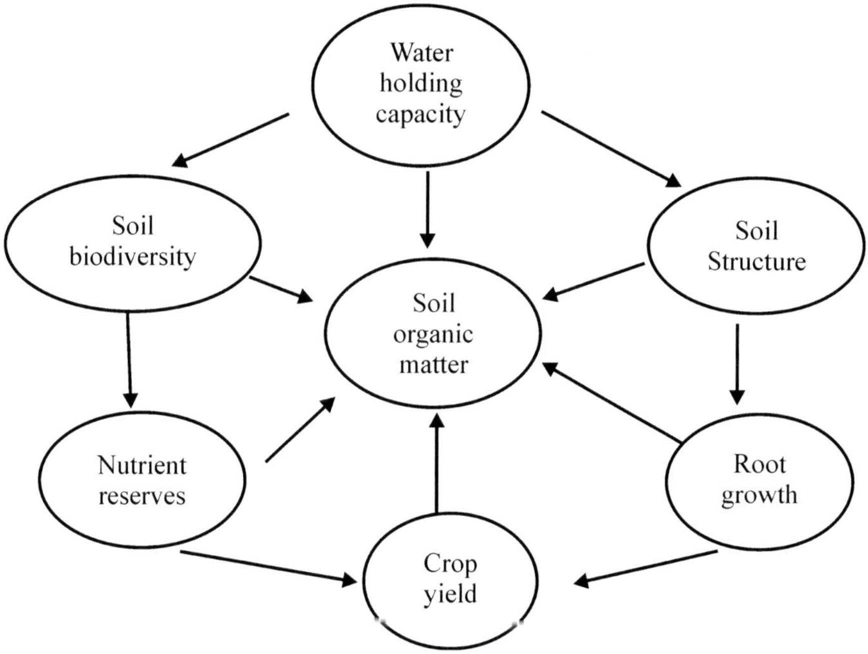

Fig. 2. A wide range of processes and technological option for C sequestration in agricultural ecosystems.

7.3 Mechanism for carbon sequestration

Through the process of photosynthesis, plants assimilate carbon and return some of it to the atmosphere through respiration. The carbon that remains as plant tissue is then consumed by animals or added to the soil as litter when plants die and decompose. The primary way that carbon is stored in the soil is as soil organic matter (SOM). SOM is a complex mixture of carbon compounds, consisting of decomposing plant and animal tissue, microbes (protozoa, nematodes, fungi, and bacteria), and carbon associated with soil minerals. Carbon can remain stored in soils for millennia, or be quickly released back into the atmosphere. Climatic conditions, natural vegetation, soil texture, and drainage all affect the amount and length of time carbon is stored.

Reducing atmospheric CO_2, that is sequestering carbon, can take place three ways:

- Carbon production or trapping carbon within plants. The more permanent vegetation that is present, the more CO_2 is required.
- Minimizing organic carbon mineralization. That means managing crops and soil to reduce conditions that break down or oxidize organic matter – letting plant material decompose more slowly and naturally.
- Reducing soil erosion and keeping carbon trapped in the soil. Eroded soil is exposed soil and exposed carbon.

7.4 Amount of carbon stored in soils

There are a whole range of SOC levels in different soils. For instance, for the surface soils, SOC ranges from about 10 per cent in the alpine soils to less than 0.5 per cent in the desert soils. The amount of SOC stored in the soil profile can be considerable. For example, if there is 1 per cent SOC over 30 cm soil depth, the amount of SOC stored over 1 hectare of land can weigh about 42 tonnes. Usually, the surface layer has the highest level of SOC which decreases with depth down the soil profile. The actual amount of SOC present in a soil is dependent on a number of factors.

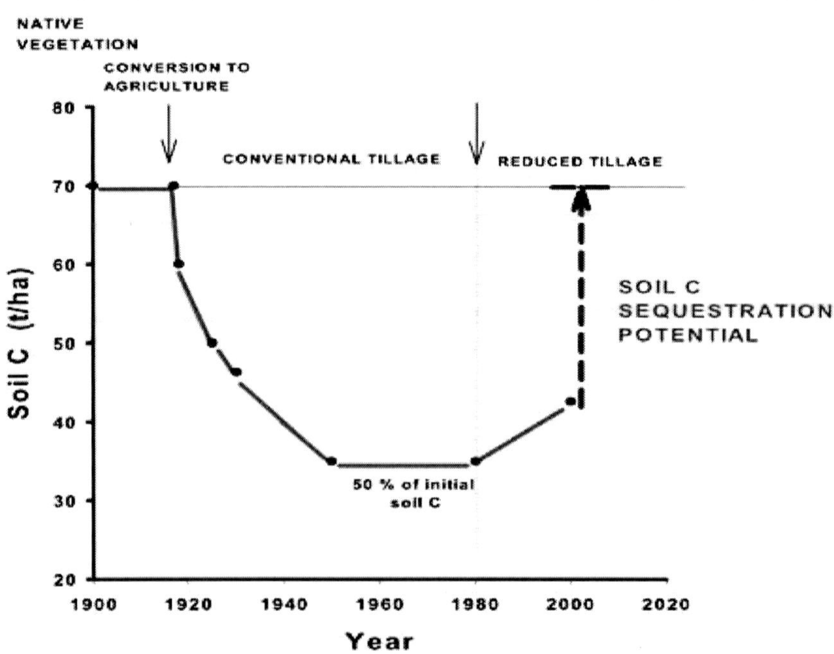

Fig. 3. Conversion of C sequestration in relation to conventional tillage and reduced tillage.

7.5 Benefits of soil sequestration of carbon

Removing CO_2 from the atmosphere is only one significant benefit of enhanced carbon storage in soils. Improved soil and water quality, decreased nutrient loss, reduced soil erosion, increased water conservation, and greater crop production may result from increasing the amount of carbon stored in agricultural soils. Management techniques, which are successful in providing a net carbon sink in soils, include the following:

- **Conservation tillage** (Fig. 3) minimizes or eliminates manipulation of the soil for crop production. It includes the practice of mulch tillage, which leaves crop residues on the soil surface. These procedures generally reduce soil erosion, improve water use efficiency, and increase carbon concentrations in the topsoil. Conservation tillage can also reduce the amount of fossil fuel consumed by farm operations. It has been estimated to have the potential to sequester a significant amount of CO_2.
- **Cover cropping** is the use of crops such as clover and small grains for protection and soil improvement between periods of regular crop production. Cover crops improve carbon sequestration by enhancing soil structure, and adding organic matter to the soil.
- **Crop rotation** is a sequence of crops grown in regularly recurring succession on the same area of land. It mimics the diversity of natural ecosystems more closely than intensive mono-cropping practices.

The amount of carbon sequestered at a site reflects the long-term balance between carbon uptake and release mechanisms. Many agronomic, forestry, and conservation practices, including best management practices, lead to a beneficial net gain in carbon fixation in soil.

8. MANAGEMENT PRACTICES THAT SEQUESTER SOIL CARBON

Practically, there are three areas of farm management that can affect soil carbon in the Great Plains: tillage, cropping intensity and fertilization. Tillage and soil carbon are inversely proportional to each other. The greater the tillage, the less soil carbon. No-till systems build soil organic matter, which is about 58 per cent carbon. Cropping intensity and soil carbon are positively related. The more frequent the cropping and greater the biomass inputs, the more soil carbon. Summer fallow reduces cropping intensity. Fertilization affects soil carbon mainly through crop biomass. However, the carbon:nitrogen ratio of soil organic matter results in stable organic matter typically within a range of about 8–10:1. If insufficient nitrogen is present to permit stable formation of soil organic matter via soil microbial degradation of crop residues, then little carbon may be sequestered.

Table 2. Possible activities sequestering C; its opportunities and barrier towards management.

Possible sinks/activities *for SOC sequestration*	**Opportunities/existing drivers**	**Barriers**
Cropping land	Currently C 'source' Conservation farming; Stubble initiative of EG Graham Centre	Heavy stubble handling and management, costs for adoption
Pasture Land	Improved pasture productivity; Environmental incentives - Move to increase perennially; Relation to dryland salinity control	SOC benefits need to be quantified for various management and geographical areas
Rangelands	Large area involved; Restoration of degraded areas	Costs; research needed on SOC and management
Environmental and rehabilitation programs	Recycling of "organics" – government legislations on waste minimisation Land care; Dryland Salinity Program Trees on Farms; Re-vegetation Erosion and nutrients Control – buffer strips, Agroforestry	Cost of transport, markets (acceptability by rural sector); SOC sequestration values needed to be quantified
New initiatives	Bioenergy crops	New industry/technology; research need on SOC benefit of biochar and land conversion
Soil Amelioration	Sodic soils; Acid soils – improved land value and overcoming land degradation problem	Yield increases; technology of amelioration known, low adoption, costs, soil C sequestration rate uncertain

8.1. Management practices that increase soil organic carbon

There are a wide range of management options and farming practices that can increase SOC levels by either increasing inputs or decreasing losses. Inputs can also be increased by direct additions of organic materials, composts, manure and other recycled organic materials.

- Conservation farming– This is rapidly gaining worldwide acceptance as a farming practice to improve soil and water conservation. In cropping, cultivation is either reduced (reduced tillage) or completely eliminated (no-tillage) and stubble (crop residue) is retained. Reduced tillage reduces carbon losses (from both reduced cultivation and reduced fossil fuel usage) and stubble retention increases carbon inputs to the soil; both of these lead to SOC increases.

- Use of organic amendments – These are manure, plant debris, composts and biosolids from sewage which are applied to agricultural soils. They are all high in organic carbon and therefore represent additional carbon inputs to the system. Some of these recycled organics also contain a high plant nutrient content and can act as organic fertilizers, reducing the use of inorganic fertilisers. They are important for organic farming systems.

Table 3. Management practices that can increase soil organic levels of agricultural soils

Crop management	Conservation tillage	Pasture management	Irrigation	Organic amendments
Soil fertility enhancement	Stubble retention	Fertiliser management	Improved grass species	Animal manure
Better rotation	Reduced tillage	Grazing management	Introduction of legumes	Green manure
Erosion control	-----	Earthworm introduction	Sown pasture	------

9. THE ROLE OF FORESTS IN REDUCING ATMOSPHERIC CARBON

Forests are capable of effective sequestration and storage of atmospheric carbon in above-ground and below-ground biomass by way of processes of photosynthesis and tree growth. Rate of biomass increase and hence rate of carbon sequestration vary through rotation of a new plantation as an integration of genetic, environmental and silvicultural determinants.

Trees in forests (including plantations), if well-stocked, typically sequester carbon at a maximum rate between, about age 10 and age 20–30. As an indication, at age 30 years about 200 to 520 tonnes CO_2-e are sequestered per ha in forests with productivity ranging from low to high (Australian Greenhouse Office, 2001). After this age, if the trees are not harvested, the sequestration rate slows gradually until maturity at about 80 to 100+ years of age.

1. **Carbon uptake and assimilation,** including immediate respiratory losses which detract from previously 'fixed' carbon in photosynthetic plant cells;
2. **Carbon transport, allocation and partitioning** of carbon for storage, structural and metabolic use in above-ground and below-ground parts of the tree; and

3. **Return of forest carbon to the atmosphere** via oxidative pathways, notably via the food chain, biological decay and combustion of forest biomass and forest products.

Human activities, especially the burning of fossil fuels such as coal, oil, and gas, have caused a substantial increase in the concentration of carbon dioxide in the atmosphere. This increase in atmospheric CO_2 – from about 280 to more than 380 parts per million (ppm) over the last 250 years – is causing measurable global warming. Potential adverse impacts include sea-level rise; increased frequency and intensity of wildfires, floods, droughts, and tropical storms; changes in the amount, timing, and distribution of rain, snow, and runoff; and disturbance of coastal marine and other ecosystems. Rising atmospheric CO_2 is also increasing the absorption of CO_2 by seawater, causing the ocean to become more acidic, with potentially disruptive effects on marine plankton and coral reefs. Technically and economically feasible strategies are needed to mitigate the consequences of increased atmospheric CO_2.

10. CONTRIBUTION OF AGRICULTURE TO GREENHOUSE GAS EMISSIONS

Agriculture activities serve as both sources and sinks for greenhouse gases. Agriculture sinks of greenhouse gases are reservoirs of carbon that have been removed from the atmosphere through the process of biological carbon sequestration. The primary sources of greenhouse gases in agriculture are the production of nitrogen based fertilizers; the combustion of fossil fuels, such as coal, gasoline, diesel fuel and natural gas; and waste management. Livestock enteric fermentation, or the fermentation that takes place in the digestive systems of ruminant animals, results in methane emissions. Carbon dioxide is removed from the atmosphere and converted to organic carbon through the process of photosynthesis. As organic carbon decomposes, it is converted back to carbon dioxide through the process of respiration. Conservation tillage, organic production, cover cropping and crop rotations can drastically increase the amount of carbon stored in soils. In 2005, agriculture accounted for from 10 to 12 per cent of total global human caused emissions of greenhouse gases, according the Intergovernmental Panel on Climate Change (IPCC, 2007).

Greenhouse gases have varying global warming potentials; therefore climate scientists use carbon dioxide equivalents to calculate a universal measurement of greenhouse gas emissions.

The U.S. Department of Agriculture released a report in May 2008 that focused on the effects of climate on agriculture, specifically on cropping systems, pasture and grazing lands and animal management. The following findings were excerpted from the report:

- With increased carbon dioxide and higher temperatures, the life cycle of grain and oilseed crops will likely progress more rapidly.

- The marketable yield of many horticultural crops, such as tomatoes, onions and fruits, is very likely to be more sensitive to climate change than grain and oilseed crops.
- Climate change is likely to lead to others migration of weeds. Many weeds respond more positively to increasing carbon dioxide than most cash crops.

11. CLIMATE CHANGE INFLUENCING AGRICULTURE

Climate change may have beneficial as well as detrimental consequences for agriculture. Some research indicates that warmer temperatures lengthen growing seasons and increased carbon dioxide in the air results in higher yields from some crops. A warming climate and decreasing soil moisture can also result in production patterns shifting northward and an increasing need for irrigation. Changes, however, will likely vary significantly by region. Benefits to agriculture might be offset by an increased likelihood of heat waves, drought, severe thunderstorms and tornadoes. An increase in climate variability makes adaptation difficult for farmers.

12. ROLE OF AGRICULTURE IN MITIGATING CLIMATE CHANGE

Several farming practices and technologies can reduce greenhouse gas emissions and prevent climate change by enhancing carbon storage in soils; preserving existing soil carbon; and reducing carbon dioxide, methane and nitrous oxide emissions (Fig. 4).

12.1. Conservation tillage and cover crops

Conservation tillage refers to a number of strategies and techniques for establishing crops in the residue of previous crops, which are purposely left on the soil surface. Reducing tillage reduces soil disturbance and helps mitigate the release of soil carbon into the atmosphere. Conservation tillage also improves the carbon sequestration capacity of the soil. Additional benefits of conservation tillage include improved water conservation, reduced soil erosion, reduced a farmer already engaged in conservation tillage would not provide additional carbon storage by maintaining that practice.

In addition it is also important because of the possibility that perverse incentives may be created that encourage farmers or ranchers to release carbon so that they can get paid to store it.

12.2. Improved cropping and organic systems

Recent reports have investigated the potential of organic agriculture to reduce greenhouse gas emissions (Rodale Institute, 2008). Organic systems of production increase soil organic matter levels through the use of composted animal manures and cover crops. Organic cropping systems also eliminate the emissions from the production and transportation of synthetic fertilizers. Components of organic agriculture could be implemented with other sustainable farming systems, such as conservation tillage, to further increase climate change mitigation potential. Generally, conservation farming practices that conserve moisture, improve yield potential and reduce erosion, and fuel costs also increase soil carbon. Examples of practices that reduce carbon dioxide emissions and increase soil carbon include direct seeding, field windbreaks, rotational grazing, perennial forage crops, reduced summer fallow and proper straw management. Using higher-yielding crops or varieties and maximizing yield potential can also increase soil carbon.

Table 4. Improvement in crop yields per ton of carbon in the root zone (Lal *et al.*, 2011)

Crop	Potential yield increase (kg/ha)
Maize	200–400
Wheat	20–70
Soybean	20–30
Cowpea	5–10
Rice	10–50
Millet	50–60

12.3. Land restoration and land use changes

Land restoration and land use changes that encourage the conservation and improvement of soil, water and air quality, typically reduce greenhouse gas emissions. Modifications to grazing practices, such as implementing sustainable stocking rates, rotational grazing and seasonal use of rangeland, can lead to greenhouse gas reductions. Converting marginal cropland to trees or grass maximizes carbon storage on land that is less suitable for crops.

12.4. Irrigation and water management

Improvements in water use efficiency, through measures such as irrigation system mechanical improvements coupled with a reduction in operating hours; drip irrigation technologies; and centre-pivot irrigation systems, can significantly reduce the amount of water and nitrogen applied to the cropping system. This reduces greenhouse emissions of nitrous oxide and water withdrawals.

12.5. Nitrogen use efficiency

Improving fertilizer efficiency through practices like precision farming using GPS tracking can reduce nitrous oxide emissions. Other strategies include the use of cover crops and manures (both green and animal); nitrogen-fixing crop rotations; composting and compost teas, and integrated pest management.

12.6. Methane capture

Large emissions of methane and nitrous oxide are attributable to livestock waste treatment, especially in dairies. Agriculture methane collection and combustion systems include covered lagoons and complete mix and plug flow digesters. Anaerobic digestion converts animal waste to energy by capturing methane and preventing it from being released into the atmosphere. The captured methane can be used to fuel a variety of on-farm applications, as well as to generate electricity. Additional benefits include reducing odours from livestock manure and reducing labour costs associated with manure removal.

13. ROLE OF CROP AND ANIMAL IN MITIGATING GHGs

Crops

- Conservation tillage and reduced field pass intensity
- Sequestration, emission reduction improves soil, water and air quality.
- Reduce soil erosion and fuel use
- Efficient nutrient management sequestration, emission reduction improves water quality. Saves expenses, time and labour.
- Crop diversity through rotations and cover crops
- Sequestration Reduces erosion and water requirements.
- Improve soil and water quality.

Animals

- Manure management emission reduction On-farm sources of biogas fuel and possibly electricity for large operations,
- Provide nutrients for crops.
- Rotational grazing and improved forage
- Sequestration, emission reduction reduces water requirements. Helps withstand drought. Increases long term grassland productivity.
- Feed management emission reduction reduces quantity of nutrients.
- Improve water quality. More efficient use of feed.

Judicious nutrient management is crucial to soil organic carbon sequestration in tropical soils (Bhattacharyya *et al.,* 2007; Mandal *et al.,* 2007). Adequate supply of nutrients in soil can enhance biomass production and soil organic carbon (SOC) content (Van Kessel and Hartley, 2000). Use of organic manure and compost enhances the SOC pool more than application of the same amount of nutrients as mineral fertilizers (Gregorich *et al.,* 2001). Besides the conventional organic sources, Indian cities generate 4.1 Million tonnes C/year from municipal solid waste (MSW) and it can be an important source of organic

C. Long-term manure application increases the SOC pool (Gilley and Risse, 2000), which not only sequesters CO_2 but also enhances productivity of soil (Swarup *et al.*, 2000). It is, however, argued that SOC sequestration is a major challenge in soils of the tropics and sub-tropics, where climate is harsh and the rate of C mineralization is high because of high temperature and low humification efficiency (Ladha *et al.*, 2003).

Tonnes carbon per ha	=	SOC (%)	x	Soil bulk density (Mg/m^3)	x	Sampling depth (cm)
	=	1.0	x	1.4	x	30
	=	42 tonnes/ha				

Fig. 4. Formulae for calculating C/ha in tonnes

14. CALCULATION OF SOIL CARBON

- SOC results are usually expressed as percentage C by weight (*i.e.* g C per 100 g of soil). SOC results can be converted to soil organic matter (SOM) level by multiplying SOC value by a conversion factor of 1.72. This assumes that SOM present in soil, on average, is made up of 58 per cent carbon.
- Very often it is more practical to express SOC on per ha basis, namely as tonnes C per ha. To perform such calculation, knowledge of the bulk density of the soil to the sampling depth and the sampling depth is needed. As an example, if SOC = 1.0% and bulk density of the soil = 1.4 Mg/m^3 to 30 cm depth (1 Mg = 1000 kg = 1 tonne), the amount of SOC present in the soil to 30 cm depth of 1 hectare of land can be worked out for soil C amount.

15. ECONOMIC AND ENVIRONMENTAL BENEFITS OF SOIL CARBON SEQUESTRATION

- Carbon credit - Carbon credit is a generic term for any tradable certificate or permit representing the right to emit one tonne of carbon dioxide or the mass of another greenhouse gas with a carbon dioxide equivalent (tCO_2e) equivalent to one tonne of carbon dioxide. Carbon credits and carbon markets are a component of national and international attempts to mitigate the growth in concentrations of greenhouse gases (GHGs). One carbon credit is equal to one tonne of carbon dioxide, or in some markets, carbon dioxide equivalent gases. Carbon trading is an application of an emission trading approach. Greenhouse gas emissions are capped and then markets are used to allocate the emissions among the group of regulated sources.

- Carbon trading - Carbon trading is a form of emissions trading that specifically targets carbon dioxide (calculated in tonnes of carbon dioxide equivalent or tCO_2e) and it currently constitutes the bulk of emissions trading. This form of permit trading is a common method countries utilize in order to meet their obligations specified by the Kyoto Protocol; namely the reduction of carbon emissions in an attempt to reduce (mitigate) future climate change. Under Carbon trading, a country having more emissions of carbon is able to purchase the right to emit more and the country having less emission trades the right to emit carbon to other countries. More carbon emitting countries, by this way try to keep the limit of carbon emission specified to them.
- Carbon markets - offer farmers financial incentives while providing binding contracts to businesses that wish to voluntarily offset their CO_2 emissions. The CC pays land managers between $2 and $3 per acre for sequestering CO_2 using certified conservation tillage practices and/or grass and tree planting. Voluntary carbon markets show great promise as a vehicle to improve stewardship and mitigate climate change.
- Carbon dioxide equivalents (CO_2-e) provide a universal standard of measurement against which the impacts of releasing (or avoiding the release of or actively sequestering) different greenhouse gases can be evaluated. Every greenhouse gas has a Global Warming Potential (GWP), a measurement of the impact that particular gas has on 'radiative forcing'; that is, the additional heat/energy which is retained in the Earth's atmosphere system through the addition of this gas to the atmosphere.
- The GWP of a given gas describes its effect on climate change relative to a similar amount of carbon dioxide. As the base unit, carbon dioxide is 1.0. This allows the greenhouse gases regulated under the Kyoto Protocol to be converted to the common unit of CO_2-e.
- Increasing SOC improves soil quality, boosting agricultural productivity while reducing atmospheric CO_2 concentrations. In addition, because carbon stored in agricultural soils can be easily quantified using well-accepted scientific practices, it can provide benefits to farmers through rental payments and to society by avoiding the cost of implementing expensive new technologies.

16. THE KYOTO PROTOCOL: CARBON SINKS, CARBON CREDITS AND EMISSIONS TRADING

In recognition of the adverse effects of unregulated GHG emissions on global climate, in Rio de Janeiro in 1992 the United Nations adopted a Framework Convention on Climate Change (UNFCCC) to develop mechanisms to stabilise atmospheric GHG concentrations. Subsequent meetings of the parties to the

UNFCCC at Kyoto in 1997 drafted a protocol (the Kyoto Protocol) to place binding GHG emission targets on signatory nations.

The Kyoto Mechanisms are:

1. The Clean Development Mechanism, a means to earn credits by investing in emission reduction projects in developing countries,
2. Joint Implementation, a means to earn credits by investing in emission reduction projects in other Kyoto-signatory countries, and
3. International Emissions Trading, in which industrialised countries can buy and sell emission (or carbon) credits amongst themselves.

17. SUMMARIZING ABOVE FACTS

1. If we want to reduce global hunger, a healthy soil biosphere is crucial. 70 per cent of poor people in rural areas depend on agriculture for their livelihoods. These rural areas comprise large numbers of smallholder farmers, who cultivate less than 2 acres of land. Lacking access to quality inputs, tools, training, and financing, smallholder farmers are often at the mercy of unproductive soil. Promoting soil health, through strategies such as agroforestry, intercropping, and composting, is one important way to increase the productivity of these small plots of land.
2. Soil is the greatest reservoir and the last frontier of biodiversity. Most known antibiotics come from organisms that were isolated from the soil. The soil biosphere controls the cycling of most major plant nutrients, such as nitrogen, phosphorus, and sulphur. In one pinch of soil, there are over one billion individual organisms. These microorganisms help in restoring fertility.
3. We can't combat climate change without focusing on soil. Climate change is driven by atmospheric carbon. Surprisingly, soil can absorb nearly twice as much carbon as can be contained within plants and the atmosphere combined. Soil's enormous capacity to absorb more carbon has additional benefits: adding carbon to the soil via plant materials (compost, green manure, animal manure, biochar) will actually reduce carbon (and thereby the amount of carbon dioxide) in the atmosphere. Increasing the amount of carbon in soil will also allow the soil to capture and hold more rainwater, which reduces the amount of soil erosion and increases the soil's ability to filter ground water, which increases water quality.
4. The C pools and SOC stock changes were strongly associated with C input rate in no tillage soils and the C losses associated with conventional treatments agree with previous findings. In the tropical sites, C increased in the upper and deeper layers, indicating the importance of the root system of the grass plants when developed in deeper layers during the dry season. These findings demonstrate the importance of

introducing species of plants with high biomass input and with high adaptation capacity in adverse environments. Thus it is crucial in tropical areas to develop cropping systems with high biomass input able to reach environmental and economic sustainability.

5. Carbon sequestration is highly related to soil and management systems. Research on the impact of tillage practices and crop rotation has demonstrated that no-till and permanent vegetation are more effective in storing carbon in the soil. The use of crop rotation and conservation tillage, in addition to effective manure and nitrogen management systems, contributed significantly to improving soil carbon status.
6. Farmers can benefit from carbon sequestration through the use of conservation tillage, crop rotation, the use of buffer strips, and permanent vegetation for highly eroded soils. These benefits include improved soil productivity, an improved environment due to less erosion, and improved physical and biological properties of soil.
7. Carbon credit is another benefit that has been explored recently by many different entities. Carbon credit is worth exploring, but needs careful consideration. The issues of market value, policy, carbon monitoring procedures, and management entities are among those that need to be addressed when considering carbon sequestration and credits.
8. SOC sequestration has a finite capacity. Soil carbon stock may increase only until the environmental equilibrium level is achieved, possibly in 50–100 years of improved management practice.

All the evidence points to substantial potential of SOC sequestration to mitigate climate change. Overall this potential contribution is equivalent to 5–14 per cent of total annual GHG emissions and is expected to last for the next 50–100 years. Furthermore, according to the latest IPCC estimates, soil carbon sequestration globally can contribute to 89 per cent of the total technical mitigation potential for agriculture whereas mitigation of soil CH_4 and N_2O emissions account for only 9 per cent and 2 per cent respectively (Smith *et al.* 2007). This further indicates the significant role that SOC sequestration can play in mitigating climate change.

Globally around 90 per cent of the technical GHG mitigation potential from agriculture will arise from improved management of crop lands, grazing lands, organic soils and degraded soils. Further small contributions can be made from (in order of significance) improved management of rice, livestock, bioenergy crops, irrigation water, land use change, agroforestry and manure management.

18. CONCLUSION

Adding organic matter to farmland is good for soil quality and crop yields, both short-term and long-term. Continuous no-till is an efficient way of doing this.

Cover crops and manure also help raise carbon levels. Soil carbon sequestration is a natural, cost-effective, and eco-friendly process. Once sequestered, carbon remains in the soil as long as restorative land use, continuous no-till, and other Best Management Practices are followed. It is a win-win option. While mitigating climate change by off-setting fossil fuel emissions, it also improves quality of soil and water resources, enhances agronomic productivity, and buys us time to identify and implement viable alternatives to fossil fuel. In addition to providing farmers with another source of income, carbon sequestration increases soil organic carbon which improves productivity, reduces soil erosion and nutrient runoff, and enhances water quantity and quality. Soil carbon sequestration is a win-win solution for agriculture and the environment.

REFERENCES

Bajpai RK, Chitale S, Upadhyay, SK and Urkarkar JS (2006) Long-term studies on soil physico-chemical properties and productivity of rice-wheat system influenced by integrated nutrient management in Inceptisols of Chhattisgarh. *J. Soc. Soil Sci.* **54:** 24–29.

Bhaskaran, Usha Pankajam (2009). Effect of organic farming on soil fertility yield and quality of crops in the tropics: http:// scholarship.org/uc/item/7k12 w04m.

Bhattacharyya R, Chandra S, Singh RD, Kundu S, Srivastva AK and Gupta HS (2007) Long-term farmyard manure application effects on soil properties in a silty clay loam soil under irrigated wheat-soybean rotation. *Soil and Tillage Research* **94:** 386–396.

Brady NC and Well RR (2007) The Nature and Properties of Soil, pp. 514–558. Thirteenth edition. Pearson Education. Dorling Kindersley (India) Pvt. Ltd.

Chan KY, Bowman A and Oates A (2001) Oxidizable organic carbon fractions and soil quality changes in an Oxic Paleustult under different pasture leys. *Soil Sci.* **166:** 61–67.

Chaudhury J, Saha MN and Kundu S (2007) The potential of cropping systems and soil amendments for carbon soil physico-chemical properties and productivity of rice-wheat system influenced by integrated nutrient management in Inceptisols of Chhattisgarh. *J. Soc. Soil Sci.* **54:** 24–29.

Climate Change (2007). The Physical Science Basis. Contribution of Working Group I to the Fourth Assessment Report of the Intergovernmental Panel on Climate Change edition. Pearson Education. Dorling Kindersley (India) Pvt. Ltd.

Gilley JE and Risse LM (2000) Runoff and soil loss as affected by the application of manure. *Transactions of the American Society of Agricultural and Biological Engineers* **43:** 1583–1588.

Gregorich EG, Drury CF and Baldock JA (2001) Changes in soil carbon under long-term maize in monoculture and legume based rotation. *Canadian Journal of Soil Sciences* **81(1):** 21–31.

IPCC (2007b) Climate Change 2007: Agriculture. Contribution of Working Group III to the Fourth Assessment Report of the Intergovernmental Panel on Climate Change.

[B. Metz, O.R. Davidson, P.R. Bosh, R. Dave and L.A. Meyer (eds.)], Cambridge University Press, Cambridge, United Kingdom and New York, NY, USA.

Jackson ML (1973) Soil Chemical Analysis. New Delhi, India: Prentice Hall of India.

Jha MN, Gupta MK, Saxena A and Kumar R (2003) Soil organic carbon store in different forests in India. *Ind. Forester* **129:** 714–724.

Kumar B, Gupta RK and Bhandari AL (2008) Soil fertility changes after long-term application of organic manures and crop residues under rice-wheat system. *J. Ind. Soc. Soil Sci.* **56:** 80–85.

Ladha JK, Dawe D and Pathak H (2003) How extensive are yield decline in long-term rice-wheat experiments in Asia. *Field Crops Res.* **81:** 159–180.

Lal R (2004a) Soil carbon sequestration in India. Climatic Change 65: 277–296.

Lal R, Follett RF and Kimble JM (2011) Achieving soil carbon sequestration in the United States: a challenge to the policy makers. *Soil Science* **168:** 827–845.

Lal R., Follett RF, Stewart BA and Kimble JM (2007) Soil carbon sequestration to mitigate climate change and advance food security. *Soil Science* **172:** 943–956.

Mandal B, Majumder B, Bandyopadhyay PK, Hazra GC, Gangopadhyay A, Samantaray RN, Mishra, AK, OR Davidson PR, Bosch R and Dave LA Meyer (2007) (eds)]. www.ipcc.ch/pdf/assessment-report.

Pathak H, Byjesh K, Chakrabarti B and Aggarwal PK (2011) Potential and cost of carbon sequestration in Indian agriculture: Estimates from long-term field experiments. *Field Crops Res.* **120:** 102–111.

Prasad R (2005) Rice–wheat cropping system. *Advances in Agronomy* **86:** 285–339

Rodale Institute (2008) Regenerative Organic sequestration in soils under long-term experiments in subtropical India. *Global Change Biology* **13:** 357–369.

Singh RK, Singh SK and Tarapdas JC (2008). Influence of cropping sequence and nutrient management on soil organic carbon and nutrient status of Typic Rhodustalfs. *J. Ind. Soil Sci.* **56:** 174–181.

Singh YV, Dhar DW and Agarwal B (2011) Influence of organic nutrient management on basmati rice (*Oryza sativa*) – Wheat (*Triticum aestivum*) greengram (*Vigna radiata*) cropping system. *Indian Journal of Agronomy* **56(3):** 169–175.

Smith P, Martino D, Cai Z, Gwary D, Janzen HH, Kumar P, McCarl B, Ogle S, O'Mara F, Rice C, Scholes RJ, Sirotenko, O, Howden M, McAllister T, Pan, G, Romanenkov V, Schneider U, Towprayoon S, Wattenbach M and Smith JU (2007) Greenhouse gas mitigation in agriculture. *Phil. Transactions of Royal Society* **363:** 789–813.

Swarup A, Manna MC and Singh GB (2000) Impact of land use and management practices on organic dynamics in soils of India. Advance inSoil Science CRC/Press, Boca Raton, New York, pp. 261–81.

Van Kessel C and Hartley C (2000) Agricultural management of grain legumes: has it led to an increase in nitrogen fixation. *Field Crops Research* **65:** 165–181.

9

Biosequestration – A Sustainable Approach Towards Rejuvenation of Soil Organic Carbon in India

Ekta Kumari, Ram Swaroop Meena, Avijit Sen, Pravin K. Upadhyay, Santosh K. Meena, Mona Nagargade and Vishal Tyagi*

Department of Agronomy, Institute of Agricultural Sciences, Banaras Hindu University, Varanasi – 221 005, India
**E-mail: ekta14.bhu@gmail.com*

ABSTRACT

The biosequestration of C often accompanied both by soil and biota is found not only to improve the agronomic/biomass productivity but had also the capability to improve water quality and mitigate climate change by decreasing the rate of enrichment of atmospheric CO_2. Biosequestration plays a considerable role in enhancing the soil organic carbon. In India, the SOC concentration of most of the soils is low, *i.e.* 5 g/kg, so the sequestration of carbon required that ends up in a stable form *i.e.* soil organic matter. Soil carbon sequestration is an important and immediate sink for removing atmospheric carbon dioxide and slowing global warming. Biosequestration is the capture and secure storage of carbon that would otherwise be emitted to or remain in the atmosphere. Therefore in this chapter we would study in detail the potential cause and ultimate solution to revive status of soil by storing the C into the soil. This improvement is directly related to maintaining the quantity of SOM, which is a critical component of soil productivity. Sequestered C boosted the SOC in the soil and could frequently be improved by adoption of management practices including the use of crop residues and organics, conservation tillage, INM including biochar, restoration of eroded and degraded soils by agroforestry systems encroaching towards sustainability of cropland and climate.

1. INTRODUCTION

Soil organic carbon (SOC) is an indicator of soil productivity. It is an important constituent of the soil's chemical, physical and biological health. Measurement of SOC will help us to assess the productivity and the sustainable fertility of the soil. Besides, quantifying the SOC content will also provide us with an idea about the potentials of the soil for sequestering carbon from the atmosphere or the emission potential when the soil is disturbed. The heat-trapping (greenhouse) gas, carbon dioxide (CO_2) is found to increase tremendously over a last 20 year. The present atmospheric CO_2 levels measured at Mauna Loa Observatory, Hawaii is 406.69 ppm (July, 2017). Soils act as the sink for the atmospheric carbon and have the potential to sequester and store in it (Lal, 2005). Soil proved to be storehouse to this converted carbon from where microbes derived energy and substantially their decay leads to increased fertility status in soil. The upper layer of the soil stores nearly three times more than the amount of carbon held in the atmosphere (UNEP, 2012). The sequestered carbon in the soil is mainly stored in the form of soil organic carbon (SOC). Plant, animal and microbial residues in all stages of decomposition contribute to SOC.

2. SOIL ORGANIC CARBON – THE BASIS OF SUSTAINABLE AGRICULTURE

Soil organic carbon (SOC) is the carbon associated with soil organic matter. Soil organic matter is the organic fraction of the soil that is made up of decomposed plant and animal materials as well as microbial organisms, but does not include fresh and un-decomposed plant materials, such as straw and litter, lying on the soil surface. Soil carbon can also be present in inorganic forms, *e.g.* lime or carbonates in some soils in the drier areas. The status of C in soils of India is low. The principal cause of decline in SOC pool in degraded soils is a reduction in biomass productivity and the low amount of crop residue and roots returned to the soil. Soil organic matter is an important component necessary to the formation of both micro- and macro-aggregates. Therefore, enhancement and management of SOC concentration are important to sustainable management of soil and water resources. Agricultural intensification implies use of recommended management practices (RMPs) on prime agricultural land to enhance food production per unit area and unit time so that marginal land can be taken out of production and converted to a restorative land use. Use of crop residues and manure to enhance soil biodiversity, especially earthworm activity, is an important strategy for increasing SOC concentration and soil quality (Lavelle *et al.*, 1998; Bhadauria and Ramakrishnan, 1996).

3. PROCESS OF SEQUESTRATION

Since some soil carbon comes from mineral sources, the vast majority of it is derived from plants. As plants grow and die, they leave behind organic, carbon-based compounds in the soil of varying size and chemical composition. Under the right conditions, soil fauna metabolize these compounds, incorporating some of the carbon in them into new chemical compounds within their own biomass, while respiring the rest to the atmosphere as CO_2 or excreting it back into the soil. This continuous movement of carbon through the soil food web means that carbon is constantly changing forms in the soil as it is incorporated into new organisms or converted into different compounds. There are general categories or pools based on how long the carbon remains in the soil, a figure often referred to as "mean residence time." The most commonly used model of these pools includes three different groupings: the fast or labile pool, the slow pool, and the stable pool (Jenkinson and Rayners, 2008). The fast pool is soil carbon that turns over and returns to the atmosphere sometime within a few days to a few years. Carbon in this pool is typically composed of recently incorporated plant residues and simple carbon compounds that are exuded by roots. This labile pool is the one most readily used by soil microbes, meaning it generates a great deal of CO_2. The slow pool is composed of more processed plant residues, microbial by-products of the fast pool, and carbon molecules that are protected from microbes by physical or biochemical soil processes (Collins *et al.*, 2000). Mean residence time of the slow pool is generally considered to be in the range of years to decades, but this range can be heavily influenced by soil texture, management, and climate. In contrast, the stable pool is more resistant to disturbances and is extremely slow to change, with mean residence times ranging from centuries to millennia. This pool is comprised of what is often called humus, a loose term for a group of carbon compounds that are extremely resistant to decomposition, and soil carbon that is very well protected from microbial decomposition (Six *et al.*, 2002). The relative size of each of these pools can vary in different soils. But in general, the size of the stable pool remains relatively constant, while the sizes of the labile and slow pools are sensitive to management. These are important in establishing a wide role in maintaining the fertility status of soil.

4. BIOSEQUESTRATION OF CARBON IN SOIL

Biosequestration in soil is the process of transferring CO_2 from the atmosphere into the soil through crop residues, tree roots and other organic solids, and in a form that is not immediately reemitted. The sequestered carbon in soil helps in reducing CO_2 present in the atmosphere, while enhances soil quality and a sustained productivity. Soil carbon sequestration can be accomplished by management practices that add high amounts of biomass to the soil, cause minimal soil disturbance, conserve soil and water, improve soil structure, and enhance soil faunal activity. Conditions for a crop or set of crops for cultivation decided the type of cropping system. However, the choice of crop species, cropping system, timing of fallowing and quality and quantity of residue returned to the soil have considerable impact on soil organic carbon stocks (Halvorson *et al.*, 2001). It is estimated that throughout the India more than 250 double cropping systems are followed. However, in each district in the country based on rational spread of crops 30 important cropping systems were identified. Looking into the CSP of cropping systems in without organic matter added scenario (NPK), it can be inferred that balanced application of NPK to crop could also promote carbon sequestration in crops; however, integrated nutrient management (INM) remained always at higher side as compared to NPK in terms of CSP of cropping system. Several reports have published showing the positive effect of no-tillage practices on soil carbon pool in rice wheat cropping systems. Based on eight years study in rice wheat cropping system, Pandey *et al.* (2014) reported 0.59 Mg C/ha/yr carbon sequestration potential of rice-wheat cropping system in no-tillage conditions.

It influences the soil in four ways:

1. By decreasing the level of soil disturbance (*i.e.* tillage) to enhance the physical protection of soil carbon in aggregates.
2. Through increasing the mass and quality of plant and animal inputs to soils.
3. It improves soil microbial diversity and abundance.
4. With maintaining continuous living plant cover on soils year-round.

Managing these processes can quickly lead to increases in soil carbon that may be highly useful in drawing down atmospheric CO_2.

4.1. SOC- Improves soil productivity

Productivity is nothing but the inherent capacity of a soil to support the growth of specified plants, plant communities, or a sequence of plant communities. It's a well-known fact that all living biota *i.e.* animal, microbe and non-living entity *i.e.* plant residues in all stages of decomposition contribute to soil organic carbon (SOC). Measurement of SOC would lead to sustainable soil health by assessing the productivity of cropland. It also contributes to the potentials of the soil for sequestering carbon from the atmosphere to mitigate the climate change. SOC is an indicator of soil productivity, such as soil texture, depth, pH, organic matter and fertility. It is an important constituent of the soil's chemical, physical and biological health. The measurement of SOC stock will help us to assess the crop productivity and the sustainable soil health. Besides, quantifying the SOC pool will also provide us with an idea about the potentials of the soil for sequestering carbon from the atmosphere or the emission potential when the soil is disturbed.

4.2. Advantage of biosequestration in soils

Soils are the largest carbon reservoir of the terrestrial carbon cycle. It stores large amount of SOC, which is originated from plants and animal tissue that exist at different stages of decomposition. According to Batjes and Sombroek (1997), soils contained 1550 Pg of organic C up to 1 m depth, 2500 Pg of organic C up to 2 m and 750 Pg of inorganic carbon at 1 m depth. This total soil C pool of 2300 Pg is three times the atmospheric pool of 770 Pg and 3.8 times the vegetation pool of 610 Pg. Carbon storage in soils is the balance between the input of dead plant material (leaf and root litter) and losses from decomposition and mineralization processes (heterotrophic respiration). Removing CO_2 from the atmosphere is only one significant benefit of enhanced carbon storage in soils. The benefits of atmospheric carbon sequestration into SOM are as follows:

i. Improved soil structure and aggregate stability,

ii. Increased water holding capacity (WHC)

iii. Improve nutrient cycling,

iv. Reduced soil erosion

v. Increased capacity to reduce the toxic elements from the soil

vi. Increase crop productivity

vii. Improve soil health and soil sustainability

viii. Increase fertilizer use efficiency (FUE),

ix. Mitigate climate change through atmospheric CO_2 fixation.

The researchers explored the impact of raised atmospheric CO_2 levels on carbon sequestration in soil by crops. The stored soil carbon is vulnerable to loss through both land management change and climate change. The important strategies of soil C sequestration include restoration of degraded soils, and adoption of improved management practices (IMPs) of agricultural and forestry soils.

5. ASSESSMENT OF SOIL CARBON

- SOC results are usually expressed as percentage C by weight (*i.e.* g C per 100 g of soil). SOC results can be converted to soil organic matter (SOM) level by multiplying SOC value by a conversion factor of 1.72. This assumes that SOM present in soil, on average, is made up of 58 per cent carbon.
- Very often it is more practical to express SOC on per ha basis, namely as tonnes C per ha. To perform such calculation, knowledge of the bulk density of the soil to the sampling depth is needed the amount of SOC present in the soil to ploughing depth (cm depth of 1 hectare) of land can be worked out as follows.

Tonnes Carbon per ha = SOC (%) × Soil bulk density × Sampling depth (cm)

6. MANAGEMENT TECHNIQUES AN ULTIMATE SOLUTION

Management technique has a crucial role on soil carbon sequestration and is highly related to the soil management system.

1. Zero/minimum tillage combined with crop residue retention in the field helps to increase water use efficiency (WUF), reduces fossil fuel utilization and sequestering carbon.
2. Crop rotation and cover crops contribute to soil carbon sequestration because they can increase the rate of accumulation of SOC stock at various depths in the soil.
3. Organic manures are the best resources of incorporating SOM to the soil and increasing the soil carbon stock. These are more resistant to

microbial decomposition than plant residues. Thus, the carbon storage is higher with manure application than with crop residue. Organic farming is a production system that sustains the soil health and quality.

4 The application of FYM/vermicompost has long been treated as a valuable source of organic matter to enhance soil nutrient cycling and carbon concentration. The animal dung management can also be done by using it in biogas production and slurry left could be used as manure in field.

5. Conservation practices, deep rooted crops/grasses should be developed through hybridization to enhance the carbon stock in the deeper layer of soil profile.

6. Biochar application can hold carbon in the soil profile for hundreds and thousands of years. It helps in improving soil fertility that stimulates plant growth which ultimately increases the biomass leading to higher CO_2 consumption.

Management techniques, which are successful in providing a net carbon stock in soils, include the following:

6.1. Conservation agriculture

CA is a set of soil management practices that minimize the disruption of the soil's structure, composition and natural biodiversity. Despite high variability in the types of crops grown and specific management regimes, all forms of conservation agriculture share three core principles. These include:

1. Maintenance of permanent or semi-permanent soil cover (using either a previous crop residue or specifically growing a cover crop for this purpose);
2. Minimum soil disturbance through tillage (just enough to get the seed into the ground)
3. Regular crop rotations to help combat the various biotic constraints;

CA also uses or promotes where possible or needed various management practices listed below:

- Utilization of green manures/cover crops (GMCC's) to produce the residue cover;
- No burning of crop residues;
- Integrated disease and pest management;
- Controlled/limited human and mechanical traffic over agricultural soils.

It is a more sustainable cultivation practice for the future than those presently practiced. Conservation agriculture has a proven potential of converting many soils from source of C to sinks of atmosphere C by sequestering it into soil (FAO/CTIC, 2008; Lal *et al.*, 1998). CA improves agriculture by improving soil surface aggregates, reducing erosion, increasing water infiltration, reducing compaction through promotion of biological tillage, carbon content, and increasing surface soil organic matter moderating soil temperatures, and suppressing weeds growth. It also helps in reducing costs of production, saves time, increases yield through more timely planting, reduces diseases and pests through stimulation of biological diversity, and reduces greenhouse gas emissions (Hobbs, 2007).

6.2. Minimum/zero tillage

Minimum tillage: These are tillage operations that leave at least 30 per cent of the soil surface covered by plant residues in order to increase water infiltration and cut down on soil erosion and runoff. Conservation tillage is an intermediate form of CA since it keeps some soil cover as residue from the previous crop. But some tillage is usually done. It developed as a management system after the "Dust Bowl" of the 1930's in the Mid-West areas of the USA. It was found to reduce erosion by protecting the soil surface from wind and rain.

Tillage and Zero Tillage are technical components used in conservation agriculture that simply involve the absence of tillage/plowing operations on the soil. Crops are planted directly into a seedbed not tilled after harvesting the previous crop. Not everyone utilizing no-till technologies adopts other important components of CA. One major difference is that NT or ZT do not necessarily leave residue mulch. Some recent research data suggest this is vital, since without the residue mulch many of the benefits of CA are lost or decreased in value.

The main purpose of this tillage is to provide favourable soil environment for plant growth. It is one of the major factors responsible for reducing carbon stocks in soil (Yadav *et al.*, 2015). SOM is oxidized when it is exposed to the air by tillage, resulting in a reduction in organic matter (OM) content, unless additional OM is returned to the soil as residues, compost, or other means. Tillage disrupts the pores left by roots and microbial activity. Changes in tillage practices could lead to C sequestration in agricultural soils. Conservation tillage includes no-tillage, reduced tillage and mulch tillage (Hati *et al.,* 2007). The effect of tillage on soil carbon sequestration could be noticeable only in long term rather than short term tillage practice (Deen and Kataki, 2003). West and Post (2002) concluded that by converting conventional tillage to no-till sequestered an average of 0.57±0.14 t C/ha/year. Higher carbon stocks were recorded in the direct seeded rice (DSR) as compared to the transplanted rice (TPR) treatments, because of the effects of puddling on aggregation/SOC losses. If only normal tillage was used (on the basis that deep tillage is not practiced) then DSR led to an increase in soil C of 2.63 t/ha compared to TPR. As a result, without puddling of soil, the annual increase in soil C was 0.375 t/ha, which can also be applied to soils of similar texture. This value was higher than that (0.34 t C/ha) found for the global average for the switch from conventional tillage to no-tillage. Similar results were reported by Roldan *et al.* (2005) and observed that substantial amounts of carbon could be sequestered in irrigated cotton-growing soil where stubble was incorporated under a minimal tillage regime. Soil aggregates physically protected soil organic matter that would otherwise decompose rapidly. Thus, increased tillage disrupted the soil aggregates and increased the availability of carbon substrates to the microbial biomass ultimately decreasing long term storage of carbon (Six *et al.*, 2000). Rochester (2011) conducted long term trial on minimum tillage in clay soil and concluded that SOC increased throughout the monitoring period. Higher rates of C sequestration were observed in systems including legume crops. It was observed that in sub-soil the sequestration of C is found to be occurred greater in the sub-soil than in the surface soil.

6.3. Cover crops

A cover crop is a type of plant grown primarily to suppress weeds, manage soil erosion, help build and improve soil fertility and quality, and control diseases and pests. Cover crops are usually grasses or legumes, but may be comprised of other green plants.

Benefits of cover crops:

1. Cover crops improve biodiversity by increasing the variety of species in a given area. They also can help to reduce the amount of water that drains off a field, protecting waterways and downstream ecosystems from erosion.
2. And because each root creates pores in the soil, they help allow water to filter deep into the ground. As a result, they help to conserve water in several ways.
3. Cover crops can help to break disease cycles by reducing the amount of bacterial and fungal diseases in a soil. If you have a soil that is infested, you can plant a cover crop in that area and see if you can eradicate the disease.
4. They are also called "green manure" and "living mulches" because they provide nutrients to the soil much like manure does. And they are called "living mulches," as they can prevent soil erosion.

The use of crops such as legumes and small grains for protection and soil improvement between periods of regular crop production is the cover crop. Cover crops improve carbon sequestration by enhancing soil structure and adding organic matter to the soil. Wang *et al.* (2011) reported in their study on fine sandy, the lowest amounts of C accumulation observed 0.597 kg/m^2 by *Vicia faba* L. and 0.149 kg/m^2 by white clover (*Trifolium repens*), respectively. Among summer cover crops, sunhemp (*Crotalaria juncea* L.) accumulated the highest quantity of C (0.481 kg/m^2), while that by castor bean (*Ricinus communis*) was 0.102 kg/m^2 at 30ºC in the GL soil. Following a whole cycle of winter and summer cover crops grown, the mean SOC increased by 13.8 and 39.1 per cent in the GL and FS soils, respectively as compared to the respective soils. Pulses add a significant amount of organic carbon in an upland maize-based cropping system in Inceptisols of Indo-Gangetic plains, were studied after seven cropping cycles (Venkatesh *et al.,* 2013). The results indicated that inclusion of pulses improved the total soil organic carbon content. It was more in surface soil (0–0.2 m) and declined with increase in soil depth. Maize-wheat-mung bean and pigeonpea-wheat systems resulted in significant increase of 11 and 10 per cent, respectively in total soil organic carbon.

6.4. Crop rotations or crop sequencing

Growing of set of crops in a regular succession over a same piece of land (field with) in a specific period of time. In crop rotation soil improving crops should be rotated in time over the entire farm in a regular sequence as permissible by soil, climatic and economic factors. In general cropping intensity in dryland is only 100 per cent. At few places on partial lands occasionally two crops are taken in favourable season (Monoculture is the rule in dryland agriculture). Increasing the cropping intensity is one of the methods for increasing crop production. Cropping intensity is increased by sequence cropping and double cropping, but intercro-pping may also prove effective measure for increasing production per unit area.

6.4.1. *Benefits of crop rotation*

1. Crop rotation maintains and improves soil fertility.
2. Prevent – build-up of pests, weeds and soil diseases.
3. Control of soil erosion.
4. Ensures balanced programme of work throughout the year.
5. Prevent or limit periods of peak (requirements of irrigation water)
6. Conserve moisture from one season to next.

Varying the type of crops grown can increase the level of soil organic matter. The main component of crop rotation is replenishment of nitrogen through the use of green manure in sequence with cereals and other crops. Continuous maize cultivation with a legume-based rotation was studied by Gregorich *et al.* (2001). After 35 years, the difference between monoculture maize and the rotation was 20 tonne C/ha. This suggested that the legume-based rotation was more efficient in converting biomass C into soil organic C than the grass based. Likewise Baker and Ritchie (2007) found that crop rotation systems along with conservation tillage accumulated about 11 t/ha of carbon after nine years. Whereas, according to the report of FAO (2001), the carbon liberation into the atmosphere was about 1.8 t/ha/ year of CO_2 under conventional tillage with monoculture systems. Most of the C sequestered due to crop rotations took place in the mineral-associated organic matter fraction, possibly because of the effects of stabilization mechanism of organic matter related to organo–mineral interactions.

6.5. Crop residue

Crop residues are materials left in an agricultural field or orchard after the crop has been harvested. These residues include stalks and stubble (stems), leaves, and seed pods. The residue can be ploughed directly into the ground, or burned first. Good management of field residues can increase efficiency of irrigation and control of erosion. Depending on the crops grown, cropping intensity and productivity in different regions of India, there is a large variability in generation and end use of these crop residues. Most CR converted into high-value manure of better quality than FYM, and its use, along with chemical fertilizers, can help to sustain or even increase the agronomic yield. The crop residues generation is the highest in Uttar Pradesh (60 Mt) and followed by Punjab (51 Mt) and Maharashtra (46 Mt). Estimated total crop residues unutilized in India is 92.81 Mt annually, which is burnt on-farm primarily to clear the fields to facilitate planting of succeeding crops (Pathak *et al.,* 2010). The problem of on-farm burning of crop residues has intensified in recent years due to use of combines for harvesting and high cost of labours in removing the crop residues by conventional methods. Burning of crop residues produces CO, CH_4, N_2O, NOx, NMHCs (nonmethane hydro carbons), SO_2 and many other gases. About 0.23 million tonnes of CH_4 and 0.006 million tonnes of N_2O were emitted from burning of crop residue in India in 2007. Burning disturbs the microbial population present in the soil, leads to moisture loss and increases the pH of soil due to production of ash, which contains Ca, Mg and K ions. Leaving crop residue on the field is another practice which will have an important impact on the sequestration of carbon. In case of soybean-wheat system in sub-tropical central India, the annual contribution of C from aboveground biomass was about 22 per cent for soybean and 32 per cent for wheat (Kundu *et al.*, 2001). This resulted in 18 per cent of the annual gross carbon input being incorporated into the SOM. Unless a root crop was harvested, all below-ground production was available for incorporation into the SOM. Below-ground residues and root turn-over represented direct inputs into the soil system, and as such had the potential to make major contributions to SOM stocks (Sanderman *et al.*, 2010). On an average, crop residue contains about 40–50 per cent of C, but N is a much more variant component.

6.6. Organic agriculture

Organic agriculture does not permit the use of synthetic chemicals to produce plant and animal products, relying instead on the management of soil organic matter (SOM) and biological processes. In some parts of the world, farms must be inspected and certified before their food products can be sold as organic, indicating that no synthetic chemicals were used in producing them. But organic farming uses the principles of CA to some extent and one objective similar to CA is to maintain and improve soil health. Unlike organic farming, CA does allow farmers to apply synthetic chemical fertilizers, fungicides, pesticides and herbicides. Many farmers rely on using these to control weed and pest problems, particularly during the early transition years. As soil physical, chemical and biological health improves over time; the use of agrochemicals can be significantly reduced or, in some cases, phased out entirely.

Organic farming is not new to the Indian farming community. India has diverse climate and different forms of organic farming are successfully practiced. In 2003, only 73,000 ha of cultivated land were certified organic. In 2010, this rose to 780,000 ha with 0.43 per cent share of total agricultural land. Kukal *et al.* (2009) conducted two long term experiments involving application of FYM and inorganic fertilizer in rice-wheat and maize-wheat cropping systems. They reported that the SOC sequestration rate was higher in FYM plots in comparison to NPK plots in both the cropping systems. Also the sequestration rate was three times higher in rice-wheat than in maize-wheat cropping system. In rice-wheat system, the FYM plots sequestered 0.31 Mg/ha/year, while the rate was 0.14 Mg/ha/year in maize-wheat cropping system. This was in consonance with the studies conducted by Parker *et al.* (2002) who obtained 7-20 per cent higher organic carbon in top soil layer (5 cm) in cotton-rye cropping system with poultry litter as compared to fertilizer. Recent studies have highlighted the substantial contribution of organic agriculture to climate change mitigation and adaptation. The main potential of organic agriculture lies in its significant capacity to sequester carbon in soils. It also reduces emission as it emits less nitrous oxide (due to lower nitrogen input) and methane from biomass waste burning (as burning is avoided). Mainly due to zero chemical fertilizer use the requirement of energy is also less. At the same time it helps in increasing the SOC stock and soil quality simultaneously reducing vulnerability to extreme weather events.

6.7. Grasses and forages

Grasses and forages have the power to rebuild and upgrade the status of marginal and wasteland soil. These pastures, grass and legume species grown for animal feed made changes in environmental conditions, soil nutrient status or grazing management can lead to rapid changes in botanical composition and thus soil productivity (Snaydon, 1987). A better understanding of the effects of grazing on soil chemical properties and their subsequent influence on nutrient cycling through the soil-plant-animal system is necessary to maintain productivity and sustainability of pastures.

Grasses sequestered more C than leguminous cover crops. Photosynthetic assimilation of atmospheric carbon and the translocation of photo-assimilate to roots by deep-rooted grasses, not only help to trap the excess CO_2 in deeper soil layer, but partly replenish the SOC in the long run (Lal *et al.*, 1999a). Lavania and Lavania (2009) studied the importance of Vetiver grass [*Vetiveria zizanioides* (L.)] in sequestering carbon. Its root penetrates deep into the soil with an initial growth rate of 3 cm per day reaching over 2m in just 6 months to 6m in three years. It has an annual biomass production potential of 100–120t/ha. Soussana *et al.* (2004) suggested that introducing legumes into pasture can promote soil C storage through increasing plant production and N supply. The improvement of soil carbon in grasslands offered a global greenhouse gas mitigation potential of 810 Mt of CO_2 up to 2030, almost all of which would be sequestered in the soil (Conant *et al.*, 2001). The main factor responsible for increased carbon storage in grassland sites was the high carbon input derived from plant roots. This high root production provided the potential to increase SOM in pastures and vegetated fallows compared to cropped land (Ravindranath and Ostwald, 2008).

6.8. Rotational grazing

Rotational grazing is the practice of moving grazing livestock between pastures (often called *paddocks*) as needed or on a regular basis. There are many approaches and types of grazing that fall under the broad umbrella of rotational grazing. The simplest is moving livestock between paddocks every set number of days *i.e.,* in days, week or one month. But well-managed rotational grazing means that you evaluate the nutritional and forage needs of your animals, assess forage quality and quantity, regulate the acreage of access and control which parts of the pasture/range that the animals have access to.

Grazing is expected to decrease the availability of residues that can be used to sequester C, especially as the quantity of C returned in manure is less than that consumed. However, if proper grazing management is followed, it will have a positive effect on the C stock of soil. For improved grasslands, high rates of sequestration can be achieved through introduction of more productive grass species and legumes. Improved nutrient management and irrigation can also increase productivity and sequester more carbon (Verchot and Singh, 2009). Souza *et al.* (2008) in a study observed that forages accumulate more C in soils as compared to grain crops, due to a higher root biomass production stimulated by grazing. This practice keeps the pastures in a vegetative state, improves grass quality and reduces the total amount of CO_2 released into the atmosphere by enhancing the soils C storage capacity. Livestock is concentrated in small paddocks for days at a time so they graze lightly, but evenly. This encourages roots to grow deeper into the soil. Rates of SOC sequestration in grazing lands were lower than those in cropland and may range from 200 to 500 kg C/ha/year (Conant *et al.*, 2001).

6.9. Growing plants on semi-arid lands

In India, mainly grasses and shrubs are found in semiarid and marginal lands. Here, that plant can be grown which do not require excess moisture and can grow easily. These regions have been suggested as a way to increase carbon storage in soils. However, the fossil fuel costs of irrigating these lands may exceed any net gain in carbon sequestration. Additionally, in many semi-arid regions surface and groundwater contain high concentrations of carbonate and bicarbonate ions of Ca, Mg and Na. As these are deposited in the soil, they release CO_2 into the atmosphere. On degraded croplands, an increase of 1 tonne of soil carbon pool may increase crop yield by 20 to 40 kg/ha of wheat, 10 to 20 kg/ha for maize, and 0.5 to 1.0 kg/ha for cowpeas. Trees could also make significant contribution in increasing C content of soils of dryland. N-fixing trees in particular leads to increased accumulation of soil *Prosopis* and *Acacia* adapted to subtropical semi-arid lands were reported to increase the level of soil C by 2 tonne/ha (Geesing *et al.,* 2000).

6.10. Biochar or black carbon

Biochar is charcoal used as a soil amendment which is stable solid, rich in carbon, and can endure in soil for thousands of years. Like most charcoal, biochar is made from biomass via pyrolysis. It is supposed to facilitate carbon sequestration in soil. Biochar is gaining importance as a viable option of storing carbon permanently. It is a solid material obtained from the carbonization of the biomass. Described bio-char as charcoal produced by pyrolysis of biomass feedstock (e.g. plant material), an excellent soil amendment that offered a cheap, easily accessible source of organic carbon (OC) and improved water retention as well as provided habitat for microbes. According to Zimmeman (2010) and Roberts *et al.* (2010), soil application of biochar, charcoal created by low-temperature pyrolysis of biomass under anaerobic conditions, is also being considered as an option to increase the SOC. The conversion of biomass carbon to biochar leads to sequestration of about 50 per cent of the initial carbon as compared to low amounts retained after burning (3%) and biological decomposition (less than 10–20% after 5–10 years) (Lehmann *et al.,* 2006). Estimates suggest that around 0.05–0.2 Pg (1015 g) C yr^{-1} of black carbon or charcoal are stored annually in soil (Lehmann *et al.,* 2005; Nguyen and Lehmann, 2009).

6.11 Land use change and soil amendments

Land use change involves the management and modification of natural environment or wilderness into built environment, such as settlements and semi-natural habitats arable fields, pastures, and managed woods. It also has been defined as "the total of arrangements, activities, and inputs that people undertake in a certain land cover type. The land use pattern of India indicates that cropland dominates and followed by forest. Based on data from Forest Survey of India (FSI, 2009) and National Remote Sensing Centre (NRSC, 2008), the land use change matrix indicated that the forest area increased marginally, whereas the net sown (cropped) area and grassland area declined. The land use change and forestry sector (LULUCF) include emissions and removals from changes mainly in forest land, crop land and grasslands, which sequesters 177 million tone of CO_2 (NAAS, 2014). Thus, this sector plays an important role in regulating the emission profile from the agriculture sector and provides avenues for increasing the sink. Lal (2001) reported that by converting degraded soils under agriculture and other land uses into forests and perennial land use can enhance the SOC pool in India. The magnitude and rate of SOC sequestration with afforestation depends on climate, soil type, species and nutrient management. Carbon emissions attributed to changes in land use and land cover, can significantly affect management strategies that are intended to enhance carbon sequestration and decrease the atmospheric CO_2 concentration. Therefore, in such cases, agroforestry may be another option of conserving soil and improving the SOC pool. Soil amendments are the left over residues from processes that have beneficial properties when added to soil.

6.12 Inorganic fertilizer

Fertilization is one of the most important crop inputs in the production system. It is the primary means of increasing plant population and crop yield. Any increase in plant biomass ultimately increases the scope of carbon sequestration. Study revealed that fertilizer application as a successful method of sequestering carbon. Nitrogen can increase soil organic matter because it often limited in agro-ecosystems. Wilson and Al-Kaisi (2008) conducted experiment to examine the short-term effects of crop rotation reported that soil CO_2 emission rate was decreased, with the increase of N application rate. It has been observed that increased N can cause a depression in soil CO_2 emissions not only in agricultural soils but also in forest soils. Kukal *et al.* (2009) conducted research on rice-wheat and maize-wheat cropping systems and concluded that application of FYM or balanced fertilization with NPK resulted in higher carbon sequestration. Rice-wheat system has greater capacity to sequester C as compared to maize wheat because of greater C input through enhanced productivity.

7. CARBON SEQUESTRATION POTENTIAL OF AGROFORESTRY IN INDIA

The CO_2 reduction in atmosphere can only be achieved by shifting from lower biomass land uses (*e.g.*, grasslands, crop fallows etc.) to tree based systems, such as agroforestry, forests, and plantation forests (Roshetko *et al.* 2007). There are ample evidences to show that the overall (biomass) productivity, soil fertility improvement, soil conservation, nutrient cycling, microclimate moderation, and carbon sequestration potential of an agroforestry system is generally greater than that of an annual system (Dhyani *et al.*, 2013). According to Pandey (2002) carbon sequestration in Indian agro forests varies from 19.56 Mg C/ha/yr in north Indian state of Uttar Pradesh to a carbon pool of 23.46–47.36 Mg C/ha/yr in tree-bearing arid agro-ecosystems of Rajasthan. Carbon sequestration in terrestrial pools include the above ground plant biomass, such as timber, fuel wood and belowground biomass like roots, soil microorganisms, and all the forms of organic and inorganic C in soils including deep root zone. In agroforestry systems, two major components *i.e.,* trees and crops are mainly responsible for CO_2 sequestration. The total amount sequestered in each component differs greatly and is dependent largely on a number of factors that includes the type of system (and the nature of components and age of plant), site quality, and previous land-use (Newaj and Dhyani, 2008).

8. CARBON SEQUESTRATION BY TREES

The carbon sequestration potential (CSP) of trees varies with species, structure, age and spatial distribution. For the most common tree density in the range of 312–800 trees/ha Agroforestry System (AFS), the CSP varied in the range of 0.25 to 76.55 Mg C/ha/yr. Nair *et al.* (2010) have also reported world scenario of carbon stored in AFSs ranged from 0.29 to 15.21 Mg C/ha/yr in above ground, and 30–300 Mg C/ha up to a depth of 1 m in the soil (the age varied from 4 to 35 years). Thus the trees in agroforestry not only improving livelihood of small and marginal farmers, but also helping in mitigating global warming by enhancing carbon sequestration potential of Indian agriculture (Ajit *et al.*, 2013). The Central Agroforestry Research Institute (CAFRI), Jhansi has been working on carbon sequestration potential of various agroforestry systems since 2000 through in house and externally aided projects. CAFRI as a partner in ICAR scheme "National Innovations on Climate Resilient Agriculture" (NICRA) is assessing the carbon sequestration potential of selected AFS in the country. The research under the NICRA scheme indicated that in agroforestry, the tree has a capacity for biomass production at least as great as that of natural vegetation. The survey (32 districts of 12 states) for existing agroforestry systems in the farmer's field revealed that agroforestry is practiced in all parts of India and is recognized as having high potential for carbon sequestration. The observed number of trees on farmers' fields in these districts varied from 2 to 204/ha. The baseline standing biomass in the tree component varied from 0.58 to 113.12 Mg DM/ha, whereas, the total biomass (tree and crop) ranged from 4.96 to 123.58 Mg DM/ha. The soil organic carbon in the baseline ranged from 4.28 to 24.13 Mg C/ha. The CSP of existing AFS at district level has been estimated to range from 0.05 to 2.78 Mg C/ha/yr. Based on this study, estimated value of trees/ha was 19.44 and average CSP of existing AFS at country level was 0.34 Mg C/ha/yr or equivalently AFS in India has the potential to mitigate 1.245 Mg CO_2 ha/yr. Thus, the trees in existing agroforestry systems on farmers' fields are estimated to mitigate more than 33 per cent of the total GHG emissions from agriculture sector annually at the country level.

9. CARBON SEQUESTRATION BY CROPS

In agroforestry systems, although trees sequester more carbon, but crops also fix and store carbon in considerable amounts. Crop improves the organic matter in soil, which is a significant component of the terrestrial C pool (Ciampitti *et al.*, 2011). An increment in carbon pool in soil can be achieved through adoption of appropriate crop rotations (Wright and Hons, 2005), integrated soil fertility management (Lal, 2010; Srinivasarao *et al.*, 2012), precise use of fertilizers and organic amendments (Schuman *et al.*, 2002), and adoption of conservation agriculture. Multiplicity of cropping systems is one of the main features of Indian agriculture.

REFERENCES

Ajit SK, Dhyani R, Handa AK, Prasad R, Alam B, Rizvi RH, Gupta G, Pandey KK, Jain A and Uma (2013) Modeling analysis of potential carbon sequestration under existing agroforestry systems in three districts of Indo-Gangetic plains in India. *Agroforestry Systems* **87:** 1129–1146.

Baker CJ and Ritchie WR (2007) No-tillage for forage production in No-tillage seeding in conservation agriculture (2nd edition). Food and Agriculture Organization of the United Nations, CABI., United Kingdom.

Batjes NH and Sombroek WG (1997) Possibilities for carbon sequestration in tropical and subtropical soils. *Global Change Biology* **3:** 161–173.

Ciampitti IA, Garcia FO, Picone LI and Rubio G (2011) Soil carbon and phosphorus pools in field crop rotations in Pampean soils of Argentina. *Soil Sci. Soc. Am. J.* **75:** 616–625.

Collins HP, Elliott ET, Paustian K, Bundy LC, Dick WA and Huggins DR (2000) Soil carbon pool and fluxes in long term corn belt agroecosystem. *Soil Boil. Biochem.* **32:** 157–168.

Conant RT, Paustian K and Elliott ET (2001) Grassland management and conversion into grassland effects on soil carbon. *Ecol. Appl.* **11:** 343–355.

Deen W and Kataki PK (2003) Carbon sequestration in a long-term conventional versus conservation tillage experiment. *Soil Till. Res.* **74:** 143–150

Dhyani SK, Handa AK and Uma G (2013) Area under agroforestry in India: An assessment for present status and future perspective. *Indian J Agrofor.* **15(1):** 1–11.

Dixon RK, Brown S, Houghton RA, Solomon AM, Trexler MC and Wisniewski J (1994) Carbon pool and flux of global flux ecosystems. *Science* **263:** 185–190

FAO (2001) Soil carbon sequestration for improved land management, by M. Robert. World Soil Resources Report No. 96. Rome.

FSI (2009) State of Forest Report. 2009. Forest Survey of India, Ministry of Environment and Forests, Dehradun.

Geesing D, Felker P and Bingham RL (2000) Influence of mesquite (*Prospis glandulosa*) on soil nitrogen and carbon development: Implications for global carbon sequestration. *J. Arid Environ.* **46:** 157–180.

Gregorich EG, Drur CF and Baldock JA (2001) Changes in soil carbon under long-term maize in monoculture and legume-based rotation. *Can J Soil Sci* **81:** 21–31.

Halvorson AD, Wienhold BJ and Black AL (2001) Tillage and nitrogen fertilization influence grain and soil nitrogen in an annual cropping system. *Agron. J.* **93:** 836–841

Hati KA, Swarup A, Dwivedi AK, Misra AK and Bandyopadhyay KK (2007) Changes in soil physical properties and organic carbon status at the topsoil horizon of a vertisol of central India after 28 years of continuous cropping, fertilization and manuring. *Agr Ecosyst Environ.* **119:** 127–134.

Hobbs PR (2007) Conservation agriculture: what is it and why is it important for future sustainable food production. *J. Agric. Sci.* **145:** 127–137.

Jenkinson D and Rayners K (2008) The turnover of organic carbon in subsoils: II. Modelling carbon turnover. *Eur. J. Soil Sci.* **59:** 400–413

Kukal SS, Rasool R and Benbi DK (2009) Soil organic carbon sequestration in relation to organic and inorganic fertilization in rice-wheat and maize-wheat systems. *Soil Till. Res.* **102:** 87-92.

Kundu S, Singh M, Saha JK, Biswas A, Tripathi AK and Acharya CL (2001) Relationship between C addition and storage in a Vertisol under soybean–wheat cropping system in sub-tropical central India. *J. Plant Nutr. Soil Sci.* **164:** 483–486.

Lal R (2001) World cropland soils as a source or sink for atmospheric carbon. *Adv. Agron.* **71:** 145–119.

Lal R (2004) Soil carbon sequestration impacts on global climate change and food security. *Science* **304:** 1623-1627.

Lal R (2005) Soil carbon sequestration in natural and managed tropical forest ecosystems. *J. Sustain. Forestry.* **21:** 1–30.

Lal R (2009a) Soil degradation as a reason for inadequate human nutrition. *Food Sec.* **1:** 45-57.

Lal R (2010) Enhancing eco-efficiency in agroecosystems through soil carbon sequestration. *Crop Sci.* **50:** S120–S131. doi:10.2135/ cropsci2010.01.0012

Lal R, Kimble JM, Follett RF and Cole CV (1998) The potential of U.S. cropland to sequester carbon and mitigate the greenhouse effect. Chelsea, Mich.: Ann Arbor Press.

Lavania UC and Lavania S (2009) Sequestration of atmospheric carbon into subsoil horizons through deep-rooted grasses- vetiver grass model. *Cur. Sci.* **97(5):** 618–619.

Lavelle P, Pashanasi B, Charpentier F, Gilot C, Rossi JP, Derourard L, Andre L, Ponge JF and Bernier N (1998) Large-scale effects of earthworms on soil organic matter and nutrient dynamics. *Earthworm Ecology* 103–122.

Lehmann, J., Gaunt, J. and Rondon, M. 2006. Biochar sequestration in terrestrial ecosystems: a review. *Mitig Adapt Strategy Global Change* **11:** 403–427.

NAAS (2014) Carbon economy in Indian agriculture. Policy Paper No. 69, National Academy of Agricultural Sciences, New Delhi: 19 p.

Nair PKR, Nair VD, Kumar BM and Showalter JM (2010) Carbon sequestration in agroforestry systems. *Adv Agron,* **108:** 237–307

Newaj R and Dhyani SK (2008) Agroforestry for carbon sequestration: Scope and present status. *Indian Journal of Agroforestry* **10:** 1–9.

Nguyen BT and Lehmann J (2009) Black carbon decomposition under varying water regimes. *Organic Geochemistry* **40:** 846–853.

NRSC (National Remote Sensing Centre). 2008. Biodiversity Characterization at Landscape Level using Remote Sensing and Geographical Information System, Project Manual, NRSC, Indian Space Research Organization, Department of Space, Government of India, Balanagar, Hyderabad.

Pandey DN (2002) Carbon sequestration in agroforestry systems. *Climate Policy* **2:** 367–377.

Pandey SS, Pandey G, Cockfield TN and Maraseni (2014) Carbon stock dynamics in different vegetation dominated community forests under REDD+: a case from *Nepal For. Ecol. Manag.* **327:** 40–47

Parker MA, Nyakatawa EZ, Reddy KC and Reeves DW (2002) Soil carbon and nitrogen as influenced by tillage and poultry litter in North Alabama. Pp. 283–287. In: E. van Santen (ed.) Making Conservation Tillage Conventional: Building a Future on 25 years of Research, Proceedings of the 25th Annual Southern Conservation Tillage Conference for Sustainable Agriculture, Auburn AL, 24–26 June 2002.

Pathak H, Jain N, Bhatia A, Patel J and Aggarwal PK (2010) Carbon footprints of Indian food items. *Agric. Ecosyst. Environ.* **1–2:** 66–73.

Ravindranath NH and Ostwald M (2008) Carbon inventory methods. Handbook for greenhouse gas inventory, carbon mitigation and round wood production projects. Netherlands, *Ecol. Appl.* **11:** 343–355.

Roberts KG, Gloy BA, Joseph S, Scott NR and Lehmann J (2010) Life cycle assessment of biochar systems: Estimating the energetic, economic and climate change potential. *Environmental Science and Technology* **44(2):** 827–833.

Rochester LJ (2011) Sequestering carbon in minimum tilled clay soils used for irrigated cotton and grain production. *Soil Till. Res.* **112:** 1–7.

Roldan A, Salinas-Garcia JR, Alguacil MM and Caravaca F (2005) Changes in soil enzyme activity, fertility, aggregation and C sequestration mediated by conservation tillage practices and water regime in a maize field. *Appl. Soil Ecol,* **30:** 11–20.

Roshetko JM, Lasco RD and Delos Angeles MD (2007) Smallholder agroforestry systems for carbon storage: mitigation and adaptation strategies for global change. **12:** 219–242.

Sanderman J, Farquharson R and Baldock JA (2010) Soil carbon sequestration potential: a review for Australian agriculture *CSIRO Land & Water Report* www.csiro.au/resources/Soil-Carbon-Sequestration-Potential-Report.html.

Six J, Conant RT, Paul EA and Paustian K (2002) Stabilization mechanisms of soil organic matter: Implications for C saturation of soils. *Plant Soil* **241:** 155–176.

Six J, Paustian K, Elliott ET and Combrink C (2000) Soil structure and soil organic matter. Distribution of aggregate size classes and aggregate associated carbon. *Soil Sci. Soc. Am. J.* **64:** 681–689.

Snaydon RW (1987) Managed Grasslands – Analytical Studies. Ecosystem of the World 17B, Elsevier, Amsterdam, *The Netherlands.* **28:** 237–238

Soussana JF, Loiseau P, Vuichard N, Ceschia E, Balesdent J, Chevallier T and Arrouays D (2004) Carbon cycling and sequestration opportunities in temperate grasslands. *Soil Use Manage.* **20:** 219–230.

Souza AP, Gaspar M, Silva EA, Ulian EC, Waclawovsky AJ and Nishiyama MY (2008) Elevated CO_2 increases photosynthesis, biomass and productivity, and modifies gene expression in sugarcane. *Plant, Cell and Environment* **31:** 1116–1127.

Srinivasarao (2012) Long-term effects of crop residues and fertility management on carbon sequestration and agronomic productivity of groundnut–finger millet rotation on an Alfisol in southern India. *Int. J. Agric. Sustain.* **10:** 1–15.

UNEP (2012) Global environment outlook 5 — Environment for the future we want, United Nations Environment Programme.

Venkatesh MS, Hazra KK, Ghosh PK, Praharaj CS and Kumar N (2013) Long-term effect of pulses and nutrient management on soil carbon sequestration in Indo-Gangetic plains of India. *Can. J. Soil Sci.* **93:** 127–136.

Verchot LV and Singh VP (2009) Carbon sequestration opportunities with smallholder communities: forestry, agriculture and agro-forestry. In: 4th World Congress on Conservation Agriculture 4–7th February, 2009 at New Delhi: 351–361.

Wang WJ, Qiu L, Zu YG, Su DX, An J, Wang HY, Zheng GY, Sun W and Chen XQ (2011), Changes in soil organic carbon, nitrogen, pH and bulk density with the development of larch (*Larix gmelinii*) plantations in China, *Global Change Biol.* **17(8):** 2657–2676.

West TO and Post WM (2002) Soil organic carbon sequestration rates by tillage and crop rotation: a global analysis. *Soil Sci Soc Am J* **66:** 1930–1946.

Wilson HM and Al-Kaisi MM (2008) Crop rotation and nitrogen fertilization effect on soil CO_2 emissions in central Iowa. *Applied Soil Ecology* **39:** 264–270.

Wright AL and Hons FM (2005) Tillage impacts on soil aggregation and carbon and nitrogen sequestration under wheat cropping sequences. *Soil Tillage Res.* **84:** 67–75.

Yadav RS, Pandya IY and Jangid MS (2015) Estimating status of soil organic carbon in tropical forests of Narmada forest division, Gujarat, India. *International Research Journal of Environment Sciences* **4(1):** 19–23.

Zimmerman AR (2010) Abiotic and microbial oxidation of laboratory-produced black carbon (biochar). *Environmental Science & Technology* **44:** 1295–1301.

■■■■

10

Polymers: A Potential Way to Enhance Agricultural Sustainability

V. Tyagi, R.K. Singh, Ram Swaroop Meena and M. Nagargade*

Department of Agronomy, Institute of Agricultural Sciences, Banaras Hindu University, Varanasi – 221005, India
**Author for correspondence: vish926@gmail.com*

ABSTRACT

The greatest challenge of today's agriculture is to feed the growing population and restore the natural resources. Global food production needs to be increased to feed this ever increasing population. Modern crop production technology has considerably raised output, but has created problems of land degradation, pesticide residue in farm produce, gene erosion, atmospheric and water pollution. The task of meeting the needs of the present generation without eroding the ecological assets of the future generation should be the prime objective. This objective can be achieved only by adopting such technologies which improve resource use efficiency in agriculture. In this context use of polymers may be a novel technology which can improve food production without threatening the environment and the natural resources. The use of polymer in agriculture is gaining popularity in science, particularly in the field of polymer chemistry. Polymers are used to improve soil permeability, density, structure, increase moisture content in soil, increase the efficiency of pesticides and herbicides, allowing lower doses to be used and to indirectly protect the environment by reducing pollution and clean-up existing pollutants.

Keywords: Biodiversity, Food security, Natural resources, Polymer, Sustainability

1. INTRODUCTION

World population is increasing at an alarming rate and is expected to reach seven billion by the end of year 2050 (Ekebafe *et al.,* 2011). Population growth and the resultant development of large high-density urban populations, together with parallel global industrialization, have placed major pressures on our environment, potentially threatening environmental sustainability and food security. This has resulted in global warming and the build-up of chemical and biological contaminants throughout the biosphere, but most notably in soils and sediments. During the 20th century, the main emphasis of agricultural development all over the world was the increasing productivity per unit area of land used for crop production to feed the ever-increasing population. This was substantially accomplished through over exploitation of natural resources such as water and plant resources and excessive use of fertilizers and pesticides. Although this practice resulted in considerable increase in crop yields in the short-term, it was not sustainable in the long-run. The widespread use of irrigation, agrochemicals and new seeds have largely increased agriculture production, but this model of agricultural growth is fatally flawed because of declining crop yields and massive environmental impacts. The practical effects of such changes from diverse low-intensity agriculture to intensive monocultures greatly improved yields from crops and livestock, but drastically reduced native habitat cover, leading to an impoverished agricultural matrix, and exposure of many wildlife species to toxic levels of pesticides.

2. ADVERSE IMPACT OF MODERN AGRICULTURE ON NATURAL RESOURCES

The greatest challenge of today's agriculture is to feed the growing population and restore the natural resources. By 2050 the world's population will reach 9.1 billion, in order to feed this larger population, food production must increase by 70 per cent (FAO, 2013). Uncontrolled population growth in developing countries accelerated the imbalance between human needs and sustainable use of land. Though by virtue of chemical fertilizers the production and productivity of crops has increased, the increased use of pesticides has posed many environmental and health problems. There is now overwhelming evidence that some of these chemicals do pose a potential risk to humans and other life forms and unwanted side effects to the environment (Akter *et al.,* 2009). Modern agricultural practices use many kinds of chemicals, such as fertilizers, pesticides, cleaners, crop preservatives to produce and keeping large amount of high-quality food. But every single of these chemicals has dangerous and unforeseen side-effects as like toxicity to non-target organisms which causes to ecological imbalance (Onder, 2011). The chemical fertilizers and pesticides used over a long period of time have adverse toxic effects on the production potential of the land and the ultimate consumers of the products.

Technology generated and implemented for improving the productivity of crops during the past three decades resulted in fast depletion of natural resources base besides creating several other crop production problems. Soil productivity is fast declining due to greater dependence on inorganic fertilizers as against organics in traditional agriculture (Fig. 1). Improved cultivars are susceptible to several insect pest and diseases that traditional varieties would have tolerated. The problem is aggravated by the widespread use of one or a few genetically uniform varieties over a large area (spatial dominance), which is considered the prerequisite for the widespread of disease and insect pests as it provides ideal condition and creates vulnerability (Keneni *et al.,* 2012). To counteract these pests and disease problems, large amounts of pesticides and fungicides are used at the cost of environmental safety. With dwindling natural resources base and increasing environmental concern, balancing food and environmental security assumes greater importance to sustainable livelihood for posterity (Reddy, 2011).

2.1. Depletion of soil fertility

There is concern that modern farming, especially, continuous cultivation of arable land, is degrading the soil on which agriculture depends, particularly by reducing concentrations of soil organic carbon (SOC) and nutrient reserves, by damaging soil structure, by increasing the burden of heavy metals, and through increased erosion risks. More amounts of nutrients is depleted from soil due to intensive cropping along with continued lack of required nutrient replenishment of nutrient depleted soils as well as nutrient losses through wind and water erosion are not only exacerbating soil degradation, but also jeopardizing agricultural sustainability in these regions (Tan *et al.*, 2005).

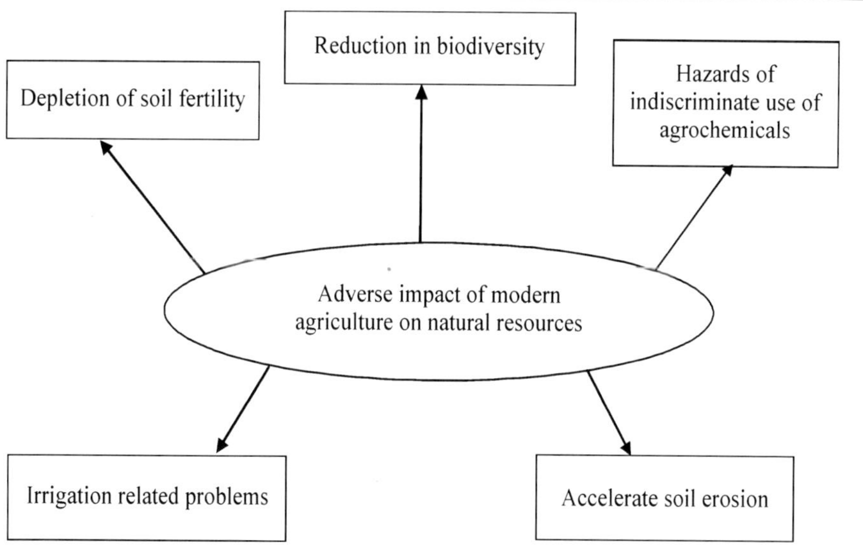

Fig. 1. Adverse impact of modern agriculture on natural resources

Conventional tillage, increase porosity on the short term but decrease stable soil aggregation over the long term, and can reduce soil organic matter content, deteriorate soil structure, lower water-holding capacity, and compact subsoil. Decline in quantity and quality of organic matter in most soil is adversely affecting soil biodiversity and biological regulation of soil process (Reddy, 2011).

2.2. Reduction in biodiversity

Current agricultural practices are based on growing monocultures or binary mixtures over large areas, with a resultant impoverishing effect on biodiversity at several trophic levels. Areas under intensive agriculture, in which applications of high doses of mineral fertilizers to monoculture crops are a prominent feature, have a negative impact on biodiversity (Reidsma *et al.,* 2006). High yielding varieties having narrow genetic base replaces several local cultivars which were rich in genetic diversity. The FAO suggests that three-quarters of the varietal genetic diversity of agricultural crops has been lost in the past 100 years (FAO 2008). Widespread use of modern cultivars raises question about the stability of production and threatens the pest and disease attack. However, maintaining acceptable levels of agricultural productivity will continue to be a high priority, mainly because of an increasing global population that needs to be fed, but also due to the need for economic sustainability of the individual farmer (Frankow-Lindberg *et al.*, 2009).

2.3. Hazards of indiscriminate use of agrochemicals

Modern cultivars are responsive to agrochemicals. These cultivars are also very much sensitive to pest, disease and weeds than traditional cultivars due to narrow genetic base. Keeping in view the productivity of high yielding cultivars, farmers used agrochemicals including the fertilizers indiscriminately. Which are mixed with soil, water, air and food, this causes problems on the agricultural foods and affect both human health and natural balance, so finally they become an environment problem. Pesticide runoff is an important contributor to surface-water contamination (Wohlfahrt *et al.,* 2010). A pesticide that specialized on a harmful doesn't kill only target, it also kills many harmless organisms.

2.4. Accelerate soil erosion

Wrong soil tillage with regards of without any concern field location, soil structure and climate conditions cause to soil moving with rain, in other words cause erosion. This situation not only cause to inefficient soils, it also pollutes streams and fills up dams with soil etc. serious environment problems (Onder, 2011). Cultivation of natural ecosystems has led to marked decline in soil C storage, such that conservation agricultural practices are widely recommended as options to increase soil C storage, thereby mitigating climate change (Luo, 2010).

2.5. Irrigation related problems

Due to increasing demand for food and fibre by ever increasing population of the country, the pressure on fresh water resources is increasing. Optimum utilization of surface and ground water resources of paramount importance to fill the gap between water demand and supply. Inadequacies in irrigation water management resulted in serious resource degradation problems. In areas of canal irrigation without adequate drainage, salinity and high water table are the serious constraints in enhancing the productivity. There are other areas where unrestricted installation of tube-wells has resulted in continuous lowering of ground water tables leading to increasing cost in lifting water for irrigation (Reddy, 2011).

3. ROLE OF POLYMERS IN MAINTAINING SUSTAINABILITY THROUGH ENHANCING INPUT USE EFFICIENCY

The use of polymer in agriculture is gaining popularity in science, particularly in the field of polymer chemistry (Fig. 2). This has provided solutions to the problems of the present day agriculture which is to maximize land and water productivity without threatening the environment and the natural resources. Superabsorbent polymer hydrogels potentially influence soil permeability, density, structure, texture, evaporation, and infiltration rates of water through the soils. Functionalized polymers are also used to increase the efficiency of pesticides and herbicides, allowing lower doses to be used and to indirectly protect the environment by reducing pollution and clean-up existing pollutants (Ekebafe *et al*., 2011).

3.1. Super absorbent polymers

The shortage of irrigation water, drought, deforestation and soil erosion are one of the major problems of agriculture around the globe. Prolonged periods of drought can cause environmental, agricultural and economic problems leading to social unrest and humanitarian crisis. The development of non-traditional new technologies to conserve water is becoming important for attaining a sustainable economic growth, especially in agricultural countries.

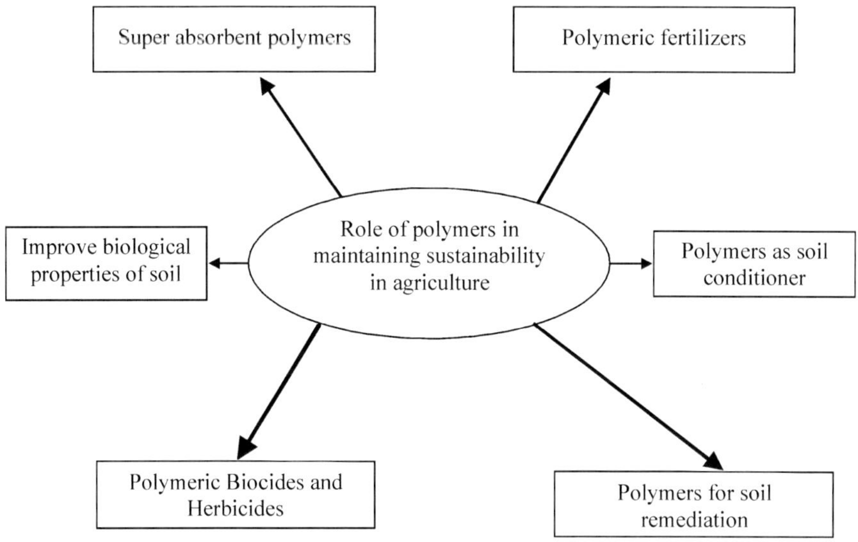

Fig. 2. Role of polymers in maintaining agricultural sustainability

In recent years, cross-linked polymers with hydrophilic nature, known as superabsorbent polymers (SAPs), have attracted much interest in this area as they can improve water and fertilizer retention in soil and thus increase plant growth (Zhang *et al.,* 2006).

Synthetic polymers in the form of crystals or tiny beads available under several trade names, such as super absorbent polymers, root watering crystals and drought crystals are collectively known as hydrogels. They have enormous capacity to absorb water and make it available to plants over time (Akhter *et al.,* 2004). Super absorbent polymers (SAPs) can absorb water up to 500 times their own weight and swell to many times their original size and weight (Farrell *et al.,* 2013). They are lightly cross-linked networks of hydrophilic polymer chains. The network can swell in water and hold a large amount of water while maintaining the physical dimension structure (Zohuriaan-Mehr *et al.,* 2007). It was known that commercially used water-absorbent polymeric materials employed are partial neutralization products of cross-linked polyacrylic acids, partial hydrolysis products of starch-acrylonitrile copolymers and starch–acrylic acid graft copolymers. At present, the material's biodegradability is an important focus of the research in this field because of the renewed attention towards environmental protection issues.

SAP hydrogels potentially influence soil permeability, density, structure, texture, evaporation and infiltration rates of water through the soils. Particularly, the hydrogels reduce irrigation frequency and compaction tendency, stop erosion and water runoff, and increase the soil aeration and microbial activity. Hydrogel addition improves water storage properties of porous soils and resulted in the delay and onset of permanent wilting per centages under intense evaporation. An increase in water holding capacity due to hydrogel significantly reduced the irrigation requirement of many plants (Taylor and Halfacre 1986). In arid areas, the use of SAP in the sandy soil (macroporous medium), to increase its water-holding capacity seems to be one of the most significant means to improve the quality of plants (Bakass *et al.,* 2002). The SAP particles may be taken as "miniature water reservoirs" in soil. Water will be removed from these reservoirs upon the root demand through osmotic pressure difference. The hydrogels also act as a controlled release system by favouring the uptake of some nutrient elements, holding them tightly, and delaying their dissolution. Consequently, the plant can still access some of the fertilizers, resulting in improved growth and performance rates (Liang *et al.,* 2007, Bowman *et al.,* 1990 and Wu *et al.,* 2008).

Mechanism of water absorption

Many hydrogels are polymers of carboxylic acids. The acid groups stick off the main chain of the polymer (Fig. 3). When these polymers are put into water, the hydrogen atoms react and come off as positive ions. This leaves negative ions along the length of the polymer chain. When polymer chains are in solution, they tend to coil up (Fig. 4). However, the hydrogel now has lots of negative charges down its length (Fig. 5).

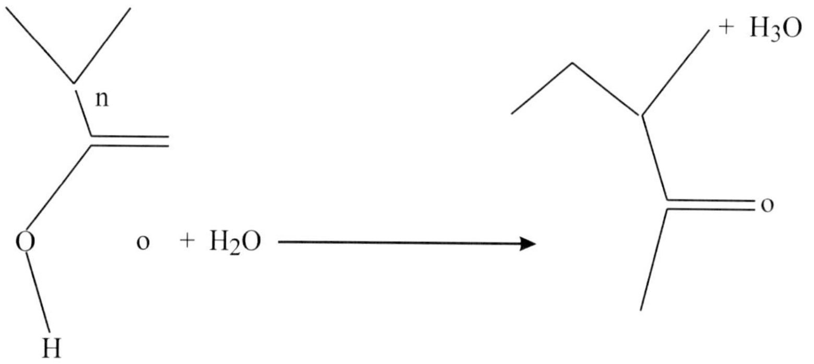

Fig. 3. The polymer chain of a hydrogel.

Note: H_3O^+ is another way of writing H^+ and shows an acid.

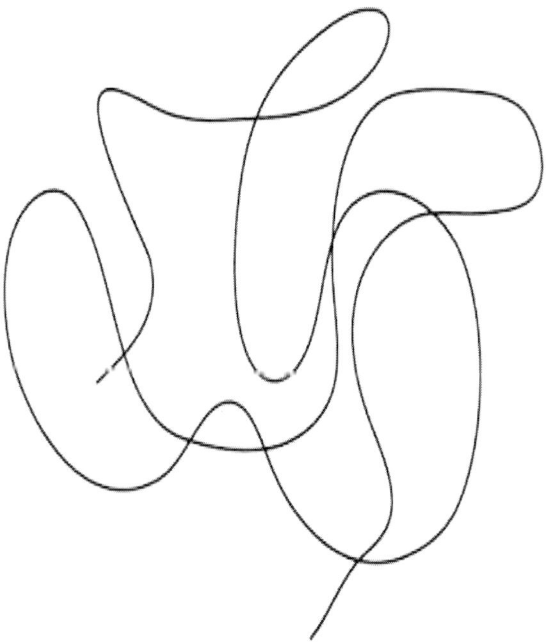

Fig. 4. A polymer chain coiled up in solution with lots of negative ions

These negative charges do two things:

- They repel each other. This forces the polymer chain to unwind and open up.
- They attract water molecules.

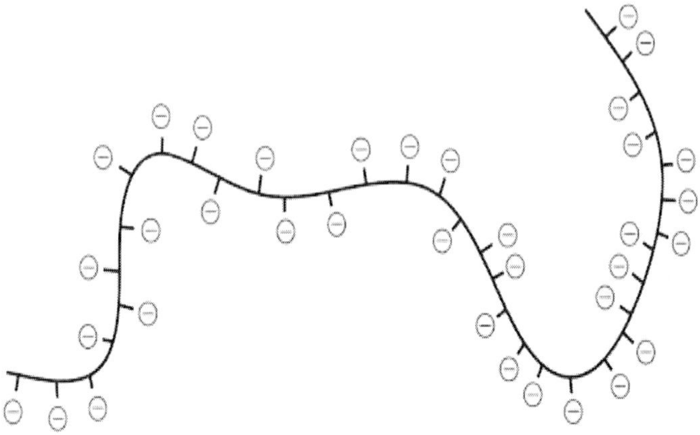

Fig. 5. A hydrogel polymer chain charges along its length

Advantages of Gel Polymers

Generally gel polymers perform well at high temperatures also (40–50ºC). Hence these are suitable for semi-arid and arid regions. These polymers absorb a minimum of 350 times of its dry weight in pure water and gradually releases it as per the need of crop plant, so minimize irrigation requirement of crops, water loss due to evaporation is significantly reduced. Help plants to withstand extended moisture stress and delays onset of permanent wilting point. Gel polymers not affect nutrient availability, soil chemical composition because these polymers have neutral pH and have ability to minimize nutrient leaching. Since these polymers are effective in soil for at least one crop season, they maintain sufficient moisture in soil during entire crop growth. These polymers improve physical properties of soils (porosity, bulk density, water holding capacity, soil permeability, infiltration rate), reduce soil compaction, increase the oxygen/air availability in the root zone, reduce soil erosion, improve seed germination and the rate of seedling emergence, improve root growth and density. Gel polymers also increase biological/microbial activity in the soil and these are biodegraded in soil within 1–3 years.

Limitations of Super Absorbent Polymers

Though the SAP amendment in soil has many beneficial uses, there are some limitations to its applications in soil.

SAPs are quite fragile and tend to break apart easily thereby losing their water retention property. Further SAPs can also dehydrate rapidly in a matter of hours thus losing their absorbed water (Kim and Nadarajah, 2008). The water absorption of SAPs greatly reduces in the soils as SAPs are under pressure and unable to swell and take in water (Bhardwaj *et al.,* 2009). The water absorption of SAPs in soils further decreases due to formation of additional crosslinks with certain ions like Ca^{2+} and Al^{3+} present in the soil (Chatzoudis and Rigas, 1999). The water absorption of the SAP also decreases with increase in salinity of irrigation water (Taban and Naeini, 2006; Lentz and Sojka, 2009). The SAPs in soils releases water with increase in temperature and this water could be potentially lost to deep percolation (Andry *et al.,* 2009). Effectiveness of SAP decreased on rewetting and can affect the hydraulic properties of soil only if applied in higher application rates (Geesing and Schmidhalter, 2004). Also the efficacy of the SAP decreases over a period of time and to compensate for these loses, higher application rates are required. This factor affects the economic value of crops grown on fields amended with SAP.

5.2. Polymeric Fertilizers

Fertilizers are one of the most important products of the agrochemical industry. They are added to the soil to release nutrients necessary for plant growth. However, the potential hazards of fertilizers to the environment have results in stringent limitation to their use. About half of the applied fertilizers, depending on the method of application and soil condition, are lost to the environment, which results in the contamination of water. One possible way to improve nutrient use efficiency while reducing the environ-mental impacts is through the use of controlled release fertilizers (CRFs) (Wu *et al.,* 2008). The CRF products are developed to release nutrients gradually trying to coincide with the nutrient demand during crop growth. Compared to the conventional type, slow-release fertilizers have many advantages, such as decreasing fertilizer loss rate, supplying nutrients sustainably, lowering application frequency and minimizing potential negative effects associated with over dosage (Guo *et al.,* 2006). These fertilizers are prepared by coating the active soluble nutrient component with a membrane that serves as a diffusion barrier. Several polymer materials have been proposed for CRF encapsulation (Gonzalez *et al.,* 2015).

Polymer coated fertilizers (PCFs) enhance longevity and nutrient use efficiency. They control the rate of nutrient release by forming semi-permeable membrane on the surface of nutrient particles. Most of PCFs release nutrients by diffusion through a semi-permeable polymer membrane and release rate can be controlled by varying the composition and thickness of the coating. New gel based controlled release fertilizers are of immense use in nutrient management. They are developed by mixing and processing N, P and K fertilizers with natural and semi-natural polymers (Hong and Zhang, 2006). These gel based fertilizers are easily produced and have lower price than coated fertilizers and expressed better physical and chemical properties. Besides, the nutrients (N, P and K) in gel based fertilisers are leached more slowly into the soil than common fertilizers. The release kinetic of these gel based fertilizers can be changed by adding inorganic particles into hydrogel (so-called hydrogel composites). The particles, when sufficiently dispersed, cause tortuosity effect that disturbs the release of nutrients. Such an effect provides longer (tortuous) pathways that makes difficult the diffusion of the nutrients toward the outside hydrogels. The particles play a role as a retardant factor for nutrient release. As a result, the release is extended for longer times (Marcos *et al.*, 2015).

In arid and desert regions water scarcity also limit fertilizer use efficiency because nutrients are taken up in the form of solution from the soil. In dry land areas, plants become unable to take nutrients from soil even they are present in sufficient quantity in soil. So to overcome this problem we should combine slow released fertilizers with superabsorbent polymers to obtain both slow-release and water-retention properties and also increase the soil's aeration and microbial activity, and mitigate at the same time the environmental impact from water-soluble fertilizers, lower frequency of irrigation (Rashidzadeh and Olad, 2014). In general, combining fertilizers with superabsorbent is done by two methods. In the first method, fertilizers are blended with superabsorbent. In the second approach, fertilizers are added to the reaction mixture and polymerized in situ whereby the fertilizers are entrapped in the superabsorbent. Recently, the preparation of polymer superabsorbent composite has attracted great attention because of their relative low production cost, excellent water-retention and their considerable applications in agriculture and horticulture. Diatomite (a kind of clay mineral), as a good substrate for superabsorbent composite materials, is a kind of lightweight sedimentary rock composed of principally silica microfossils of aquatic unicellular algae (Wu *et al.*, 2008).

5.3. Polymers as Soil Conditioner

Soil management is aimed at effectively maintaining or increasing agriculture production for the benefit of society and preserving or improving the environment. Soil factors include soil type, thickness, compaction of soil layers, and ground water, conditions. Soil provides a medium to support plants, and is a reservoir for water, and plant chemical nutrients made up of a mixture of solids, liquids, and gaseous materials. A destiny of present and future generations mostly depends on how they manage soil and its resources. The main problem of the world's land is the degradation of agricultural soils (Zezin *et al.*, 2015). Several factors such as excessive tillage, monocropping, excessive use of inorganic fertilizers without adding organic manure, faulty irrigation practices etc. all are responsible for soil degradation. Intensive agriculture practiced with insufficient attention to ecological principles has harmed soil health, depleted groundwater resources and reduced agrobiodiversity (Kesavan and Swaminathan, 2008). In clayey soils, crust formations cause problems for seedling emergence. Thus, there is a great interest in soil reclamation to overcome these problems. Polymeric materials are being added to soils for reclamation and to improve soil composition and structure. These polymeric materials improve the soil grain structure by forming cloddy soil suitable for vegetation for improving plant growth. They reduce water demand especially in sandy soils via increase of water-holding capacity, reduction of water stress, preventing soil erosion by altering soil mechanical structure, improving friability, enhance porosity

(Paluszek and Zembrowski, 2008), enhancing the establishment of seedlings, and increasing crop yields. During the last two decades, water-soluble polymer emulsion has been developed and applied as a soil stabilizer in many countries (Liu *et al.*, 2009). The mode of action of these polymers can be targeted to a particular physical property of the soil: (1) *Surfactants* affect the surface tension of soils to water and are most commonly used to enhance the wetting and infiltration of treated soils. (2) *Flocculants* enhance the cohesive attraction among dispersed fine particulates and lead to formation of aggregates (flocs) in aqueous media that achieve sufficient size and weight. These materials enhance the existing structural stability of the soil and increase shear strength and reduce detachment.

Advantages

Hence, when we used polymers as a soil conditioner these polymers reduce soil runoff and save water in sub soils by decreasing crust formation so maintain sufficient moisture in the root zones during plantation. These polymers also improve fertilizer use efficiency by minimizing nutrient leaching.

5.4. Polymers for Soil Remediation

The release and persistence of toxic heavy metals into the natural environment is a serious concern especially in urban areas. The problem of heavy metal pollution is gaining momentum from year to year as more and more amounts of heavy metals are extracted from their ores and released into the environmental segments (water, air and soil) during processing or afterwards. Heavy metals are essentially nonbiodegradable and therefore, accumulate in the environment and subsequently find their way into the food chains (Ali *et al.*, 2012). Soil may become contaminated with metals from a variety of anthropogenic sources. Various potentially toxic elements, including heavy metals, are present in industrial waste water. Ideally, a contaminated soil should be restored to regain its original potential, but this can be a very expensive process, and thus depends not only on the expected benefit of the cleanup and future value of the soil, but also on political and public awareness of the problem (Ekebafe *et al.*, 2011). Conventional remedial approaches to severely metal-contaminated soils involve removal and replacement of soil with clean materials or capping the soil with an impermeable layer to reduce exposure to contaminants (Brown *et al.*, 2006), chelating agent ethylenediaminetetraacetic acid (EDTA) has been used extensively for heavy metals extraction from soil (Cline and Reed, 1995).

Water-soluble polymers (WSP) have been used extensively for the removal or recovery of metal ions from aqueous solutions. However, this technology has only recently been used to remove metals such as lead from solid surfaces. Three distinct advantages of using WSP are: (1) The metal bound to the polymer in solution can be easily concentrated by ultrafiltration (2) the metal subsequently released from the WSP can be easily segregated by ultrafiltration to allow for recycle of the extraction agent and disposal of the metal; and (3) commercially available polymers can be modified to selectively bind the target metal ions. Polyethylenimine (PEI), a highly branched aliphatic polyamine, was chosen as the backbone polymer for the studies on lead extraction from soil (Ekebafe *et al.*, 2011).

5.5. Polymeric Biocides and Herbicides

Pesticides provide substantial agronomic and economic benefits. However, pesticide migration may cause soil, surface and ground-water contamination (Werf and van D 1996) because applied conventional agrochemicals (90%) reach their objective to produce desirable response at the precise time and in precise quantities required, due to nonspecific and periodic application of active agents (Dubey *et al.*, 2011). New technique has recently emerged for the controlled release formulations designed to avoid or reduce the possible side effects accompanying the use of biologically active agents. Controlled released formulations technology is based on interaction of pesticides with polymer. The purpose of this technique include to protect the supply of the agent, to allow the automatic release of the agent to the target at controlled rates, and to maintain its concentration within the optimum limits over a specified period of time, thereby producing a great specificity and persistence (Ekebafe *et al.*, 2011).

Advantages of Polymeric Biocides

Polymers allow much lower amounts than conventional biocides to be used as it releases the required amount of active agent over a long period. Reduce number of applications through achieving a long period of duration of activity by a single application. Polymers eliminates the time and cost of repeated applications because less active materials are needed. Reduce mammalian toxicity. These polymers convert liquids to solids; hence it results in easily handled transported materials with the reduction of their flammability. It eliminates the need for widespread distribution of large amount of biocides at one time; thereby it reduces the excess of toxic biocide levels in the surrounding environment, thus reduces environmental pollution. Reduction of evaporation and degradation losses by environmental forces or leaching by rain into the soil or waterways due to the macromolecular nature. Polymers extend the duration of activity of less or non-persistent biocides which are unstable under an aquatic environment by protecting them from environmental degradation and hence enhance the practical applicability of these materials. Reduction of phytotoxicity by lowering the high mobility of the biocides in the soil and hence reduces its residues in the food web. These polymers extend herbicide selectivity to additional crops by providing a continuous amount of herbicide at a level sufficient to control weeds, but without injuring the crop.

There are two different approaches in combining the biological agents with polymeric materials: either by physical combination (encapsulation, or heterogeneous dispersion) to act as a rate controlling device, or by chemical combination to act as carrier for the agent (Ahmed, 1990).

5.6. Polymers for Improving Biological Properties of Soil

Micro-organisms such as bacteria, fungi and actinomycetes influence profoundly the physical, chemical and especially the biological properties of soils. Activities of such organisms include the decomposition of plant residues and other organic materials and the formation of humus, the most chemically and physically active group of compounds in the soil. One result of these processes of decay is the release from organic forms of essential plant nutrients, such as nitrogen, phosphorus and sulfur. Subsequently, still other micro-organisms can oxidize, reduce and otherwise change the state of the nutrients elements in the soil (El-Hady *et al.,* 2009). Polymers improve physical, chemical as well as biological properties of soil. Soil biological parameters are positively affected by changing soil physical and chemical properties, *e.g.* the addition of hydrogels to sandy soils can change the water holding capacity to that of silty clay or loam (Huttermann *et al.,* 1999) and influence infiltration rates, density and soil structure, and reduce the negative effect of nutrient losses which available for plant growth (Sarapatka *et al.,* 2006). Acrylamide based hydrogels increase the bacteria, fungi and actinomycetes counts in the soil.

Several polymers have been reported for improving activity of various enzymes such as catalase enzyme activity (CEA). Soil enzymes have a close relationship with soil biological properties and play a significant role in soil mineralization. Polymers improve soil aeration in root zone hence increase enzymatic activity and soil fertility (Ozturk *et al.,* 2005). Superabsorbent polymer also improves nodulation hence nitrogen fixing microbes can also benefited from hydrogel applications. Polymers have been attributed to the continuous supply of nitrogen, increasing nitrification or nitrate content in the soil, enhance soil bacterial growth and microbial populations and impart friability which results in increasing rooting

6. CONCLUSION

Throughout human history, agriculture has been a source of food, fuel and fibre. Opportunities have arisen through external events and trends that impacted patterns of production and utilization. Numerous publications describe the increase in yield of various plants as a result of better soil conditions.

SAPs have created a very attractive area in the viewpoint of super-swelling behaviour, chemistry, and designing the variety of final applications. When working in this field, we always deal with water, aqueous media and bio-related systems. Thus, we increasingly walk in a green area becoming greener via replacing the synthetics with the bio-based materials, *e.g.,* polysaccharides and polypeptides. Considering the high cost and increasing prices of crude oil, the necessity of preparing natural based SAPs seems more obvious. This paves the way for further developments in this area in the mid and far future ahead. Key opportunities exist to build biodegradable polymers from annually renewable crops and agro industrial waste-streams. The production of monomers and polymers with enzymes, microbes, or plants represents a cleaner and safer way of doing chemistry. However, polymers have provided solution for the need to develop cost-effective techniques that would contribute to phytostabilization of severely metal contaminated soils.

REFERENCES

Ahmed A (1990) Applications of Functionalized polymers in agriculture. *J of Islam Acad Sci* **3(1):** 49–61

Akhter J, Mahmood K, Malik K, Mardan A, Ahmad M and Iqbal M (2004) Effects of hydrogel amendment on water storage of sandy loam and loam soils and seedling growth of barley, wheat and chickpea. *Plant Soil Env* **50:** 463–469.

Aktar W, Sengupta D and Chowdhury A (2009) Impact of pesticides use in agriculture: their benefits and hazards. *Interdisc Toxicol* **2(1):** 1–12.

Ali H, Naseer M and Sajad M A (2012) Phytoremediation of heavy metals by *Trifolium alexandrinum. I J Envt Sci* **2 (3):** 1459–1469.

Andry H, Yamamoto T, Irie T, Moritani S, Inoue, M and Fujiyama H (2009) Water retention, hydraulic conductivity of hydrophilic polymers in sandy soil as affected by temperature and water quality. *J of Hydrology* **373:** 177–183.

Bakass M, Mokhlisse A and Lallemant M (2002) Absorption and desorption of liquid water by a superabsorbent polymer: Effect of polymer in the drying of the soil and the quality of certain plants. *J Appl Poly Sci* **83:** 234–243.

Bhardwaj A K, McLaughlin R A, Shainberg I and Levy G J (2009) Hydraulic characteristics of depositional seals as affected by exchangeable cations, clay mineralogy and polyacrylamide. *Soil Sci Society of America J* **73:** 910–918.

Bowman DC, Evans RY and Paul JL (1990) Fertilizer salts reduce hydration of polyacryamide gels and physical properties of gel-amended container. *J Amer Soc Hort Sci* **115:** 382–386.

Brown S L, Sprenger M, Maxemchuk A and Compton H (2005) Ecosystem function in alluvial tailings after biosolids and lime application. *J Environ Qual* **34(1):** 139–148.

Chatzoudis GK and Rigas F (1999) Soil salts reduce hydration of polymeric gels and affect moisture characteristics of soil. *Comm in Soil Sci and Plant Analysis* **30(17):** 2465–2474.

Cline SR and Reed BE (1995) Lead removal from soils via bench-scale soil washing techniques. *J Env Eng* **121(10):** 700–705.

Dubey S, Jhelum V and Patanjali VK (2011) Controlled release agrochemical formulations. *J Scientific and I Res* **70:** 105–112.

Ekebafe LO, Ogbeifun DE and Okieimen FE (2011) Polymer Applications in Agriculture. *Biokem* **23(2):** 81–89.

El-Hady OA, El-Kader Abd A A and Shafi A M (2009) Physico-bio-chemical Properties of Sandy Soil Conditioned with Acrylamide Hydrogels after Cucumber Plantation. *Aus J Basic and Applied Sci* **3(4):** 3145–3151.

FAO (2008) Biodiversity to curb world's food insecurity. Food and Agriculture Organisation, Rome. http://www.fao.org/Newsroom/en/news/2008/1000841/ index. html. Accessed 12 Oct 2016.

FAO (2013) How to Feed the World in 2050. Food and Agriculture Organisation, Rome. http://www.fao.org/fileadmin/templates/wsfs/docs/expert paper/How to Feed the World in 2050. Accessed 2 Oct 2016.

Farrell C, Qi X, Ang, John P and Rayner (2013) Water-retention additives increase plant available water in green roof substrates. *Eco Eng* **52:** 112–118.

Frankow-Lindberg B E, Brophy VR, Collins P and Connolly J (2009) Biodiversity effects on yield and unsown species invasion in a temperate forage ecosystem. *Annals Bot* **103:** 913–921.

Geesing D and Schmidhalter U (2004) Influence of sodium polyacrylate on the water-holding capacity of three different soils and effects on growth of wheat. *Soil Use and Mgt* **20:** 207–209.

Gonzalez M E, Cea M, Medina J, Gonzalez A, Diez M C, Cartes P, Monreal C and Navia R (2015) Evaluation of biodegradable polymers as encapsulating agents for the development of a urea controlled-release fertilizer using biochar as support material. *Sci of the Total Env* **505:** 446–453.

Guo M Y, Liu MZ, Liang R and Niu AZ (2006) Granular urea-formaldehyde slow-release fertilizer with superabsorbent and moisture preservation. *J Appl Poly Sci* **99:** 3230–3235.

Hong D and Zhang Y S (2006) Effect of gel base controlled release fertilizers on crop yield and nutrient use efficiency. 18th world congress of soil science, held during 9-15 July, Philadelphia, USA.

Huttermann A, Zommorodi M and Reise K (1999) Addition of hydrogels to soil for prolonging the survival of *Pinus halepensis* seedlings subjected to drought. *Soil Tillage Res* **50:** 295–304.

Keneni G, Bekele E, Imtiaz M and Dagne K (2012) Genetic vulnerability of modern crop cultivars: causes, mechanism and remedies. *Int J Plant Res* **2(3):** 69–79.

Kesavan PC and Swaminathan MS (2008) Strategies and models for agricultural sustainability in developing Asian countries. *Philos Trans R Soc B* **363:** 877–891.

Kim S and Nadaraja A (2008) Development of double-Layer hydrogels for agricultural applications. In AICHE Annual Meeting.

Lentz R D and Sojka R E (2009) Long-term polyacrylamide formulation effects on soil erosion, water infiltration, and yields of furrow-irrigated crops. *Agronomy J* **101(2):** 305–314.

Liang M, Zhan R, Liu F and Niu A (2007) Preparation of superabsorbent slow release nitrogen fertilizer by inverse suspension polymerization. *Polym Int* **56:** 729–737.

Liu J, Shi B, Jiang H, Bae S and Huang H (2009) Improvement of water-stability of clay aggregates admixed with aqueous polymer soil stabilizers. *Catena* **77:** 175–179.

Luo, Z, Wang E and Sun O J (2010) Soil carbon change and its responses to agricultural practices in Australian agroecosystems: A review and synthesis. *Geoderma* **155:** 211–223.

Marcos R, Guilherme, Fauze A, Aouada, Andre R, Fajardo, Alessandro F, Martins, Alexandre T, Paulino, Magali F T, Davi, Adley F, Rubira, Edvani C and Muniz (2015) Superabsorbent hydrogels based on polysaccharides for application in agriculture as soil conditioner and nutrient carrier. *European Poly J.* Article in press.

Onder M, Ceyhan E and Kahraman A (2011) Effects of Agricultural Practices on Environment. International Conference on Biology, Environment and Chemistry IPCBEE 24, IACSIT Press, Singapoore.

Ozturk H S, Turkmen C, Erdogan E, Baskan O, Dengiz O and Parlak M (2005) Effects of a soil conditioner on some physical and biological features of soils: results from a greenhouse study. *Bioresource Tech* **96:** 1950–1954.

Paluszek J and Zembrowski (2008) Improvement of water-air properties of eroded soils in a loess landscape after the application of agrohydrogel. *Annals of Warsaw University of Life Sciences-SGGW Land Reclam* **39:** 85–93.

Rashidzadeh A and Olad R (2014) Slow-released NPK fertilizer encapsulated by NaAlg-g-poly(AA-co-AAm)/MMT superabsorbent nanocomposite. *Carbo Poly* **114:** 269–278.

Reddy S R (2011) Principles of Agronomy. Fourth edition. Pp. 623. Kalyani publishers.

Reidsma P, Tekelenburg T, van DBM and Alkemade R (2006) Impacts of land-use change on biodiversity: an assessment of agricultural biodiversity in the European Union. *Ag Eco and Env* **114:** 86–102.

Sarapatka B, Rak L and Bubenikova I (2006) The effect of hydroabsorbent on Selected soil biological and biochemical Characteristics and its possible use in Revitalization. *Ekologia (Bratislava)* **25(4):** 422–429.

Steven J, Kohls, Dwight D, Baker, Douglas A, Kremer and Jeffrey OD (1999) Water-retentive polymers increase nodulation of actinorhizal plants inoculated with Frankia. *Plant and Soil* **214:** 105–115.

Taban M and Naeini SARM (2006) Effect of aquasorb and organic compost amendments on soil water retention and evaporation with different evaporation potentials and soil textures. *Comm in Soil Sci and Plant Analysis* **37:** 2031–2055.

Tan ZX, Lal R and Wiebe KD (2005) Global soil nutrient depletion and yield reduction. *J Sustain Ag* **26(1):** 123–146.

Taylor KC and Halfacre RG (1986) The effect of hydrophilic polymer on media water retention and nutrient availability to *Ligustrum lucidum*. *Hort Sci* **21:** 1159–1161.

Werf H and van D (1996) Assessing the impact of pesticides on the environment. *Ag Eco Env* **60:** 81–96.

Wohlfahrt J, Colin F, Assaghir Z and Bockstaller C (2010) Assessing the impact of the spatial arrangement of agricultural practices on pesticide runoff in small catchments: Combining hydrological modeling and supervised learning. *Eco Indicat* **10:** 826–839.

Wu L, Liu M and Liang L (2008) Preparation and properties of a double-coated slow-release NPK compound fertilizer with superabsorbent and water-retention. *Bioresour Tech* **99:** 547–554.

Xiaozhao H B, Sensen C, Xianguo H (2009) Controlled-release fertilizer encapsulated by starch/polyvinyl alcohol coating. *Desal* **240:** 21–26.

Zezin AB, Mikheikin SV, Rogacheva VB, Zansokhova MF, Sybachin AV and Yaroslavov AA (2015) Polymeric stabilizers for protection of soil and ground against wind and water erosion. Adv in Colloid and Interface Sci. Article in press

Zhang J, Chen H, Li P and Wang A (2006) Study on superabsorbent composite, Preparation of poly(acrylic acid)/organo-attapulgite composite hydrogels and swelling behaviours in aqueous electrolyte solution. *Macromol Mat Eng* 291: 1529–1538.

Zohuriaan-Mehr J and Kabiri K (2008) Superabsorbent Polymer Materials. *Iranian Poly J* **17:** 451–477.

■■■■

11 Adaptation Strategies for Enhancing Agricultural and Environmental Sustainability under Current Climate

Sandeep Kumar*, Ram Swaroop Meena[1], Shish Ram Jakhar[2], Chetan Kumar Jangir[3], Anshul Gupta[4] and B.L. Meena[5]

Department of Agronomy, CCS Haryana Agricultural University, Hisar – 125005 (Haryana), India
[1]Department of Agronomy, Institute of Agricultural Sciences, Banaras Hindu University, Varanasi – 221005 (U.P.), India
[2]Department of Soil Science & Agricultural Chemistry, J.N. Krishi Vishwa Vidyalaya, Jabalpur – 48004 (M.P.), India
[3]Department of Soil Science, CCS Haryana Agricultural University, Hisar – 125005 (Haryana), India
[4]Department of Agronomy, SKNAU, Jobner – 303 329
[5]ICAR-Central Soil Salinity Research Institute, Karnal – 132 001
Author for correspondence: sandeepsihag1992@gmail.com

ABSTRACT

Climate change is putting global agriculture under significant stress and, if no adaptation is made, it will be severely impacted in the course of upcoming decades. Therefore, revision in agricultural production techniques under existing climate era is vital towards ensuring good agricultural production, sustainably on long term basis. To implement adaptations strategies in such situation there is current necessity to well recognize the connection among potential adaptation tactics, decision-making policies of government, prevailing farm levels processes and risk bearing activities. Thus, to achieve the larger extent of sustainable strategies, it needs to refine the crop intensification techniques. Likewise, the scope of sustainability need not be narrowed only to the aspects of sustainability of natural resources (*e.g.*, water, land, forest, energy and biological-diversity). To overcome the harmful impacts of changing climatic stimuli and to exploit potential positive impacts, numerous short and long term adaptation approaches for sustainable agriculture are proposed here. The short-term adaptation strategies contain attempts to enhance crop cultivation devoid of major changes in system *i.e.,* minimum/no tillage, legume based crop rotation, manuring and balanced fertilizer application, cover/mulch crops, adapting resilient in varieties, fallows period management, sowing dates, cropping pattern, irrigation technique, water harvesting, agroforestry, erosion control etc. The reduced/no-tillage along with inclusion of legume in crop rotation, application of on-farm inputs i.e. FYM, compost and vermicompost, intercropping of leguminous crop with commercial exhaustive crops like cereals and fallow period management improves soil physical (*e.g.,* structure, WHC), chemical (*e.g.,* soil fertility, nutrient balance, organic matter) and biological (diversity and functionaries of macro and micro fauna-flora). The conservation tillage, agroforestry system and organic based fertility management have a great ability towards building and maintaining the soil carbon storage by sequestering the atmospheric carbon-dioxide and enhance the capacities of the aboveground plant parts to capture more carbon from atmosphere and thereafter, minimizing the warming of atmosphere caused by increasing carbon-dioxide levels. Moreover, these practices enhance the resilience of cultivated soils from enlarged treats arising from change in climatic stimuli *e.g.* silent and severe soil erosion, flooding and drought like conditions. Conservation agriculture (CA) seeks to improves the small farmer's capability to go for crop production and ensures the per day food-intake, alleviation of poverty, biodiversity conservation and safeguards for ecological services. The CA contributes for making agricultural production system more resilient to climate change by reducing the vulnerability of the system to biotic and abiotic stresses and by enriching the local natural resource base. The long-term adaptation options are those which cause major structural changes in the direction of overcome adversity arisen from climate change *viz.* breeding of cultivars, changes in farming system and land use pattern and new innovative land improvement skills, timely and precise forecasting, expansion of irrigation schemes, governmental policies and funding etc. to stabilize sustainable crop production. In this respect, we propose a wide variety of comprehensive review of literature related to the adaptation options those have potential to reduce sensitivity of agricultural systems and encourage good production in face of adverse climatic conditions besides safeguard to environmental sustainability. The biophysical adaptation returns consequently increases directly with increase in the level of adaptation actions.

***Key word*:** Adaptation strategies, Climate change, Environment, Sustainability

1. INTRODUCTION

Climate is usually known as a vital biophysical factor influencing vegetation growth and development. Nowadays, climate change is a global truth which has already begun (Kumar *et al.,* 2017). It referred in the observational record of climate occurs because of internal changes within the climate system or in the interaction among its components, or because of changes in external forcing, persisting for decades or longer general more than 30 years, arising from anthropogenic factors (high confidence) or due to natural phenomena (IPCC, 2007a; Hassan *et al.*, 2013). Climate change endangers all sectors of human life, but agriculture is characterized by means of most sensitive system to climatic stresses (Anonymous, 2009; NAAS, 2013; Huang and Wang, 2014). A number of scientific studies noticed the negative effects of already changed climatic stimuli on global agricultural sectors (Neumann *et al.*, 2010; Clements *et al.*, 2011; Venkateswarlu and Shanker, 2012; Ray *et al.*, 2015). Every year one and another region of the world got affected through extreme events like cyclones, frost, hailstorm, floods, droughts, and so on. The significant changing trends of climatic parameters, including annual temperature variability (low and high temperature regimes), spatial and temporal rainfall variability, and increasing oceanic levels which in turn cause warming of the planet, decline coastal freshwater availability and increased frequency and intensity of floods and more unpredictable weather events (Lu *et al.*, 2015; NAAS, 2013). These can also result in originating the new threats which did not seen till now in a specific region or crop like for outbreak of new pest and diseases favouring changed atmospheric temperature (Clements *et al.*, 2011). Average global temperatures and their impacts on agricultural production system expected to rise continuously in upcoming centuries. Globally, the earth surface's mean temperature is likely to raise 2 to 5°C over the subsequent decades, assuming GHGs concentration will continue to increase in the atmosphere at the existing rates (IPCC, 2007b). The increased temperature besides shrinking water availability for agricultural purpose as a result of climate change tends to augmenting weed and pest proliferation consequently lessen the yields of plants of economic importance (Venkateswarlu and Shanker, 2012; NAAS, 2013). The grain yield may reduce as much as 5 per cent with each 1°C increase in global warming (Olive Heffernan, 2013). Even though, some crops exploits positive response in temperate regions in higher latitudes at globally from increased temperature as these already facing low temperature (NAAS, 2013).

The increasing oceanic salinization, coastline inundation, and erosion are the consequences of sea level rise and anthropogenic causes and reduce the size of productive agricultural lands, thereby threatening farmer's livelihood and food security (Faroque *et al.*, 2013). In several areas, change in climate is projected to upsurge the land menace from drought and heat are generally arable areas. For example, in absence of rainfall a severe drought in Australia in 1901 resulted in a huge failure of the wheat crop and caused the Darling River in New South Wales to dry up. Likewise, a severe dust bowl due to the decreased rainfall, poor land management as well in Central Great Plains and United States reduced agricultural production by 17 per cent (Olive Heffernan, 2013).

Climate alteration affects on agriculture in a numeral of ways, *viz.* rainfall, temperatures, and climate extremes (*e.g.,* cold and heat waves); increase in plant pests and diseases; alarming change in biosphere environment; changes in the food quality (Meena *et al.*, 2016). The connection between climate change and crop yield variability displays discrete spatial patterns in the comparative influences of temperature, precipitation and their interaction within and across regions. The amount of yield reduction associated with climate change is highly location and crop-specific (Ray *et al.*, 2015). However, the yield variability of all the crop growing regions do not significantly influenced from inter-annual climate change. Various studies found huge yield differences across the world, even among regions having same growing conditions, telling the existence of yield gaps (Neumann *et al*., 2010). The influences of climate variability on crop yield variability are higher in poorer regions as in northeastern Brazil for maize and Central India (Madhya Pradesh, Chhattisgarh and Karnataka) for rice. South Asia is categorized as one of the most vulnerable regions of the world to climate change impacts. The climate change impacts are being witnessed all over the world, but developing countries like India are more vulnerable in view of the huge population dependent on agriculture, excessive pressure on natural resources and poor coping mechanisms (NAAS, 2013). The agriculture in the Mediterranean region seems to be more vulnerable than in other European regions. The most productive regions like maize in the United States Midwest and wheat in Western Europe highly affected by climatic variability and in some cases over 75 per cent yield variability is noticed (Ray *et al*., 2015). Smallholder farmers are in the front line of climate change impacts. Global poor farmers, especially those living in marginal areas of low agricultural productivity are highly affected by climate change disasters (NAAS, 2013). The ecosystems on which they rely are increasingly degraded and their access to suitable agricultural land is declining. Many smallholder farmers produce on marginal rainfed land that is affected by increased water scarcity and erosion.

The precise implications of climate change remain unclear: predictions of rainfall rates, amounts and patterns, the likely frequency of extreme weather events, and regional changes in weather patterns cannot be made with certainty (Clements *et al*., 2011). If risks in agricultural production from changing climate from the local to the global and links between them are evaluated, farmers can pick effective actions that build resilience. Tracing the pattern of yields change with time make available an honest method for evaluating the impacts of climate change variables on agricultural productivity. However, the measuring an increase in climate change impact is more problematical as adaptation aids are influenced by a number of distinct or interacting causes that become conscious over time, and are extremely context-specific (Price Waterhouse Coopers, 2011). Thus, it needs such a framework that can trace the performance of adaptation measures, measure breaks and regulate any increase in resilience (Adger *et al*., 2005; De Nijs *et al*., 2014). They found the efficacy of the modeled adaptation actions and established a fact that with no-adaptation strategies the values of vulnerability to climate change that was medium to somewhat high, whereas, somewhat low to low vulnerability when adaptation actions employed and this variation was directly associated with the magnitude of the alterations in site-specific biophysical environments like for erosion proneness, slope or type of soil (De Nijs *et al*., 2014). Recent studies at European level have revealed the need to take account of climatic changes and non-climate changes, such as technological, socio-economic, etc. for assessing the changes in crop yield and suitability. Regional climate models are becoming more accurate, but unfortunately too little effort has been invested into research on regional climate models in developing countries and in low latitude regions (Clements *et al.*, 2011). Currently, the impact of uncertainty can be seen most clearly in the failure of climate models to provide, good agreement at the regional scale, and in particular on future levels of precipitation. In West Africa, for example, the impact of climate change on rainfall is unclear. Climate models sometimes fail to agree as to whether precipitation will increase or decrease in any season (Christensen *et al.*, 2007). To make it more reliable, it necessitate to develop a database on climate, soil and water use and crop yields to assess, map and monitor land use performance under given technology conditions. Besides this, assessment of how vulnerable our food system is and how we can adapt agriculture to future climate related disasters (Faroque *et al.*, 2013).

Climate change adaptation refers as an adjustment in management strategies in response to actual or expected climatic stimuli or their effects, which moderates harm or exploits beneficial opportunities (IPCC, 2007b; Smit *et al.*, 2000). According to Clements *et al.* (2011) adaptation to climate change in the agriculture sector describes what policy makers, development planners, agriculture experts and other stakeholders in countries should consider while determining a technology development path in agriculture. Adaptations to climate change are not just discrete technical measures, but are modifications to farm practices with respect to multiple (climatic and non-climatic) stimuli and conditions (Smit *et al.*, 2000).

It is not surprising that most developing countries are interested in sustainable technologies for adapting agriculture to climate change. Technologies and practices do exist, or have been developed in different parts of the world, to facilitate sustainable adaptation to climate change in the agriculture sector (Clements *et al.*, 2011). Sustainable agriculture strategies are not just a set of practices but a process requiring skills of adaptability (Milestad and Darnhofer, 2003). The use of sustainable adaptation strategies, reduce excessive use of external inorganic inputs, enhance the diversity of production, while tailoring production intensity to the capacity of the landscape, and use a mix of traditional and new technologies. The focus of agricultural production system should performs major improvements in food production or environmental performance at the least effort and cost through existing or new improved practices and technologies under climatic hazards. Although, yields are only part of a range of ecological, social and economic benefits delivered by farming systems, it is broadly established fact that high yields are crucial to sustainable food security on a limited land basis. Thus to increase productivity, the underperforming land whose yields are presently beneath average may be the best places to expand crop yields. To accomplish sustainable food security probably farmers must follow the different production techniques *i.e.* organic, conventional and imaginable 'hybrid' technology in judicious pathway instead in distinct way.

Before planning for sustainable crop production one must take a great deal of attention how the past and future climate variability affects crop yields, how climate change and crop yields are interrelated over time and also make sure for future food security (Ray *et al.*, 2015; Kumar *et al.*, 2017a). Agriculture production system must encounter the dual challenges; first it needs to enhance food production sustainably to feed growing more challenging global population with least external inputs, both in terms of quality and quantity, second, at the same time lessening its global environmental influences, all under more erratic and risky climatic conditions in the future while charting a more environmentally sustainable path (Wilson and Tyrchniewicz, 1995; Tilman *et al.*, 2002; Godfray and Garnett, 2010; Godfray *et al.*, 2010; Cameron *et al.*, 2015). However, the international agricultural research have abundant potential for wider adoption than has been reached up to now at farmer's field and their adoption at agricultural land will build the resilience of millions of global population to change in climate variables and uncertainty at present, imminent time as well (Anonymous, 2009).

The capacity of a farming system to adapt the changing climate and weather conditions is based on its natural resource endowment and associated economic, social, cultural and political conditions (Wall and Smit, 2005). Adaptation to climate risks and opportunities is then reviewed and discussed in terms of adaptive capacity, a concept that ties directly into sustainability. Examples from Canadian producers in the Prairie region, who are currently meeting challenges associated with climate change, provide concrete illustrations of the linkages between sustainable agriculture practices and climate change adaptation (Wall and Smit, 2005; Anonymous, 2009). European agricultural research and the agro-industry have considerable skills and know-how, which through technology transfer and promotion of innovation could be used for improving adaptive capacity in less well developed regions of the world, which will generally be more severely affected by climate change (Jorgen Olesen, no date).

The approach must build sufficient resilience of food systems to avoid enormous future economic losses in agriculture. Evolving comprehensive climate resilience strategies comprising risk assessment, developing of varieties that can perform well in stressful and adverse conditions, better land and water management and bringing about specific changes in agricultural practices that can respond to climate change strongly and effectively. As whole globe is facing a challenge of producing adequate food from shrinking natural resource base for the ever-increasing population, therefore, intensification of agricultural activities through enhanced productivity and efficient resource use is the only option available as competition for land and water is increasing from non-farm sectors under a variable climate (NAAS, 2013). The promotion of scientific agroforestry forms a key component in the war against climate change (NAAS, 2013). Agroforestry can both sequester carbon and produce a range of economic, environmental, and socioeconomic benefits. For example, trees in agroforestry systems improve soil quality through nitrogen fixation, extraction of nutrients from deep soil horizons, promotion of more nutrients cycling, erosion control, maintenance of soil organic matter and physical properties, and creating a more favourable microclimate (IFAD, 2012; NAAS, 2013). Drought, heat, salt and flood tolerant crop genotypes, new ways to irrigate crops and better ways to manage soils are certain the approaches that has great adaptive capacity to benefit present farmers, and into the future (Anonymous, 2009).

Other timely access of information through early warning systems, provision of reliable weather information, scenario-based land-use planning and index-based weather risk insurance are important components of adaptation strategies. Availability of information from climate monitoring systems, and in many areas with no weather stations, gathering information from local population knowledge, is very important for the process of selecting technologies for climate change adaptation (Clements *et al.,* 2011). Appropriate environmental policies and the use of subsidies incorporating climate issues into policies for poverty reduction, agriculture and food is highly feasible, platforms for dialogue among different ministerial and non-ministerial actors, such as in the IFAD/Global Environment Facility (GEF) Participatory Control of Desertification and Poverty Reduction in the Arid and Semi-Arid High Plateau Ecosystems of Eastern Morocco project, or integrating climate vulnerability assessment into programme interventions, such as in the Rural Growth Programme in Yemen. Many countries are starting to engage with the National Adaptation Programme (NAP) process of the United Nations Framework Convention on Climate Change (UNFCCC), aiming to integrate climate change adaptation measures into medium- and long-term national planning processes. This means identifying entry points for adaptation planning in agriculture and taking action to strengthen country systems to improve management of the risks and opportunities of climate change. The NAP process represents a potential scaling-up mechanism for IFAD. By making climate finance available to agricultural investment programmes, IFAD can reorient these programmes to make them climate-resilient and better aligned with national priorities in climate change adaptation. As the Himalayan glaciers melting continue, in response to the expected changing river flows is of great concern as these rivers are the major crop production belts *i.e.,* Gangetic plains of Asia. Therefore, South Asian countries instantly must to initiate cooperative funding and policies, in collaboration with international, environmental and development organizations, to assess and monitoring of climate change taking place (Javaid, 2013).

These emerging threat to agriculture from climate change in the form of unpredictable weather, floods, and other disastrous events make the task of providing enough food for the global population even more challenging (Clements *et al.*, 2011). Apart from the challenges of coping with patterns of climate variability, there is also concern about how future climate changes due to global warming may influence the global agriculture. There is an urgent need to identify adaptation strategies that are needed for the sustainable development at various levels (NAAS, 2013). Thus the focus has been on improving the yields with the existing technology and timely availability of water.

The purpose of this chapter is to illustrate how sustainable agriculture practices and climate change adaptation strategies are mutually supportive and identifies where priorities for adaptation are greatest, and discusses the importance of ensuring that the research and development of agricultural technologies are relevant to the livelihood needs of vulnerable farmers. An assessment of relevant strategies, suggests that the two are closely aligned. Many climate and weather risk management strategies fit squarely into sustainable agriculture practices and can, therefore, be promoted with several of the programs and policies targeting environmentally responsible production (Wall and Smit, 2005).

2. VULNERABILITY OF AGRICULTURE SYSTEM TO CLIMATE CHANGE

Vulnerability is the degree to which a system is susceptible to, and unable to cope with, adverse effects of climate change, including climate variability and extremes (Smit and Wandel, 2006; IPCC, 2007b). Agricultural system is considered as the most vulnerable system to changing climate, the reasons behind this are its seasonality means the contracted range of weather over which crop cultivation can take place, the presence of significant non-climatic stressors that influence sensitivity to changes in climatic conditions, and the endemic poverty often associated with food production in the developing world, which compounds vulnerability (World Bank, 2009). The vulnerability in agricultural sector to climate change depends on three constituents *i.e.* exposure, sensitivity and adaptive capacity of system (Mahendra Dev, 2006; Tubiello and Rosenzweig, 2008; Clements *et al.*, 2011) (**Fig. 1**).

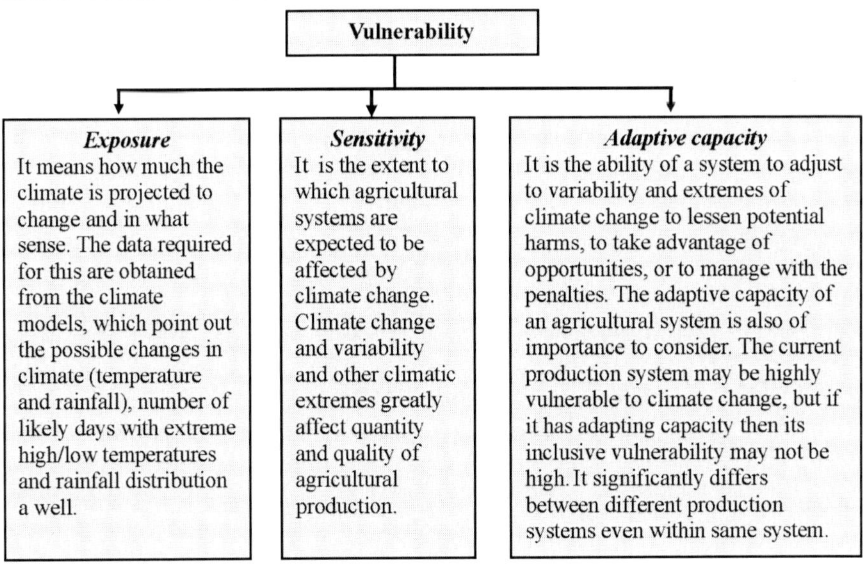

Fig. 1. Vulnerability of agricultural system to climate change and variability

Vulnerability of agricultural production system of arid, semi-arid, and dry sub-humid regions to changing climate is considered to be high in lower latitudes *i.e.,* developing nations because of theirs prevailing social, economic, and environmental circumstances that strengthen exposure to harmful bearings and add to low capacity to cope with and adapt to changing climate threats. The exciting adverse effects of increased climate variability like pervasive flooding and drought on agriculture resonate through economies where agriculture is a key giver to total gross domestic product (GDP), generally in rain-fed production (World Bank, 2009). For example in Sub Saharan Africa, India etc. recurrent precipitation/monsoon, agricultural, and national GDP are intricately connected. In India mid-season monsoon break followed by drought in 2002 reduced total GDP by 3 per cent (Gadgil and Gadgil, 2006). Like this, uninterrupted La Nina/El Nino events from 1997 through 2000 drop GDP more than 10 per cent in East Africa. Sub-Saharan Africa has recognized as 'hotspots of vulnerability' (Anonymous, 2009). Vulnerability of agricultural system to drought often results from the conjunction of manifold climate and non-climate risk issues (World Bank, 2009; Hassan *et al.*, 2013). The negative impact of changing climate is higher where natural systems are brutally corrupted coupled with land degradation and desertification, and the human systems flop to take action to recover it, specifically in poor nations (Hassan *et al.*, 2013). The most vulnerable people will suffer higher from change in climate.

ADB and IFPRI (2009) have recognized vulnerable nations to climate change in Asia-Pacific and described into three categories according to the agenda of adaptive capacity, exposure and sensitivity. The first, *highly exposed* where the temperature is likely to increase as a minimum 2°C or annual rainfall are expected to change by greater than 20 per cent. According to them 26 countries of the Asia-Pacific region falls under this category. Second, *highly sensitive* includes countries where above 40 per cent of labours are employed in agriculture and which covers 12 countries in the region. Third one, *low adaptive capacity* country having a poverty ratio of 30 per cent is considered in this category and 5 countries of Asia-Pacific region comes in low adaptive category.

Adaptation is succeeded by enhancing adaptive capacity, or by plummeting exposure and sensitivity to antagonistic changing climate ill-effects (UKCIP, 2013). Thus, initially climatic scientist's research must focus on diagnosing vulnerability and analyzing opportunities thereafter on follow a better recognizing the bad effect of projected climate change and system's sensitivity at site of question to implement such action (De Nijs *et al.*, 2014). For the reason that of the high level of vulnerability, there is a burning requisite in the developing nations to recognize the threats from climate change and make plans against climate change that decreases climatic menaces and to take the bull by the horns (Hassan *et al.*, 2013). There is also need to develop locally new and better agricultural adaptation strategies against increasing climate variability and climate change, especially in vulnerable regions where food and fiber production are most sensitive to climatic fluctuations.

3. IMPACT OF CLIMATE CHANGE ON GLOBAL AGRICULTURE

The impacts of climate change on agriculture are expected to be widespread across the globe. Climate change reduces yields of crops in lower latitudes, while it increases yields at middle and high-middle latitudes (Mahendra Dev, 2006). Climate change is likely to impact crop productivity with reference to quantity and quality of crops directly through changes in the growing environment, but also indirectly through shifts in the geography and prevalence of agricultural pests and diseases, associated impacts on soil fertility and biological function, and associated agricultural biodiversity (NAAS, 2013). Climate change is therefore expected to alter the biophysical environment of growing crops (Rosenzweig and Parry, 1994) coupled with agricultural practices like water use and application of fertilizers, insecticides, and herbicides etc. and environmental influences, particularly in relation to the frequency and intensity of soil drainage which may lead to loss of nitrogen through leaching, increased risks of soil degradation caused by erosion and possible decline in soil organic carbon concentration and reduction of crop diversity (NAAS, 2013).

Studies have indicated that the direct impacts of climate changes would be small on *kharif* crops but *kharif* agriculture will become vulnerable due to increased incidence of weather extremes such as changes in rainy days, rainfall intensity, duration and frequency of drought and floods, diurnal asymmetry of temperature, change in humidity, and pest incidence and virulence. Rabi crop production may become comparatively more vulnerable due to larger increase in temperature, asymmetry of day and night temperature and higher uncertainties in rainfall (Shetty *et al.*, 2013).

Ray *et al.* (2015) clarifies the spatial patterns in the relationship between climate variability (temperature, precipitation and interaction of both these) and crop yield variability, the global climate variability shares for approximately 32–39 per cent of the observed yield variability of rice, wheat, maize and soybean. This converts into annual production instabilities due to climate variability of approximately 22 M tons, 9 M tons, 3 M tons and 2 M tons for maize, wheat, rice and soybean, respectively. However, the key crop harvesting nations observed the effect of climate change on crop yields: approximately 70 per cent, 53 per cent, 79 per cent and 67 per cent of maize, rice, wheat and soybean harvesting regions, respectively and the per centage of global total mean production harvested over these regions and thus influenced by climate variability was approximately 78 per cent, 52 per cent, 75 per cent and 67 per cent of maize, rice, wheat and soybean (Ray *et al.*, 2015).

In keeping with a report of the International Rice Research Institute (IRRI) and partners, the Mekong and Red River Deltas in Vietnam, Irrawaddy Delta in Myanmar and Ganges-Brahmaputra Delta in Bangladesh and India are threatening due to increased flooding and salinity as a result of climate change (Anonymous, 2009). The frequency of happening of climate encouraged disease and heat stress in Central, East, South and South-East Asia has increased with continued mounting temperatures and rainfall pattern (Peng *et al.*, 2004). The cereal production potential in South-Asia could expect to decline by 2050 up to 30 per cent (IPCC, 2007b) and up to 20 per cent yield increment in East and South-East Asia by the reason of climate change (Rosenzweig *et al.*, 2001). The same decrease yields of wheat, rice and maize had also reported by NAAS (2013) but neutral to positive in yield of groundnut, soybean and chickpea. In South Asia, climate change leads to a 14 per cent, 44–49 per cent, 9–19 per cent and 12.2–19.6 per cent decline rice, wheat, maize and sorghum production, respectively as compared to no climate change scenario. Whereas rice production declines by about 10 per cent and slight increases in wheat in East Asia and Pacific sub-region (Mahendra Dev, 2006). In developing countries, irrigated wheat yields will be reduced by around 30 per cent and irrigated rice yields by 15 per cent (Anonymous, 2009), while in Bangladesh it may decline by 32 per cent and 8 per cent, respectively up to 2050 (Faisal and Parveen, 2004). Mall *et al.* (2006) using CROPGRO-soybean model forecasted the decreased soybean production in different regions of India due to climate change. However, decline in average soybean yield is reported to be less in west and south India over other parts of the India. Climate change is anticipated to strengthen drought and heat stress in groundnut production zones. Area under groundnut is likely to decrease by 5 per cent and 4 per cent in medium and long-term, respectively while groundnut yields are projected to drop by 7 per cent and 5 per cent in medium and long-term scenario, respectively (Soora *et al.*, 2012). Over Sub-Saharan Africa and Latin America, the aggregate yield decline in smallholder rainfed maize is projected by 10 per cent up to 2055 equivalent to an annual economic loss of US$2 billion (Anonymous, 2009).

The major effect on crop is due to shortening of crop duration through hastening senescence and development, by this means reduced time to gather light and produce biomass which is related to the thermal environment. These effects differ with species and characteristic of the crop, the timing of heat stress in relation to crop development and with growing environment. Annual crops are more susceptible to increased temperature and affected more than perennial crops (Easterling *et al.*, 2007). Reproduction, pollination and fertilization processes of crops such as rice, wheat, sunflower etc. are highly sensitive to temperature. At mid-day temperature hastens crop maturity, consequentially reduced grain filling, whereas higher minimum night time temperatures increase respiration losses (World Bank, 2009). Episodic heat waves also have a strong negative impact on yields, particularly at reproductive stage causing increased sterility or during seedling emergence, which influence crop establishment (World Bank, 2009). In India, climate change is projected to reduce the production of timely sown irrigated wheat, late sown irrigated wheat, irrigated rice and rain-fed rice by about 6 per cent, 18 per cent, 4 per cent and 6 per cent, respectively by 2020 (Shetty *et al.*, 2013). This significant decline in production is likely to be caused by shortening of growing period, which will have negative impact on reproduction and grain filling particularly due to terminal heat stress and decreased water availability.

The production of rain-fed rice and wheat in South-Asia is projected to fall with an increase in temperature beyond 2.5^{0}C experiencing 9–25 per cent loss in farm level net revenue (Lal, 2009). With the change in temperature by 2°C and 4°C, the prospect of growing wheat and potato would be severely impaired (Islam *et al.*, 2008). Simulation models suggest that wheat yields in south-east Australia may decrease by about 29 per cent while India could lose 4–5 million tonnes of wheat with every rise of 1°C in mean temperature (Aggarwal, 2008) and wheat productivity would reduce by 0.45 tons per hectare in India with a 0.5^{0}C rise in winter temperature. Night time temperatures above some threshold can have stronger negative impacts on rice yields compared to radiation and day time temperatures over India and other parts of south Asia (Peng *et al.*, 2004). In the Philippines the irrigated rice yields decrease by 10 per cent for each 1ºC increase in the minimum night-time temperature although the maximum temperature has no effect (Peng *et al.*, 2004).

A study of International Maize and Wheat Improvement Center (CIMMYT) stated of an expected climate shift in Indo-Gangetic Plains of wheat of South Asia spread out from Pakistan through northern India, Nepal to Bangladesh in 13 million hectares area because of increasing temperature which would make wheat area hot while wheat favours temperate climate. On the word of them by 2050 as much as 51 per cent of wheat area may turn out to be heat-stressed for wheat, with a significantly shorter season for the crop (Ortiz *et al.*, 2008; Anonymous, 2009). This shift would also result a significant reduction in wheat yields, unless appropriate cultivars and crop management practices are innovated and adopted by farmers (NAAS, 2013). Simulation studies with two commonly used process-based crop models indicated that existing models under-estimate the effects of heat on wheat, with several models underestimating yield losses for +2°C by as much as 50 per cent for some planting dates (Lobell *et al.*, 2012). Similarly, in Mid Indo-Gangetic Plains and Southern Plateau of India the grain yield of winter maize is expected to fall with raising temperature by 55 per cent and 35 per cent, respectively (Byjesh *et al.*, 2010). Simulation models study indicated that growth and yield of maize was greatly affected by means of temperature changes allied with sowing dates (Singh *et al.*, 2010). They also reported that the yield was reduced in late-sown crops due to the harmful effect of chilling temperature and the reduction in irrigated wet-maize yield is expected up to 18 per cent in 2020 and 2050 and about 23 per cent in 2080 scenarios.

Changes in rainfall cycles are impacting on agricultural yields as water availability is decreasing in already arid zones and water excesses (floods) are being experienced in other areas. This will affect agricultural production more in large parts of Latin America, Asia and Africa. Some agricultural land may no longer be cultivatable, growing seasons will change and productivity will decrease (Clements *et al.*, 2011). Statistical analyses of state-level data confirm that drought and extreme rainfall adversely affected rice yield in predominantly rain-fed areas during 1966–2002 (NAAS, 2013). Many studies confirm that as the glaciers melt, the seven major river systems in Asia will be faced in the short and medium term with a sharp increase in the influx of water and therefore with floods, followed by periods of drought in the long term (Passenier and Lak, 2009). Sea-level rises impact on coasts and mega-deltas, monsoon changes and extreme floods can occur in mega-rivers, while droughts and high inter annual variability can affect the Sub-Saharan Africa and Asian highlands (SCAR, 2011). Numerous researches on climate change confirms the impact of perceived changes in climate pattern, variability and extreme events show that the crop yield in many nations of Asia Pacific has dropped owing to decreased availability of agricultural water, partially due to increases in average temperatures, increase extreme events like drought and floods, increasing frequency of El Nino and reduction in the number of rainy days (Aggarwal *et al.,* 2000; Fischer *et al.*, 2002; Tao *et al.,* 2004). In Europe, contrasts are going to be amplified between Southern and Northern regions (Soussana, 2010). In Southern Europe, climate change is projected to increase stress by high temperatures and drought, so that in a region already vulnerable to climate variability, water availability and crop productivity will further decline.

Where excess rain occurs, extreme rainfall events will increase leading to flooding and soil erosion. Flood has most deleterious effect on crop production of Bangladesh. The 1988 flood caused reduction of agricultural production by 45 per cent (Karim *et al.,* 1996). Prolonged floods would tend to delay *Aman* plantation, resulting in significant loss of potential *Aman* production, as observed during the floods of 1998. Loss of *Boro* rice crop from flash floods has become a regular phenomenon in the *haor* areas over the recent years in Bangladesh, due to which in 2010 loss of about 150,000 metric tonnes of rice was accounted (Faroque *et al.*, 2013).

In low rainfall, drought-prone areas, there is general acceptance in the science community of more frequent moisture stress because of failed rainfall patterns and increased evaporation caused by higher temperatures (Cooper *et al.*, 2008). The frequent and severe droughts in the Sahel region since the 1970s and 1980s were mainly caused by recent ocean warming due to climate change (Giannini *et al.*, 2003). In Western Africa, the impact of future climate change is very uncertain, but might be very radical: possible future range from the complete disappearance of the African monsoon to its amplification. Understanding and projecting rainfall patterns in West Africa and their agronomic consequences has led to a strong international research initiative for African Monsoon Multidisciplinary Analysis (AMMA, 2010). However, studies suggest that African agriculture is likely to be most affected due to heavy reliance on low-input rain-fed agriculture. In Bangladesh, drought is generally affects crops in pre-monsoon and post-monsoon periods. During the last 50 years, Bangladesh suffered about 20 drought conditions. During 1981 and 1982 droughts affected the production of the monsoon crops only. The drought condition in north-western Bangladesh in recent decades had led to a shortfall of rice production of 3.5 million tons in the 1990s (Faroque *et al.*, 2013).

Declining availability of water resource for global agricultural use is the most critical challenges that agriculture is facing and also expected in coming time because of climate changes in accordance with the scenario (Anonymous, 2009). In addition to, shifting and less probable rainfall pattern and river flow; water scarcity is going to become the future milestone for global agriculture (Anonymous, 2009). Changing amounts and patterns of precipitation is already causing farmers to struggle. In Asia, changes in the monsoon pattern and in glacier melt are presumably the chief intimidations (Anonymous, 2009). The similar problem of changes in rainfall pattern in Africa is also exists, where the hunger population is fighting for their food and livelihoods that depend on the rainfall. In East Asia, for 1ºC rise in surface air temperature expected by 2020s, water demand for agricultural irrigation would increase 6–10 per cent or more (IPCC, 2007b). Growing evidence suggests that changes in the hydrological cycles can bring longer droughts and more intense rains making wet regions even wetter and arid areas drier (GHF, 2009). Temperature rise of this magnitude could significantly reduce the dry-season supply of glacial melt water, an important water source for irrigated agriculture in South and Central Asia, the Andean Region, and western China. In coastal areas of Asia, USA and Mediterranean Countries, intensive agricultural activities always demand greater amounts of water to support their economy. Climate change may particularly worsen in the Mediterranean areas, due to the combined effects of reduced recharge and consequent increase of discharge. These problems highlight the importance of a serious reflection on water resources management to guarantee agricultural sustainability and better fresh water supply of the coastal zones (Romanazzi *et al.*, no date). With a 1 m rise in sea level, the loss of 2,500 km^2 of mangroves in Asia are to be expected; Bangladesh would be expected to highly affected by this in terms of loss of cultivated land with approximately 1,000 km^2 of cultivated land (IPCC, 2007b). This in turn will curtail the ability of irrigated agriculture to respond to the expanding food requirements of tomorrow's globe (Rosegrant *et al.*, 2002).

Copious numbers of diseases pests of crops are vulnerable to climate change and can expected to change in currently capricious ways. Some diseases pests will become rampant in region where they were not occurred before, after climate turn out to be favourable in that region. Simultaneously, in that region where disease pest found currently, usually poor knowledge of pest or disease management and low immunity to a disease, enhances threatens. Insects possess very high extent of adaptability to climate change. Climate change would expect to augment the confrontational impacts of insects-pests. Water stressed conditions enhance plant vulnerability and generates more auspicious atmosphere for insects-pests (Anonymous, 2010).

4. ADAPTATION STRATEGIES FOR PRODUCTION SUSTAINABILITY

4.1 Farmers Level Approaches

The farm is typically the entity at which adaptation to climate change and climatic variability must take place through introduction of new management approaches and technologies (Kumar *et al.*, 2016). The most effective approach to adapt to climate change at farm level is to adopt a more sustainable environmental safely developed framing practice which are economically viable and conserve natural resources by shifting to pre-existing less effective environmental threatening system (Mathur, 2009). To achieve the long-term sustainability of crop production, a number of prevailing farm-level management practices are by this time available for developing climate change response strategies needed in near future (Tubiello and van der Velde, no date) which have potential to reduce losses occurring at farm level and also focused on conservation agriculture operations.

Some of the viable sustainable strategies are selection of good qualitative seed of high-yielding adapted varieties according to growing situations, improved soil management practices, needs based nutrient management to sustain soil health, efficient water management to improve water use efficiency, sustainable pest management with minimal possible use of agrochemicals, crop rotation, mixed farming, changing in sowing time, crop diversification, intercropping, cover crops, sensible labour and energy use and enlarged carbon accumulation in soil. Beside this, adaptation strategies also include rescheduling of farm management practices such as irrigation and nutrient application to better match critical growth stages; introducing and implementing of efficient technologies that conserve water and soil, etc. (Tubiello and van der Velde, no date). All these sustainable resource conserving technologies provide opportunities to build-up a strong base for long term higher crop productivity, adaptation and reduced sensitivity to climate change, enhanced environmental functioning and services, and reductions in agriculture's greenhouse gas emissions and "carbon footprint" (Anonymous, 2010; Shetty *et al.*, 2013).

The potential aids of on farm management adaptation for wheat in temperate and tropical climate are comparable *i.e.* 17.9 per cent and 18.6 per cent, respectively. However, the on farm adaptation management benefits in rice and maize was only 10 per cent as compared to no use of adaptation strategies and this was less than wheat. The use of adaptation strategies at farm level avoid the yield loss which occurs due to increase in temperature by 1–2°C in temperature region and by 1.5–3°C in tropical regions in wheat, rice and maize (Tubiello and van der Velde, no date). Intercropping system after that legume fallow and alley cropping was the best adaptation technique with greatest reduction in vulnerability by assuaging the negative impacts of climate change of decreased water availability and increased pests (Meena *et al.*, 2017). Intercropping can also help to protect the soil from excessive heat, exposure to wind, and moisture loss (Recha *et al.*, 2014). Adaptation benefits by adopting legume fallow, alley cropping, crop rotation, mulch cover, water management practices, late maturing genotypes and low/no tillage was also observed (De Nijs *et al.*, 2014). Crop rotation improves the fertility of the soil, which helps with adapting to change the climate.

The important strategies that needed to alter to deal with projected climatic changes are as follow (Tubiello and van der Velde, no date) –

- Shifting fertilizer rate, time, dose and method of application to maintain crop yield quality.
- Changing irrigation method, time and amount to enhance water use efficiency in conditions of limited water availability.
- Implementation of crop varieties resistant to biotic (*i.e.*, disease, insect, attack of parasitic weeds etc.) and abiotic stresses (*i.e.*, drought, flood, extreme temperature, cold and hot waves, frost and soil, water salinity etc.).
- Diversifying crops in the direction of legume based rotation and including cover crops for enhancing the soil capacity to retain moisture for longer time and reduced erosion.

4.1.1. *Varieties*

Farmers will need a diverse selection of improved crop varieties that are better adapted locally to ecologically based production practices and suited to resilient to climate change (Shetty *et al.*, 2013). The introduction of adapted and recognized crop varieties can potentially empower cropping systems by aggregate crop yields with building crop resilience to biotic and abiotic stresses in a sustainable way likewise by means of taking new market opportunities besides increasing crop productivity, quality, health and nutritional value (Recha *et al.*, 2014; Clements *et al.*, 2011) which will help to make sure continuous improved agricultural production even at uncertainties about future climate change impacts and help to create eco-friendly sustainable environment (Shetty *et al.*, 2013).

There are many crop varieties better adapt, better to changing climatic parameters (Sinclair and Muchow, 2001). Farmers of Meso-America, the Andes, Africa and parts of Asia are using new and improved varieties in response to environmental stress conditions. Adaptation strategies of improved and tolerant cultivar managed under improved inputs with additional nitrogen fertilizer can enhance the irrigated maize net production by about 21 per cent in 2020, 10 per cent in 2050 and 4 per cent in 2080 scenarios (Soora *et al.*, 2012). Improved varieties according the prevailing or expected stress could be an advantageous adaptation strategy for minimizing the vulnerability of maize production in India (Byjesh *et al.*, 2010). Research and developments of stress tolerant rice and wheat varieties can ensure food security by an increase in yield of up to 20 per cent (Faroque *et al.*, 2013). Bangladesh Rice Research Institute (BRRI) has developed salt-tolerant rice varieties *i.e.* BRRI dhan 40, BRRI dhan 41, BRRI dhan 53 and BRRI dhan 54, Submerge-tolerant varieties BRRI dhan 51 and BRRI dhan 52, Drought tolerant BRRI dhan 42, BRRI dhan 43, and Salt-cold-drought tolerant BRRI dhan 56, BRRI dhan 57 (Faroque *et al.*, 2013). Seeds of BRRI dhan 47 varieties are multiplied and distributed to the farmers for cultivation in the saline areas of southern districts of Bangladesh. Similarly, short duration rice varieties BR -33 BRRI and BINA -7 is being successfully cultivated against *monga* situation in the northern Bangladesh (Faroque *et al.,* 2013). Similarly, submergence tolerant *indica* type rice varieties, such as FR13A, can survive up to two weeks (Xu *et al.,* 2006). The first marker assisted selection (MAS) drought tolerant aerobic rice variety released in India was MAS946-1. Consequently, IRRI-India drought breeding network has yielded a drought tolerant genotype Sahbhagi Dhan which was released in India during 2009. After that drought tolerance was also inserted in rice varieties such as IR64, Swarna and BPT 5204 through marker assisted backcross breeding. Other salt tolerant *indica* rice varieties being used in India are Nona Bokra, *Saltol,* FL478, and Pokkali (Lin *et al.,* 2004).

4.1.2 *Crop diversification*

Crop diversification refers to the addition of new crops or cropping systems to agricultural production on a particular farm taking into account the different returns from value-added crops with complementary marketing opportunities (Clements *et al.,* 2011; Lakhran *et al.,* 2017) (Fig 2). Today farmers grow same crop again and again on same piece of land; they are exposed to high climatic risks in the event of unexpected climate changes that could severely impact agricultural production, such as arrival of new insect-pests which is not seen till now and the sudden onset of frost or drought. In the same time, crop diversification aids to agricultural production as it protects natural biodiversity, strengthening the ability of the agro-ecosystem to respond to these stresses, minimizing environmental pollution, reducing the risk of total crop failure, reducing incidence of insect pests, diseases and weed problems and secure food supply opportunities and also providing producers with alternative means of generating income (Clements *et al.,* 2011). This diversification from the monoculture of traditional staples can have important nutritional benefits for farmers in developing countries and can support a country to become more self-reliant in terms of food production. Compared to producing monocultures, management techniques for diversified crops generally consist of more sustainable natural resource practices. Crop diversification provides better conditions for food security and enables farmers to grow surplus products for sale at market and thus obtain increased income to meet other needs related to household well-being.

Altering in farming systems may also a vital role in the adaptation of European agriculture to climate change (Jorgen Olesen, no date). Berry *et al.* (2006) interpreted four IPCC-SRES scenarios and advocate that different types of adaptation of farming systems, alteration may be suited for specific scenarios and regions *i.e.,* high latitude, altitude, marginal areas etc. Maize is the main alternative to soybean in the Plains, and its yields are very sensitive to drought-related stresses. The availability of drought-tolerant maize may allow cropping in a marginal area that becomes even drier. Furthermore, rotations including drought-tolerant maize would become more attractive in a drier climate and partially alleviate concerns about soybean mono-cropping. Cultivation of bioenergy crops for energy production is also a good alternative of food production (Tuck *et al.*, 2006). Though, raising temperature and reduced precipitation may be severely reduced the choice of energy crops in future in Southern Europe (Jorgen Olesen, no date). Increasing climate change threats needed to develop the full potential of salt tolerant crops like barley, wheat and halophytes (World Bank, 2009). The halophytes have greatest potential in the production of oilseed and forage and agroforestry products. For example canola which produces high quality fatty acids (Weber *et al.*, 2007) and well suited in extreme temperature, saline condition and low moisture environments and therefore appropriate options for crop diversification in water scare areas resulted from climate change and variability (World Bank, 2009). Besides this, some halophytic grasses and shrubs also in existence which could be a good option for "bio saline" forage and livestock production along with restoration of waterlogged and salinity degraded soils (Barrett-Lennard, 2002; Ravindran *et al.*, 2007).

Fig. 2. Crop diversification with maize

Some of the other examples of crop diversification are as follow –

- Growing of forages as they have better drought tolerance and also improve soil quality and moisture retention capacity of soil (Wall and Smit, 2005).
- Introduction of native grasses for pasturing as they are drought resistant once rotational grazing is practiced on them (C-CIARN Agriculture, 2003).
- In some areas shifting wheat production by growing a more drought resistant crops *i.e.*, pulses (Wall and Smit, 2005).

- Some farmers also stagger their seeding and thus, harvesting time by selecting a crop variety that need varied growing conditions, so that crops are at different stages means more/less vulnerable if climate conditions start having a negative effect (C-CIARN Agriculture, 2003).
- Involve livestock in existing farming system *i.e.,* mixed farming to ensure better utilization of increased forage production on the farm (SSAPRI, no date).

4.1.3 *Mixed farming*

Mixed farming is an agricultural system which involves integrated crops, livestock and other possible household enterprises to increase income through different enterprises and to complement land, and labour demands across the year. This system of farming pays to climatic change adaptation for the reason that the diversification of crops and livestock provides higher chances to withstand against extreme weather phenomena which was previously not known, resulted from climate change and variability. In addition, mixed farming ensures consistent production and farmer's income since if one crop or crop variety flops to compensate, then another crop and variety may reimbursements. The second component of mixed farming *i.e.,* livestock, by which farmers can get money whenever they want by selling their animal during the periods of need, such as if crops fail due to drought or flooding (Clements *et al.,* 2011). The byproducts of livestock like crop residues as animal feed and dung as manure can be used efficiently on farm with dual advantage of its better utilization as well as reducing external inputs use i.e., fertilizers and other agrochemicals that enhance environmental sustainability by reducing fertilizer's residue buildup in soil body, eutrophication and air pollution (Meena *et al.,* 2016). In another way, bullocks can be used for tillage operations that save the energy by the way of reducing diesel consumption which is not only economically flexible, but also environmental friendly. To make sure the accomplishment of mixed farming, it is essential to increase the livestock productivity and generate a sustainable land use plan according to the soil fertility and domestic requirements. As livestock needs a high amount of fodder, unfertile land can be used for fodder production with higher success (Shetty *et al.,* 2013). Mixed farming allows the introduction and use of rotations between various crops and forage legumes and trees, or for land to remain fallow and grasses and shrubs which is also adventitious for the environments as it maintains soil fertility by recycling soil nutrients. This system also sustains soil biodiversity, minimizes soil erosion and helps to conserve water, ensures the best use of crop residues otherwise they will not be used as feed and then stalks may be incorporated directly into the soil, which may temporarily trap nitrogen, creating nitrogen deficiency (FAO, 1995).

4.1.4 *Sustainable water management*

According to IPCC (2007b) for the duration of the next decades, billions of people, mostly in developing countries, will face life-threatening freshwater scarcities or flooding, consequentially negative impacts on agriculture due to shifting in rainfall patterns by climatic variations. Several studies (Seckler *et al.*, 1998; Rosegrant *et al.*, 2002) predicated standing problem of complete water scarcity in over a third of the global population by 2025. Therefore, adaptation technologies for sustainable water use and management are very crucial for improving agricultural productivity and guaranteeing food security for global hungers.

Sustainable adaptation strategies call for smarter, precision technologies for irrigation that increase the water productivity, and farming practices that use eco-friendly approaches to conserve water resources (Shetty *et al.*, 2013; Tubiello and van der Velde, no date). In addition, for enhancing water productivity along with soil and water conservation associated adaptation strategies, such as rain water harvesting in dryland areas, watershed development through tree-based farming, use of short duration drought tolerant crops and varieties, enhancing use of organic manures to improve soil fertility levels, efficient use of available water resources through micro irrigation (i.e., drip, sprinkler/trickle irrigation etc.), moisture conservation technologies (i.e., conservation tillage, mulching, manuring etc.), better crop pest management etc. (World Bank, 2009; Shetty *et al.*, 2013). In addition, some others sustainable approaches include alternate wetting and drying irrigation technique for rice, provision of supplementary irrigation in rain-feds and appropriate water management to avert waterlogging, erosion and leaching (Tubiello and van der Velde, no date). Therefore, it needed the judicious use of such large number of approaches and application along with demand and supply side management solutions to improve water productivity. Although, certain approaches are for direct implementation for short term. Other phenomena, like recharging of aquifers, conjunctive use of good and poor quality water etc. that want alliance and capacity building before sustainable use (Anonymous, 2010). However, up to now, a small number of simulation studies predicated the importance of irrigation water as adaptation strategies to condense the climate change impact. On the whole, the greatest comparative adaptation assistance is to be gained under low-to-moderate warming conditions (IPCC, 2007b; Tubiello and van der Velde (no date)).

4.1.5 *Drip irrigation technology*

Drip technology is more fitted where water availability is limited, presently or expected to become limited in near future for agricultural use due to climate change. Thus, it sustains farmers towards adapting to climate change by efficient use of water and enhanced crop productivity (Clements *et al.*, 2011). In spite of its high crop and water productivity, irrigation by this pressured system reduces its environmental impact *i.e.* soil salinization and nitrate contamination of aquifers. Such type of precise irrigation ensures consistent and malleable water application which is a major platform for sustainable water management (Shetty *et al.*, 2013). Mainly in regions which are more prone to climate change hazards like drought; this technology lessens water demand and decreases water evaporation losses, as it enables farmers to apply just enough water to ensure good harvests at lower investment (Anonymous, 2009). Fertilizer is also applied more efficiently, as it can be applied directly through the pipes along with water (Clements *et al.*, 2011). IWMI and ICRISAT are helping to introduce drip irrigation in southern and West Africa for the production of high-value vegetables and fruits (Anonymous, 2009). The Figure 3 (a) and (b) illustrate a good example of successful sub-surface drip technology in maize and wheat.

Fig. 3. Sub-surface Drip Technology in (a) maize and (b) wheat

4.1.6. *Sprinkler/trickle irrigation technology*

Sprinkler irrigation technology is predominantly pertinent where water availability is limited presently or expected to become limited in near future for consistent supply in agricultural because of climate change. This system enables more efficient use of water for irrigation as it eliminate water loss at conveyance channels, well and more evenly distributed water across crops which avoid water wastage, increased crop yields, by reducing the risk of crops freezing due to colder than usual temperatures as frequent frosts are by now impacting on crops. As a result of climate change, most suiting for most of agronomic and horticultural crops that are grown closely together, like for cereals, oilseeds, pulses, sugarcane, cotton, vegetables, fruits, and flowers etc. (Narayanmoorthy, no date; Clements *et al.*, 2011; Snyder and Melo-Abreu, 2005). Unlike other surface irrigation system, this system confines soil disturbance which helps in eliminating the risk of soil erosion (Clements *et al.*, 2011). This adaptation approach can be used in combination with other adaptation approaches like for the, fertilizer management and multiple cropping. Nowadays, globally sprinkler irrigation is being used in America, Europe, Asia, India, Africa and Oceania on 13.3 M ha, 10.1 M ha, 6.8 M ha, 3.5 M ha, 1.9 M ha and 0.9 M ha area, respectively (Kulkarni *et al.,* 2006).

4.1.7. Rainwater harvesting and recycling

Climate change is disrupting global rainfall patterns, meaning some parts of the world are suffering from a drastic drop in precipitation leading to a fall in water levels in many reservoirs and rivers (Clements *et al.,* 2011). In such situations water harvesting techniques can help switch excess water runoff, making them important strategies for adapting to climate change. Conservation and gathering of rainwater help to lessen the risks created by rainfall scarcity (Recha *et al.*, 2014). Therefore, promoting the efficient capture, storage and utilization of rainfall through the adoption of appropriate water harvesting strategies is a vital strategy for managing climate risks in rain-fed agriculture to enhance production (UNEP and SEI, 2009; Recha *et al.*, 2014). Rainwater harvesting methods range from simple technologies that increase the capture of surface water flow and concentrate runoff to more complex measures for storage of water in macro catchments which is important for helping agriculture to adapt future climate risks (World Bank, 2009). As the water quantity to be harvested is limited; therefore, it must apply to crops at critical period of growth which ensures the more yield per drop. Subsequently, supplemental irrigation is a crucial approach which can increase yields in excess of 100 per cent. A small investment providing between 50 and 200 mm of extra water per hectare per season for supplemental irrigation, in combination with improved agronomic management, can double water productivity and yields in small-scale rain-fed agriculture (UNEP and SEI, 2009). The agriculture in Sub-Saharan Africa are suffering from the water scarcity where two-thirds of the region is desert and dryland, and in such cases there is urgent need for improving water management. However, nowadays, the World Fish Centre in sub-Saharan Africa is trying to develop governance systems to promote water harvesting at the watershed level (Anonymous, 2009).

In some parts of the word, there is not enough water for every use. However, smart utilization of water that has already been used can be very valuable. Water re-use has benefits for all. Let's see an example of how things are done today and they could improve if done differently. Cities need water for their people. Water that is used in toilets, to clean house and cook food. Consequently, it is contaminated and becomes waste water. Sometimes, wastewater is not properly treated and is disposed of further downstream. This freshwater disposal can pollute rivers, and harm while ecosystems including fish and plants, and limit human water use downstream. At the same time farms near the cities need water and fertilizers to produce vegetables. Farmers often purchase costly mineral fertilizers containing nitrogen, phosphorus and potassium from factories that may pollute the environment and use-up nonrenewable resources. Freshwater is taken from the nearby river to irrigate the field. After a few months of irrigation and fertilization farms' produce are ripe and ready to be picked. After harvest they are sent to the cities where they are sold and consumed. With this approach, rivers and air has been polluted and resources have been wasted, hence making this practice not sustainable in the long term. But, there is another way, water can be used in the cities and re-used in agriculture with benefits for all (Cavallini and Young, 2002). Even if urban waste water contains harmful contaminations, it also includes nutrients that plants need to grow (FAO, 2010a). With the proper treatments most of the contaminants can be removed from water leaving only the nutrients in (World Bank, 2010). The very same nutrients nitrogen, phosphorus and potassium, that are used in fertilizers. With the right agricultural and irrigation practices contact between treated wastewater and crops is minimized. This ensures that fruits and vegetables produced are not harmful to our health. Following this practice provides farmers with a reliable source of water and nutrients all year around (Amerasinghe *et al.*, 2013). Agriculture crop uptake the nutrients that cities produce, while preventing the pollution of rivers and lakes. With this practice, water has been re-used and nutrients have been recycled (Jhansi ad Mishra, 2013).

4.1.8. *Sustainable soil management*

Some major and extensive soil changes are likely because of global climate change that would impact positively or negatively on the fertility and soil physical conditions, even though the exact nature of these changes are not well understood yet (Clements *et al.*, 2011). Despite these uncertainties, a range of soil management strategies can improve soil health and resilience in contradiction of negative impacts of climate change (Brinkman and Sombroek, 1996). Plant nutrient management to improve soil nutrients and therefore boost crop productivity is a major technological challenge to enhance the sustainability of agricultural production under changing climatic conditions along with sustaining climate-friendly environment. Thus, to maintain the quality of soil to encounter the food demand plant nutrient management becomes essential (Anonymous, 2010) even though at the same time protecting soils from erosion (World Bank, 2009).

Implementation of integrated nutrient management (INM) and soil forming terrace approach increases nutrient use efficiency by integrating and balancing the nutrient dose as per the soil nutrient status and crop needs, in addition to protects the soil from erosion threats (Anonymous, 2010). These practices also enhance the resilience of the soil health to more intense storms and extreme events, which are expected to be increasingly prominent under climate change (World Bank, 2009).

4.1.8.1 Integrated Nutrient Management (INM)

Integrated nutrient management (INM) refers to a range of practices that combine organic and inorganic sources of nutrients. INM approach can sustain the soil fertility and plant health through adequate supply of major plant nutrients in right proportion significantly influences the resistance or susceptibility of plants to a number of pest and diseases *i.e.,* excessive nitrogen application makes plant succulents and thus attracts disease-pest attack which in contrast, potassium enhance plant resistance/tolerance against pests and diseases by knowing nutrient status of soil and plant through soil testing, plant tissue analysis and observing visual signs of plants, assessment of constraints and opportunities in the existing soil fertility management strategies (Clements *et al.*, 2011). The maximum yield will be produced when nutrients applied through a mix of fertilizers and natural sources like manure and nitrogen-fixing crops under changing climatic parameters (Shetty *et al.*, 2013).

Low external input sustainable farming system (LEISA) in addition to developments in the scientifically-informed system of organic farming systems display robust socio-technical possibilities and potential for reducing energy use and new human N fixation. Therefore, farmers must more focus on use of organic fertilizers *i.e.,* compost, FYMs, green manure and bio-fertilizers in its place of chemical fertilizers towards maintenance and productivity by arresting nutrient mining and sustaining soil fertility (Recha *et al.*, 2014; World Bank, 2009). Along with agronomic practices, such as intercropping with legumes, reduced/no tillage and leguminous mulching, in addition to strategies that condense the dangers of salinization and acidification to uphold the soil fertility (Faroque *et al.,* 2013). The application of manure and composting can improve soil fertility and productivity, and produce greater crop yield along with increase soil moisture and soil cover, as well as reduce soil loss which is important for adapting to climate change (Recha *et al.*, 2014). Mulching, improved fallow can help to prevent erosion and increase soil fertility through protecting the soil from excessive heat, exposure to wind, and moisture loss. The planting of fast-growing species of leguminous trees or crops into a short-term fallow for one or more years to improve soil fertility. INM also promotes changes in land use, crop rotation and interactions between forestry, livestock and cropping systems as part of agricultural intensification and diversification (Clements *et al.,* 2011). Assessing the suitability of green manure and agroforestry trees used in INM for heat, submergence, and drought tolerance (*i.e.*, canavalia, crotalaria, mucuna and lablab) will be an important approach for adaptation to climate change (World Bank, 2009). Crop rotation with these green manures help to build up the resilience to climate change, particularly in small and marginal farming as a consequence of their value in restocking soil fertility and organic matter reserves. In the same time, nitrogen leaching rates in organic arable fields are reduced by 35 per cent to 65 per cent over conventional fields (Brunori *et al.,* 2008).

4.1.8.2 Soil forming terraces

A terrace is a leveled surface used in farming to cultivate sloping, hilly or mountainous terrain. They can be used on relatively flat land in cases where soil and climate conditions are favourable to erosion. Terraces mainly focused on improving the natural conditions for agricultural production, decreasing soil erosion, maintaining or increasing soil moisture and create positive environmental benefits. Terraced fields are effective for growing a wide range of crops, such as rice, potatoes, and maize. This technology facilitates adaptation to climate change by optimizing water use by crops. In the respect of landslides and mudslides, slow-formation terraces reduce soil erosion and, therefore, the threats of landslides. Terraces also provide a method for regulating the micro-climate for agricultural production. By capturing the sun's heat in the rock walls, terraces absorb heat during the daytime and release this nightly and helping to create a slightly warmer internal micro-climate which can protect crops from frosts, prolong the growing season and allow for crop diversification (Mars, 2005).

4.1.9 *Conservation Agriculture*

To benefit global endangering agriculture due to climate change, conservation agriculture holds much promise as 'an approach to managing agro-ecosystems for improved and sustained productivity, increased profits and food security while preserving and enhancing the resource base and the environment' (Friedrich *et al.*, 2012). Towards the efforts of increasing agricultural productivity by the way of enhanced agricultural intensification without spreading out global cultivated area and thereby environmental degradation, conservation agriculture has been recognized as an auspicious strategies (Godfray *et al.*, 2010; Grassini and Cassman, 2012). Conservation agriculture consists of three crop management principles *i.e.,* minimal soil disturbance (no/zero-tillage), crop residue retention on soil surface, and sensible legume based crop rotation (Hobbs *et al.*, 2008; FAO, 2011; Milder *et al.*, 2011; Cameron *et al.*, 2015) which helps to develop the sustainable agricultural production system (Derpsch, 2001).

Conservation tillage involves change in tillage practices from repetitive annual tillage operations to minimal or zero tillage practices for establishing crops in a previous crop's residues (typically about one-third of crop residue), which are purposely left on the soil surface to augment management of climate change threats through better protection of soils from extreme events and improve crop productivity through increased water and nutrient holding (World Bank, 2009; Clements *et al.*, 2011). Therefore, this reducing soil disturbance helps to stabilize soil structures, increase water infiltration and retention, reducing soil compaction and prevent erosion because the crop residues placed along the contour lines can slow down surface runoff and thus recover resilience to climatic change impacts of drought and floods (Smith, 2005; Recha *et al.*, 2014). The combination of reduced soil disturbance and increased retention of crop residues also results in increased carbon sequestration (Anonymous, 2009). Conservation tillage benefits farming to climate change adaptation by saves water, labour and diesel, helps early sowing, improves soil organic carbon, increases fertilizer use efficiency and soil–water dynamics, reduced and improving yield (Anonymous, 2009; Clements *et al.*, 2011;Recha *et al.*, 2014). Enhanced soil nutrient recycling might also help fight crop pests and diseases. A study of Cameron *et al.* (2015) concluded that generally, no-till system reduces yields; however this response was variable and sometimes no-till recorded equal or superior yields than conventional tillage. But the important point to consider is, when no-till is combined with residue retention on surface soil and sensible legume based crop rotation, its negative impacts are overcome as this strategy improves the crop yield considerably in rain-fed dry climates probably owing to improved water infiltration and better soil moisture conservation (Serraj and Siddique, 2012), and suggesting that it may become an imperative climate-change eco-friendly adaptation approach for ever-drier regions of planet. Moreover, this approach is recognized suitable for sustainable intensification for millions of hectares areas in dry climates of sub-Saharan Africa and South Asia. Conservation agriculture proves more adventitious in such areas as crop production is becoming difficult gradually in these areas (Cameron *et al.*, 2015) and currently struggling with food security disputes and have a high possibility of experiencing yield drops due to climate change in the upcoming times (Lobell *et al.*, 2008).

Presently, zero-tillage is being practiced roughly in 9 per cent of global arable land corresponding to over approximately 125 million hectare areas, with unpredictable degrees of use of two other principles of conservation agriculture (FAO, 2011; Friedrich *et al.*, 2012). Besides, the technology has been rapidly taken up so that close to half a million farmers in India, Pakistan and other countries of the region. Now apply this technology on more than 3.2 million hectares, with economic benefits so far estimated at US$147 million (Anonymous, 2009). In semi-arid to dry sub-humid rice-wheat areas of Pakistan and the Indian Punjab, where the environmental and economic costs of irrigation are increasing, zero tillage systems with retention of straw has increased *in situ* water retention, resulting in significantly improved economic returns for wheat production, through reductions in fuel costs and irrigation demand (Gupta and Seth, 2006) (Fig. 4a & b).

Fig. 4. (a) Sowing of wheat in anchored rice stubbles (b) Wheat in standing rice stubbles

Conservation tillage is a popular technology in the Americas, with approximately 44 per cent practiced in Latin America. Studies suggest there is great potential to bring this technology to Africa, Asia and Eastern Europe, although limiting factors have to be taken into account (GTZ, 1998; Derpsch, 2001). In western Kenya, the adoption of such practices was found to increase soil moisture storage by 18 to 50 per cent, resulting in combined yield increases of 30 to 150 per cent for maize, beans, and wheat (Ngigi *et al.*, 2006). According to Sorrenson *et al.* (1998), total economic benefits arising from adoption of the no-tillage technique in small farms of generally less than 20 ha in Paraguay have reached around $941 million. Zhang and Liu (2005) estimated that under climate change scenario of increased precipitation in the Loess Plateau, the use of conservation tillage practices in a wheat maize rotation could reduce soil erosion by more than half compared with conventional tillage. Likewise, Zhang and Nearing (2005) estimated no change in soil erosion under no-till wheat systems compared with increased erosion rates as great as 67 to 82 per cent under conventional tillage, using estimates of erosion under a range of climate change scenarios compared with current erosion rates.

Specific efforts should be made to scale up the results of applying conservation agriculture in pilot districts, adopt and mainstream conservation agriculture practices in government planning (especially locally), train farmers and extension officers on the principles and application of conservation agriculture, promote the use of various simple and low-cost techniques (such as application of compost and organic manures or construction of ditches, terraces or trenches), and encourage farmers to use locally available organic inputs (cow dung and mulch) and develop better links between grazers and cultivators in pastoral areas (UNDP, 2013).

Fig. 5. Relay seeded green gram in anchored wheat stubbles

4.1.10. *Ecological pest management*

An increase in extreme events, changes in moisture conditions, temperature rise, and elevated CO_2 concentrations are anticipated to amplify the pest attack on crops either by expansion of existing pests, increased pressure from invasive alien species or by interruption of the chronological and topographical synchronization of pests and beneficial insects (World Bank, 2009). Sustainable climate-friendly approaches to lessen the crop losses to due insects, include growing resistant varieties, diversified farming system, conserving predators and managing crop nutrient levels to reduce insect reproduction. Whereas, against plant diseases include use of clean planting material, crop rotations to suppress pathogens, and eliminating infected host plants, weeding, minimized tillage and the use of surface residues (Shetty *et al.*, 2013). Adaptation strategies must focus on establishment of decision and information support systems for pest and disease surveillance, demonstration of best practices and quick response mechanism to deal with climatic disasters (Anonymous, 2010).

Integrated pest management (IPM) is identified as a key element in sustainable agricultural development (Frison *et al.*, 1998) which provides a better guarantee of good harvesting even in difficult conditions of pests and diseases that emerge with the temperature and water level changes (increase of relative atmospheric humidity and runoff) typical of climate change (Clements *et al.*, 2011). IPM builds resilience in agricultural production to potential risks posed by climate change, such as damage to crop yields caused by newly emerging pests and diseases. It can be promoted through, local production of bio-control agents, strict pesticide regulations and removal of pesticide subsidies. As an ecosystem-based strategy, integrated pest management affects the prospects for the effective management of pests through soil management. It applies an ecosystem approach, such as mulching, because water stress can increase the susceptibility of crops to disease, crop varietal resistance essential for managing plant diseases and many insect pests and timing, and spatial arrangement of crops to influence the dynamics of pest and natural enemy populations (Shetty *et al.*, 2013). IPM contributes to climate change adaptation by providing a healthy crop with least disruption of agro-ecosystems in which the vulnerability of plants to pests and diseases is decreased (Frison *et al.*, 1998; LEISA, 2007). Decision to apply pesticides should be made, based on the extent of the attack or infestation of disease-pest and the condition of the crop at minimal possible rate.

For better management of crop pest and diseases, sustainably under climate change includes (Clements *et al.*, 2011) –

- Grow pest resistant-tolerant and well adapted input responsive crop varieties.
- Follow legume based crop rotation to enhance biological nitrogen fixation and thus increase nutrient cycling in addition to soil fertility status and favourable conditions for robust plants that better face pests and diseases.
- Grow either cover crop (*e.g.*, green manure, leguminous) to enhance crop weed competition in favour of crop and reduce pest-diseases or keep soil covered with crop residues.
- Integration of intercropping and agro-forestry technology.
- Maintain crop spacing and intercropping to create opposed conditions to pests.

- Add organic manures to help in maintaining balanced pH and nutrient levels. Microbes in the compost will improve water absorption and air exchange.

4.1.11. *Agroforestry system*

Agroforestry is a collective name for land use systems and practices in which woody perennials are deliberately integrated spatially or temporally with crops and/or animals on the same land management unit (Recha *et al.*, 2014) (Fig. 6). Recently, agroforestry has been identified as a preferred adaptation technology against climate change (Thomas *et al.*, 2005). It helps to improve the resilience of agricultural production to present and future climate change and variability by means of using trees for intensification, diversification and buffering of farming systems (Clements *et al.*, 2011). Trees considerably enhance the adaptive capacity of system to climate change risks and thus resilience of farming systems and defending agricultural production as trees are deep rooted are less vulnerable as compared to annual crops to extreme events like droughts or floods (Clements *et al.*, 2011). However, the capability of agroforestry as an adaptation measure depends on sustaining farmer access to information, support and training for tailoring agroforestry to variable environments, appropriate policy conditions and access to necessary inputs (World Bank, 2009). The potential of agroforestry system to adapt against climate change can be understood by the below mentioned points.

- Increases diversity of on-farm trees, crops and tree cover to cushion farmers (Recha *et al.*, 2014).
- Reduce dependency of animals on a single staple crop. For example, if annual crop fails due to drought, trees will still provide fodder (Clements *et al.*, 2011).
- Enhances or maintains wildlife habitat and biodiversity (Recha *et al.*, 2014).
- Improves food security through enhancing diversification of production systems such as production of fruits, nuts and edible oils (Clements *et al.*, 2011; Recha *et al.*, 2014).
- Improve soil quality and fertility to lessen the impacts of hunger and chronic illnesses, and the trees provide microclimate conditions that are suitable for crop growth by guarding the soil from excessive heat, exposure to wind and water stress during drought (Clements *et al.*, 2011; Recha *et al.*, 2014).
- Tree-based systems also have higher rates of evapotranspiration than row crops/pastures and thus maintain soil aeration by pumping excess water out of the soil profile more rapidly than other production systems if there is sufficient soil moisture (Martin and Sherman, 1992).

- Trees manage to risk of extreme climates, such as droughts or excess rain. For example, a combination of Napier Grass and leguminous shrubs in contour hedgerows reduced erosion by up to 70 per cent on slopes beyond 10 per cent inclination without affecting maize yield in central Kenya (Mutegi *et al.*, 2008). Like this in semi-arid environments of Sahel regions of South Africa found significant recovery of agricultural lands from extreme drought, by managing natural regeneration.
- Trees enhance carbon sequestration and improve the resilience of the agro-ecosystem to climatic hazards. The high carbon sequestration potential of agroforestry stems from the potentially large land areas to which land-use modifications to accommodate agroforestry are possible (World Bank, 2009).
- Trees stabilize the soil against landslides and increase the rate of infiltration (Ma *et al.*, 2009) and thus decrease runoff of rain water and increases ground water recharge.
- Better utilization of degraded lands and proven critical for restoring degraded lands. For example, in Southeast Asia, an estimated 35 million hectares of *Imperata cylendrica* could be put into more productive use with agroforestry (Roshetko *et al.*, 2007).

Fig. 6. Agroforestry system with poplar between maize crops

4.2. Researchers level strategies

4.2.1. *Biotechnology*

Biotechnology has been the fundamental basis for the development of agriculture across the world (Clements *et al.*, 2011) and today it proves a chief technology for the improvement of genetic resources with larger adaptive capacity to cope with climate change and its variability (Anonymous, 2010). It means not to just produce the transgenic organisms; it has greatest relevance to date in developing biotic and abiotic resistant crop varieties, whose impacts are fiercely debated today (Clements *et al.*, 2011). It has the huge potential to produce more food with less environmental degradation (Wang, 2015). However, traditional crop breeding for tolerance abiotic stresses had significant achievement, but it is slow and highly restricted to exploiting the prevailing genetic variation in plants and their wild relatives, while in the meantime, biotechnology gives us the vision of creating more intense and more rapid changes to crop comebacks to stress than traditional breeding (Clements *et al.*, 2011).

To plan for adaptation, it is important to anticipate the technological changes that may shape agriculture over the next decades. One of the most anticipated developments in agricultural biotechnology is the introduction of genes to enhance drought tolerance in plants (Babu *et al.*, 2003, Masle *et al.*, 2005). There is a great deal of activity within the major biotechnology life sciences companies and the agricultural research institutes, and academic institutions on transgenic research for drought-prone environments (Ortiz *et al.*, 2007; Varshney *et al.*, 2011). The biotech drought tolerant crops are "very promising for the millions of poor in the more marginal rain-fed agricultural areas of developing countries" (Kostandini *et al.*, 2009). Drought tolerant rice, maize and tobacco, salt tolerant rice and flood tolerant rice are some of the examples of crops produced by biotechnology. HARDY (HRD) gene from *Arabidopsis* had successfully introduced in rice to produce transgenic drought tolerant rice which has increased water use efficiency and biomass/water use ratio through reduced transpiration and enhanced photosynthetic assimilation leading to increased adaptive capacity of root mass under water stress/drought conditions created by climate change (Karaba *et al.*, 2007). In addition, scientists of CIMMYT in collaboration with International Institute of Tropical Agriculture (IITA) had developed over 50 drought tolerant maize varieties so far in sub-Saharan Africa and are being grown on about one million hectares area (Anonymous, 2009). Enhanced expression of glutamate dehydrogenase (gdhA) gene introduced from *E. coli* to increase seed germination and grain yield under drought of maize (Lightfoot *et al.*, 2007). Besides this, expression of phosphatidylinositol-specific phospholipase by ZmNFYB2 reducing stomatal conductance and consequently leaf temperature, water loss, reduced wilting and maintenance of photosynthesis of maize plant under drought conditions. This was the result of a unique breeding method, in which hundreds of small farmers take part in testing the new varieties (Anonymous, 2009). Drought tolerant tobacco possess delayed drought-encouraged leaf senescence and have higher photosynthesis and reduces 30 per cent water requirements leading to less yield losses under drought conditions (Rivero *et al.*, 2007). Barley, millet and sorghum are most naturally hardy food crops, grown in dry climates which contain useful genes of drought tolerance. Scientists of ICARDA and ICRISAT have sequestered drought escaping genes of barley, millet and sorghum, known as called 'stay-green' trait and are developing early maturating drought escaping varieties (Anonymous, 2009). Salt tolerant rice A QTL is associated with drought resistance that produce normal yield under high salinity conditions in Bangladesh.

Flooding of coastal farmlands presents a real risk for farmers. To adapt under such environments, the need food tolerant rice varieties were and therefore, IRRI researchers and collaborators identified a rice gene called Sub1A and some flood tolerant varieties have been developed which allow plants to survive completely submerged for up to 2 weeks, which give high yields while protecting against flooding (Anonymous, 2009). Along with this another rice gene was identified, responsible for rapid stem elongation in deep water rice, *i.e.* 'snorkel rice' may contribute to rice breeding in lowland areas and ultimately lead to reducing the devastation of rice crops caused by monsoon flooding in Asia and boost rice production in flood-prone areas.

4.2.2. *Breeding*

The potential for developing new crop varieties by means of higher stress tolerance is promising, given recent advances in molecular biology and genomics, with implications of both conventional and transgenic breeding (Word Bank, 2009). A better scientific consideration is required of the yield restrictive practices to target future genetic and agronomic improvements. To evade the unnecessary use of inputs during those parts of the crop lifecycle that is not critical for crop growth and yield formation, and are more affected by changing climate threats and can vary with countries and agro-climatic regions (SSAPRI, no date). Though, the long-term success in breeding crops adapted to climate change may be controlled by a potential loss of genetic diversity caused by temperature rise and shifts in precipitation. This leads to range reduction or extinction of crop wild relatives. To counter this threat, maintenance is desired for policies that initiate defining the scope of the climate change risk to crop wild relatives, policies that engage local communities in their conservation, and the formulation of priority-determining mechanisms (World Bank, 2009). Adapting to climate variability and change in seed-insecure farming environments require that current emergency seed relief policies are better addressed. This can be done through seed voucher and fair programms, as well as through promoting rural credit programs, improved seed storage facilities, and measures for increased productivity that improve supply of and access to seeds (World Bank, 2009).

4.3. Governmental level approaches

4.3.1. *Information technology (IT)*

In dryland areas which are generally marginal environmentally or economically, the adoption of new improved agricultural technology that could support well climate risk management and adaptation has been not there resulted in partially failure of the technology due to poor dissemination pathway that prevent technology spillover effects (World Bank, 2009). As per the Kandlikar and

Risbey (2000) "Farmers in poor countries face high downside risks from failure of improved innovation, mainly if information technology (IT) is limited or else missing and thus they carry out such options that have been well tested in the earlier. Studies of climate change adaptation need to pay greater attention to these issues to be truly relevant in a global sense." Enhancing improved network systems that spread information on innovative agricultural research and technology, and encourage information exchange among farmers need to promote (Jørgen Olesen, no date). For that it is very important to provide the right information by right dissemination mechanism (*e.g.,* rural radio, internet, information and communication technology, mobile phones) at right time to farmers. Thus, such fast and effective information communication and exchange to farmers using the new technologies will accelerate the adaptation rate to climatic changes as weather becomes more erratic and less predictable (Jørgen Olesen, no date). It needs to implement at all levels of agricultural production system as all these are inter-related. At crop level, there is need to encourage the efforts to linkage the public and private partners with research institutions to get laboratory results quickly. At farm level, focus needs to encouraging awareness of farmers and developing agencies with scientific research and policy ingenuities to empower to take informed decisions (Anonymous, 2010).

4.3.2. *Reliable forecasting*

Strengthening the network of weather stations to provide more reliable weather information for farmers; putting in place early warning systems to reduce losses and damage from climate hazards, such as flash floods; and establishing a salinity monitoring system to track the effects of sea-level rise on rice production. These forecasts provide information on rainfall expected during the rainy season and provide an important source of information to help famers respond to climate variability. Other ways of using the information have included making decisions about the type of crop to grow, how to manage water resources and how to allocate agricultural and household resources. The information has also enabled farmers to change planting and investment activities if a drought or good rains are forecast (Clements *et al.*, 2011). If farmers could have advance warning of what is going to happen the next year or next planting season, will bear a resemblance to, they could get ready for it by altering the production strategies like for planting of different crop or variety, altering sowing or irrigation method (Neil Savage, 2013). Farmers would change the orientation of the furrows with furrows across slopes in dry years to retain water and down slope in wet years to shed water. If high rainfall is predicted farmers would plant rice or if conditions are drier upland crop may be planted *i.e.,* maize, cotton. There is also an interaction with labour availability if rains are heavy and weeds a problem they prefer to plant sorghum which is less susceptible to weeds than maize (Clements *et al.*, 2011). According to the information coming from forecasting, farmers can change the planting and harvesting weeks that coincide with new seasonal shifts. They can sow crop opposed microclimatic conditions, selecting more resistant/tolerant varieties to biotic and abiotic stresses. This is an advantage to deal with extreme climatic changes, such as low temperatures or a limited supply of water or expected pest attack (Clements *et al.*, 2011).

Seasonal climate forecasting can benefits farmers' decision making to augment climate risk management. Though, realizing the full potential of weather forecasting will involve considerable investments in education and outreach to build trust in the forecasts, improving the linkages between hydro meteorological services and agriculture research and extension, integrating climate forecast information into existing knowledge dissemination platforms, and linking forecasts with timely access to agricultural inputs (World Bank, 2009). Modeling can also aid farmers cope with climate change by making suggestions based on probable upcoming circumstances. Asseng *et al.* (2013) combined 27 different wheat models and defined how well they allocated with increasing atmospheric temperature and CO_2 concentration. Then measured how many different models would have to run together to overcome these errors and observed that by increasing temperature 3°C and CO_2 540 ppm, it would take as a minimum five different wheat models to arise with a reliable prediction.

The information should be disseminated in Hindi or local languages rather than English. Timely dissemination of forecasts provide enough time to farmers to make decisions. In addition extension workers should be well trained to communicate information timely and efficiently to farmers. Farmers have indicated a preference for receiving the information at community meetings (Ziervogel, 2001).

4.3.3. *Sustainable climate adaptation policies*

Globally, improvements in agricultural development policies and institutions are an important vehicles for encouraging farmers to adopt sustainable crop production under climate change (Olesen and Bindi, 2002). For achieving effective and equitable adaptation outcomes by realizing the current and future impacts of changing climatic threats on sustainable agricultural production would require major changes in agricultural policies globally because without large change in climate and agricultural policies, the necessary adoption of sustainable agricultural production cannot be achieved (Jim Kleinschmit, 2009; World Bank, 2009).

First we should know that what incentives and policies could lead to the adoption of sustainable farming practices. Improving the sustainable growth of agricultural sector and its role towards the environment and rural society require research into agricultural trade, production economics, and environmental policy analysis (SSAPRI, no date). Policy makers needs to better exploit climatic survey data and different modeling scenarios to plan for the future and to formulate the most apposite policies to safeguard the competitiveness of the agricultural sector and the most auspicious ways without hampering environmental health (SSAPRI, no date). International and national climate policies need to provide appropriate and sufficient resources and policy direction to ensure that farming is a major part of the climate solution, without sacrificing its other necessary functions (Jim Kleinschmit, 2009). Like fore, European agricultural policy progressively focuses on multi functionality as its goal and its organizing principle which necessitates diverse interpretation and balance among the environmental, social and economic functions as per the specific regions (Tait, 2001).

Currently, the agricultural research is facing four main challenges – i) generating the evidence to escort policy, ii) engaging with the policy world to make sure that policy responds towards the evidence, iii) recognizing the tangible impacts of policies and their change and iv) providing a fast enough response policy makers to encounter the insistence of the situation (Anonymous, 2009). To overcome all these, governmental policies must consider about the adaptation of integrated and coherent technologies that acknowledge and support the environmental friendly agricultural practices and technologies like conservation agriculture, integrated nutrient management system dominated by organic sources of nutrients, ecological pest management, stress resistant genotypes, drip and sprinkler irrigation technologies. In addition, all other practices which enhance agricultural production sustainability without compromising with environmental system because the agriculture has a unique and considerable role to play as an agent of natural resources and ecosystems (Jim Kleinschmit, 2009; Tilman *et al.*, 2002). Besides, these some of the ingredients of better sustainable adaptation policies to encounter climatic hazards are as follow.

- Empowering farmers through proper education, skill and capacity building is important to ensure that they acquire knowledge, participate in the development process, and adopt innovation and entrepreneurship having expanded their economic choices (Shetty *et al.*, 2013).
- Resource based policies that focus to encourage viable land use to support adaptation to climate change (Olesen and Bindi, 2002).
- Reforming water markets and raising water productivity to encourage more judicious water use as in case of drought prone areas (Olesen and Bindi, 2002).
- Energy and trade policies would need to revise to make them more supporting this shift in agricultural development and food, and energy distribution systems without obstructing (Jim Kleinschmit, 2009).
- Refining nutrient use efficiencies through altering cropping systems, adoption of improved nutrient management strategies and reduce nutrient mining (Olesen and Bindi, 2002).
- Considering the taxes on agrochemicals like pesticides, fertilizers etc. exclusion of subsidies for these and carrying out of new guidelines for reducing their indiscriminate application (Tilman *et al.*, 2002).
- Provision for integrated pest management systems should be increased through a combination of education, regulation and taxation (Olesen and Bindi, 2002).

- Strengthening collecting, conserving and characterizing plant genes sustainably and conducting research on alternative crops (Olesen and Bindi, 2002).

Numerous countries such as Australia, Canada, Japan, Norway, Switzerland, United States and European Union countries have inaugurated several forms of 'green payments'; a payment given to farmers who adopt sustainable or eco-friendly farming practices (Tilman *et al.*, 2002). The United States' Conservation Reserve Program provides liberal payments to farmers to take land out of production for a definite period. Some other nations have also founded 'environmental cross-compliance' conditions as an essential for farmers to get agricultural support payments. Norway and Switzerland pays farmers for 'landscape maintenance' (Tilman *et al.*, 2002).

4.3.4. *Funding and investments*

Lacking investment in agriculture is one of most important reasons of dropping agricultural productivity since the 1980s (World Bank, 2007). One of the most important ways to sustain crop production under climate change would be to prioritize free, united and timely funding as well as credit support for climate change adaptive agricultural sector that address one of these equally-important goals especially in developing countries challenging falls in crop yields because of climate change (Jim Kleinschmit, 2009; Anonymous, 2010). According to Nelson *et al.* (2010) for considering annual agricultural adaptation funding needs about US$7 billion. African Development Bank and the World Bank have agreed to provide US$ 155 million through the African Centre of Meteorology Applications for Development to enhancement weather and climate monitoring systems in Africa. In Cameroon, a National Observatory on Climate Change has been set up with US$ 6 million of funding aimed at providing climate data monitoring the impacts of climate change on agriculture and ecosystems, and guiding work on climate deed (Clements *et al.*, 2011). A special Adaptation Fund has been set up by UNFCCC, receiving funding directly from the sale of 2 per cent of CDM carbon credits. Prominently, the 'Green Climate Fund' is planned as a fundamental, supplementary UNFCCC financing mechanism, which initiates to pay about half of its US$30 billion in the period 2010–2012 in the direction of implementation of adaptation projects, increasing to US$100 billion annually by 2020 (FAO, 2008; Tubiello and van der Velde, no date). Under the United Nations Framework Convention on Climate Change (UNFCCC), there are provisions of three multifaceted funds *i.e.,* Adaptation Funds, Least Developed Countries Fund and Special Climate Change Fund to address the needs associated with climate change and are regulated by the Global Environment Facility (GEF) (World Bank, 2007). Issues being addressed by Parties under the various Convention bodies, include: enhanced action on adaptation as part of the Bali Action Plan under the UNFCCC Adhoc Working Group on Long-Term Cooperative Action; Nairobi work programme on impacts, vulnerability and adaptation to climate change; development and transfer of technologies, research and observation under the Subsidiary Body for Scientific and Technological Advice (SBSTA); National Adaptation Programmes of Action (NAPAs); and supporting adaptation through finance, technology and capacity-building under the Subsidiary Body for Implementation (SBI) (Tubiello and van der Velde, no date).

Easy and timely financial support and credit packages provided to farmers can help in adoption of improved management practices including resource conservation technologies, agro diversification, post-harvest value addition processes, etc. which would pay to dipping perils and enhancing farmers' returns (Anonymous, 2010). According to Huang and Wang (2014), four major investment areas are (i) investing on research to ensure better understanding of climate change impacts and vulnerability on agriculture, (ii) developing and distributing improved crop varieties that could be well adapted to climate alteration, (iii) enabling international technological transfer and agreement to new approaches, and (iv) others like innovation and implementation of biotechnology and water conservation strategies in developing nations in addition to ecological farm besides organic agriculture.

In spite of current climate change funds, they are still far-off from the funds needed to effectively adapt to climate change (World Bank, 2009a). Though, current climate-related financial currents to agriculture adaptation in developing countries cover only a minute portion of the entire climate change funds (Climate Focus, 2011). In addition, water conservation is one of a prime tool to expand agricultural productivity and adaptation towards climate change in the developing countries (Huang and Wang, 2014). The Drought Tolerant Maize for Africa is a good example of investment in agricultural science (Huang and Wang, 2014) as the yield of drought tolerant maize varieties have been increased by 3–34 per cent over normal maize varieties which has significantly enhanced farmers' returns and their food security. According to the report of la Rovere *et al.* (2010), drought tolerant maize can generate US$0.53 billion from amplified maize yield and reduced risk over the study period.

The areas for investments are urgent to enhance sustainable agricultural adaptation strategies under climate change (NAAS, 2013) –

- To make precise weather forecasting *i.e.*, establish rain gauges at block/ village level and enable access to and efficient management of weather related information by modern techniques like remote sensing and GIS.
- Financing in research and development of locally-adaptable crops, management practices, input sources etc., decision support system and modeling to analyze the impact of climate change, specifically in arid and semi-arid regions bearing in mind their more vulnerability.
- Invest on development and distribution of biotic and abiotic stress resistant/tolerant crop varieties as per the region specific.
- Provision of investment on water saving/conservation techniques *i.e.,* rainwater harvesting and storage, micro-irrigation.

- Investment on affordable weather insurance to deal with climate variability.
- Special attention on regions more vulnerable to climate change particularly arid and semi-arid regions.
- Agricultural and enterprise subsidies, rural finances, poverty reduction programs, technology adoption support etc. on part of the vulnerable sections of the society.
- Much of the corporate social responsibilities funds from leading private sector players in the country should be channelized to promote Climate Resilient Agriculture in vulnerable regions.
- Implement the recommendation of the National Commission on Farmers (NCF) for establishment of a National Agricultural Risk Management Fund and its need-based timely distribution.
- Investment in human resource development and capacity building through awareness and training among officials, extension agents and farmers, incentives to farmers for adoption of natural resource conservation practices and support to improve the existing indigenous technologies that are eco-friendly and sustainable in long-run.
- Revise and restructure institutionalize teaching and research facilities in concerned educational institutions to develop graduates and post-graduates suitably trained in climate change management.
- Encourage role of the NGOs, Civil Societies, and Public and Philanthropic organizations for improving adaptation readiness among the farmers community.
- Forge international/regional partnerships, especially with FAO, CCAFS of CGIAR, UNEP, IPCC for developing tools and technologies suited to local requirement through pooling financial and intellectual resources, and help judiciously implement the international agreements.
- Provision of concessional credit to poor farmers for adoption of climate resilient agricultural strategies.
- Establish efficient co-operatives, community organizations and associations/groups to tackle critical needs of farmers like resource mobilization, custom hiring, marketing their outputs and efficient natural resource management. Suitably empower Gram Panchayats to play a leading role at farm level in developing and implementing concerned programs.
- Build on priority infrastructure like markets and information gateways and create opportunities in the non-farm sector in vulnerable regions to help the farmers to diversify their incomes.
- Identify, critically document and share most successful stories in climate change management and organize pilot demonstrations of most

successful modules, and in a partnership mode scale-up and scale-out the selected ones to establish climate smart villages throughout the country.

5. FUTURE RESEARCH NEEDS

A better scientific understanding of the influence of inter-annual climate changes and variability on crop yields remains indefinable as per the specific location and region (Ray *et al.*, 2015). Moving onward, we need to find more sustainable way for crop intensification that increases crop production gradually with less consumption of water, nutrient and agrochemicals that helps to sustain the environment. This calls for a considerable scientific effort in research and development of autonomous long-term adaptation to deal with the climate changes to ruminate all the agronomic decisions made at the farm level (Kaiser *et al.*, 1993), regional level (Jorgen Olesen, no date), national and international levels as well (Clement *et al.*, 2011). Considering climate change, it is expected that much more additional efforts are needed in the coming decades. Therefore, encouraging the research on climate change adaptation, developing improved farming systems and alternative foods are needed. There is a considerable need for an increased focus on regional studies of impacts and adaptation to climate change in agriculture, since effects and responses are likely to be regionally specific, depending on interactions with soils, current climate and cropping systems (Olesen *et al.*, 2006). Such studies are now becoming more realistic with the arrival of detailed regional scale climate change scenarios (Christensen and Christensen, 2006). Besides this, need to improve technology transfer from more developed countries, and increasing the local innovation in cropping and farming systems to improve adaptation, while improving sustainability and farmer livelihoods.

6. CONCLUSION

Climate change and its impact on agriculture at global level is a reality at current and expected in future to which whole globe is facing since the beginning of the 21th century. In contrast, agriculture system on which whole global population depends coupled with economics of several particularly developing countries. Agriculture production system is facing the dual challenges; first it needs to enhance food production sustainably to feed a growing global population with least external inputs, second, at the same time lessening its global environmental influences, all under more erratic and risky climatic conditions in the future, while charting a more environmentally sustainable path. To encounter these challenges, the adaptation of sustainable agricultural approaches are need of hour to reduce negative effects and exploit possible positive effects of present and future expected climate change. These approaches include adaptation of high yielding biotic and abiotic stress resistant/tolerant crop and its varieties, appropriate crop diversification involving less vulnerable component to climate change, mixed farming and intercropping, rainwater harvesting especially in water deficit regions and efficient utilization of available water resources through micro irrigation technologies, sustainable soil management through integrated nutrient management and soil forming terraces to improve soil health and reduce soil erosion, use of principles of conservation agriculture, eco-friendly pest management, agroforestry system and some others agronomic practices, *i.e.,* crop rotation with legumes, mulching, adjusting sowing time and in other input application like irrigation, fertilizers etc. Besides this, at broad level, sustainable adaptive governmental policies, funding and investments, information technologies and reliable weather forecasting help and support the farmers to implement better eco-friendly approaches at farm level to fight climatic menaces. These technologies ensure the plenty agricultural production at reasonable costs, better living status of farmers, and condense the environmental costs of agriculture.

REFERENCES

ADB and IFPRI (2009) Building climate resilience in the agricultural sector of Asia and the Pacific, Manila: Asian Development Bank.

Adger WN, Arnell NW and Tompkins EL (2005) Successful adaptation across scales. *Global Environ. Change* **15:** 77–86.

Aggarwal PK (2008) Global climate change and Indian agriculture: Impacts, adaptation and mitigation. *Indian Journal of Agricultural Sciences* **78:** 10–16.

Aggarwal PK, Bandyopadhyay SK, Pathak H, Kalra N, Chander S and Kumar S (2000) Analysis of yield trends of the rice-wheat system in north-western India. *Outlook Agriculture* **29:** 259–268.

Alexandrov V, Eitzinger J, Cajic V and Oberforster M (2002) Potential impact of climate change on selected agricultural crops in north-eastern Austria. *Global Change Biol* **8:** 372–389.

Amerasinghe P, Bhardwaj RM, Scott C, Jella K and Marshall F (2013) Urban wastewater and agricultural reuse challenges in India. Colombo, Sri Lanka: International Water Management Institute (IWMI). 36p. (IWMI Research Report 147). doi:10.5337/2013.200.

AMMA (2010) The International Science Plan for AMMA 2010-2020. Executive Summary, ISP Phase 2 Version 2 of 15/12/2010, http://www.ammainternational.org/IMG/pdf/ISP2_ Summary -EN-December2010.pdf.

Anonymous (2009) Alliance of the CGIAR Centers. Climate, agriculture and food security: A strategy for change.

Anonymous (2010). Department of Agriculture and Cooperation Ministry of Agriculture New Delhi. National Mission For Sustainable Agriculture: Strategies for Meeting the Challenges of Climate Change.

Asseng S *et al* (2013) *Nature Climate Change* **3:** 827–832.

Babu RC, Nguyen BD, Chamarerk V, Shanmugasundaram P, *others* (2003) Genetic analysis of drought resistance in rice by molecular markers: association between secondary traits and field performance. *Crop Sci* **43:** 1457–1469.

Barrett-Lennard EG, 2002. Restoration of saline land through revegetation. *Agricultural Water Management* **53:** 213–226.

Berry PM, Rounsevell MDA, Harrison PA and Audsley E (2006). Assessing the vulnerability of agricultural land use and species to climate change and the role of policy in facilitating adaptation. *Environ Sci Pol* **9:** 189–204.

Brinkman R and Sombroek WG (1996) The effects of global change on soil conditions in relation to plant growth and food production. In F. and W.G. Sombroek eds. Global Climate Change and Agricultural Production. Bazzaz FAO and Wiley Chichester UK 49–63.

Brunori G, Jiggins J, Gallardo R and Schmidt O (2008). New challenges for agricultural research: climate change, food security, rural development, agricultural knowledge systems. EU Commission, Standing Committee on Agricultural Research, Brussels.

Byjesh K, Kumar SN and Aggarwal PK (2010) Simulating impacts, potential adaptation and vulnerability of maize to climate change in India. *Mitigation and Adaptation Strategies for Global Change* **15(5):** 413–431.

Cameron M Pittelkow, Liang X, Linquis BA, Groenigen KJV, Lee J, Lundy ME, Gestel NV, Six J, Venterea RT and Kessel CV (2015). Productivity limits and potentials of the principles of conservation agriculture. Nature, Macmillan Publishers Limited vol 517: 365-368. doi:10.1038/nature13809.

Cavallini JM and Young LE (2002) Integrated systems for the treatment and recycling of wastewater in Latin America: Reality and Potential. Pan American Center for Sanitary Engineering and Environmental Sciences, IDRCPAHO/ HEP/ CEPIS Agreement. Lima.

C-CIARN Agriculture (Canadian Climate Impacts and Adaptation Research Network for Agriculture) (2003) Meeting the Challenges of Climate Change. Summary Report for C-CIARN Agriculture and C-CIARN Prairies Roundtable Session at *Grain World 2003*, February 25, Winnipeg, Manitoba, Canada.

Chen CC and McCarl BA (2001) An investigation of the relationship between pesticide usage and climate change. *Clim Change* **50:** 475–487.

Christensen JH and Christensen OB (2006) A summary of PRUDENCE model projections of changes in European climate by the end of this century. Climate Change (In Press).

Christensen JH, Hewitson B, Busuioc A, Chen A, Gao X, Held I, *et al* (eds) (2007) Climate Change 2007: The Physical Science Basis. Contribution of Working Group I to the Fourth Assessment Report of the Intergovernmental Panel on Climate Change pp 848–940, Cambridge University Press, Cambridge, UK: p854.

Clements R, Haggar J, Quezada A and Torres J (2011) Technologies for climate change adaptation– Agriculture sector. X. Zhu (Ed.) UNEP Risø Centre Roskilde.

Climate Focus (2011) Carbon market and climate finance for agriculture in developing countries. In: Report for Agriculture & Carbon Market Assessment, DFID Climate Change, Agriculture and Food. Climate Focus North America Inc., Washington, D.C., USA.

Cooper PJM, Dimes J, Rao KPC, Shapiro B, Shiferaw B and Twomlow S (2008) Coping better with current climatic variability in the rain-fed farming systems of sub-

Saharan Africa: An essential first step in adapting to future climate change? *Agriculture, Ecosystems, and Environment* **126:** 24–35.

Costanza R, Norton BG and Haskell BD (eds) (1992) Ecosystem Health: New goals for environmental management. Washington DC Island Press pp 269.

deNijs PJ, Berry NJ, Wells GJ and Reay DS (2014) Quantification of biophysical adaptation benefits from Climate-Smart Agriculture using a Bayesian Belief Network. *Sci Rep* 4: 6682; DOI:10.1038/srep06682.

Derpsch R (2001) Keynote: Frontiers in conservation tillage and advances in conservation practice, in Stott DE, Mo Htar, RH, Steinhart GC (Eds) Sustaining the global farm. Selected papers from the 10th International Soil Conservation Organization Meeting held May 24–29, 1999 at Purdue University and the USSA-ARS National Soil Erosion Research Laboratory.

Easterling WE, Aggarwal PK, Batima P, Brander KM, Erda L, Howden SM and Kirilenko A *et al* (2007) Food, fibre and forest products. Climate Change 2007: Impacts, adaptation and vulnerability. Contribution of Working Group II to the *Fourth Assessment Report of the Intergovernmental Panel on Climate Change,* ed ML Parry, OF Canziani, JP Palutikof, PJ van der Linden, CE Hanson 273–313. Cambridge: Cambridge University Press.

Faisal IM and Parveen S (2004) Food security in the face of climate change, population growth and resource constraints: implications for Bangladesh. *Environ. Manage* **34:** 487–498.

FAO (1995) Dimensions of Need – An atlas of food and agriculture, FAO, Rome.

FAO (2008) Enabling agriculture to contribute to climate change mitigation. FAO, Rome.

FAO (2010) "Climate-Smart" Agriculture – Policies, practices and financing for food security, adaptation and mitigation, FAO, Rome.

FAO (2010a) The wealth of waste: The economics of wastewater use in agriculture. By Winpenny J, Heinz I, Koo-Oshima S, and Case Study by Salgot M, Collado J, Hernandez F and Torricelli R. ISBN 978-92-5-106578-5.

FAO (2011) Save and Grow: A Policymaker's Guide to the Sustainable Intensification of Smallholder Crop Production 1–37.

Faroque MAA, Asaduzamman M and Hossain M (2013) Sustainable Agricultural Development under Climate Change in Bangladesh. *Journal of Science Foundation* **11(1)** ISSN 1728–7855.

Fischer G, Shah M and Velthuizen H van (2002) Climate change and agricultural vulnerability. Preprints World Summit on Sustainable Development Johannesburg 160.

Friedrich T, Derpsch R and Kassam A (2012) Overview of the global spread of conservation agriculture. *Field Actions Sci Rep* **6:** 1941.

Frison EA, Gold CS, Karamura EB and Sikora RA (1998) Mobilizing IPM for sustainable banana production in Africa Proceedings of a workshop on banana IPM held in Nelspruit, South Africa 23-28 November 1998, INIBAP.

Gadgil S and Gadgil S (2006) The Indian monsoon, GDP and agriculture. Economic and Political Weekly, November 25, 4887–4895.

GHF (Global Humanitarian Forum) (2009) Human Impact Report: Climate Change – The Anatomy of a Silent Crisis. GHF, Geneva.

Giannini A, Saravanan R and Chang P (2003). Oceanic forcing of Sahel rainfall on inter annual to inter decadal time scales. *Science* **7:** 1027–1030.

Godfray HCJ and Garnett (2010) Food security: the challenge of feeding 9 billion people. *Science* **327:** 812–818.

Godfray HCJ and Garnett T (2010) Food security and sustainable intensification. *Phil Trans R Soc B* **369:** 20120273.

Grassini P and Cassman KG (2012) High-yield maize with large net energy yield and small global warming intensity. *Proc Natl Acad Sci USA* **109:** 1074–1079.

GTZ (1998) Conserving natural resources and enhancing food security by adopting no-tillage. An assessment for the potential for soil-conserving production systems in various agro-ecological zones of Africa, GTZ Eschborn, Tropical Ecology Support Programme, TÖB Publication.

Gupta R, Seth A (2006) A review of resource conserving technologies for sustainable management of the rice-wheat cropping systems of the Indo-Gangetic plains (IGP). Crop Protection 26(3): 436–47.

Hassan R El-Ramad, Samia M El-Marsafawy and Lowell N Lewis (2013) Sustainable Agriculture and Climate Changes in Egypt. E. Lichtfouse (ed.), *Sustainable Agriculture Reviews* **12:** 41 DOI 10.1007/978-94-007-5961-9_2, © Springer Science+Business Media Dordrecht 2013.

Hobbs PR, Sayre K and Gupta R (2008) The role of conservation agriculture in sustainable agriculture. *Phil Trans R Soc B* **363**: 543–55.

Huang J and Wang Y (2014) Financing Sustainable Agriculture under Climate Change. *Journal of Integrative Agriculture* **13(4):** 698–712.

IFAD (2012) Climate-smart smallholder agriculture: What's different? IFAD Occasional Paper No. 3. Rome. http://www.ifad.org/pub/op/3.pdf.

IPCC (2007) Climate Change 2007: impacts, adaptation and vulnerability. Contribution of Working Group II to the Fourth Assessment Report of the Intergovernmental Panel on Climate Change, ML Parry, OF Canziani, JP Palutikof, PJ van der Linden & CE Hanson (eds), Cambridge, UK, Cambridge University Press, 273-313.

Islam MN, Baten MA, Hossain MS and Islam MT (2008) Impact of few important Climatic Parameters on Aman Rice Production in Mymensingh District. *J Environ Sci & Natural Resources* **1(2):** 49-54.

Javaid R Laghari (2013) Melting glaciers bring energy uncertainty Nature Macmillan Publishers Limited **502:** 617–618.

Jhansi SC and Mishra SK (2013) Wastewater Treatment and Reuse: Sustainability Options: *The Journal of Sustainable Development* **10(1):** 1 – 15.

Jim Kleinschmit (2009) Agriculture and Climate - The Critical Connection. Institute for Agriculture and Trade Policy. 2105 First Avenue South Minneapolis, Minnesota 55404 USA.

Jorgen Olesen E (no date) Climate change as a driver for European agriculture. Danish Institute of Agricultural Sciences.

Kaiser HM, Riha SJ, Wilks DS, Rossiter DG and Sampath R (1993) A farm-level analysis of economic and agronomic impacts of gradual climate warming. *Am J Agric Econ* **75:** 387–398.

Kandlikar M and Risbey J (2000) Agricultural impacts of climate change: If adaptation is the answer, what is the question? *Climatic Change* **45:** 529–539.

Karaba AS, Dixit R, Greco KR, Trijatmiko N, Marsch-Martinez A, Krishnan KN, Nataraja M, Udayakumar and Pereira A (2007) Improvement of water use efficiency in rice by expression of HARDY and Arabidopsis drought and salt tolerance gene. *Proceedings of the National Academy of Sciences (USA)* **104:** 15270–15275.

Karim Z, Hussain SG and Ahmed M (1996) Assessing impacts of climate variations on food grain production in Bangladesh *Water, Air, and Soil Pollution* **92:** 53–62.

Kostandini G, Mills BF, Omamo SW and Wood S (2009). *Ex ante* analysis of the benefits of transgenic drought tolerance research on cereal crops in low-income countries. *Agricultural Economics* **40:** 477–492.

Kulkarni SA, Reinders FB and Ligetvari F (2006) Global Scenario of Sprinkler in Micro-Irrigated Areas. Sept 10 – 16 2006, PWTC, Kuala Lumpur 7th International Micro Irrigation Congress.

Kumar S, Jakhar SR, Jangir CK, Dahiya S, Meena RS, Kumar P and Panghaal P (2017) Cereal production under changing climatic scenario. In: Agricultural, Allied Sciences & Biotechnology for Sustainability of Agriculture, Nutrition & Food Security, edited by Ratnesh Kumar Rao, P.K. Sharma, M. Raghuraman and J.K. Singh, Mahima Publications, ISBN: 978-81-926935-8-3, pp.112-114.

Kumar S, Lakhran H, Meena RS and Jangir CK (2017a) Current need of sustainable food and forage production to eliminate food and forage insecurity under current climatic era. *Forage Research* **44(1):** 165–173.

Kumar S, Sheoran S, Kakraliya SK, Kumar P and Meena RS (2016) Drought: A challenge for Indian farmers in context to climate change and variability. *Progressive Research – An International Journal,* **11(Special-IX)**: 6243–6246.

la Rovere R, Kostandini G, Abdoulaye T, Dixon J, Mwangi W, Guo Z and Bänziger M (2010) Potential impact of investments in drought tolerant maize in Africa. CIMMYT, Addis Ababa, Ethiopia.

Lakhran H, Kumar S and Bajiya R (2017) Crop diversification: An option for climate change resilience. *Trends in Biosciences* **10(2):** 516–518.

Lal R (2009) The potential for soil carbon sequestration, Focus 16, Brief 5, An Agenda for Negotiation in Copenhagen 2020 vision for food, agriculture and the environment. Washington DC: International Food Policy Research Institute.

LEISA (2007) Ecological Pest Management, LEISA Magazine, 23(4).

Lightfoot DA, Mungur R, Ameziane R, Nolte S, Long L, Bernhard K, Colter A, Jones K, Iqbal MJ, Varsa E and Young B (2007) Improved drought tolerance of transgenic maize *Zea mays* plants that express the glutamate dehydrogenase gene (gdh4) of *E. coli*. *Euphytica* **156:** 103–116.

Lin HX, Zhu MZ, Yano M, Gao JP, Liang ZW, Su WA, Hu XH, Ren ZH and Chao DY (2004) QTLs for Na+ and K+ uptake of the shoots and roots controlling rice salt tolerance. *Theor Appl Genet* **108:** 253–260.

Lobell DB, Burke M, Tebaldi C, Mastrandrea M, Falcon W and Naylor R (2008) Prioritizing climate change adaptation needs for food security in 2030. *Science* **319:** 607–610.

Lobell DB, Sibley A and Ortiz-Monasterio JI (2012) Extreme heat effects on wheat senescence in India. *Nature Climate Change* **2(3):** 186–189.

Ma X, Xu JC, Luo Y, Aggrawal SP and Li JT (2009) Response of hydrological processes to land cover and climate changes in Kejie watershed, south-west China. Hydrological Processes Online. DOI: 10.1002/hyp.7233.

Mahendra Dev S (2006) Climate Change, Rural Livelihoods and Agriculture (focus on Food Security) in Asia-Pacific Region. WP-2011-014 http://www.igidr.ac.in/pdf/publication/WP-2011-014.pdf

Mall RK, Singh R, Gupta A, Srinivasan G and Rathore LS (2006) Impact of Climate change on Indian agriculture: A review, *Climatic Change*, **78:** 445–478.

Mars R (2005) The Basics of Permaculture, Chelsea Green Publishing Company.

Martin FW and Sherman S (1992) Agroforestry Principles. ECHO (Educational Concerns for Hunger Organization).

Masle J, Gilmore SR and Farquhar GD (2005) The *ERECTA* gene regulates plant transpiration efficiency in *Arabidopsis*. Nature **436:** 866–870.

Mathur A (2009) "Ajay Mathur on Energy Efficiency in India", an interview with "The Climate Group", available at http://www.theclimategroup.org.

McCarthy JJ, Canziani OF, Leary NA, Dokken DJ and White KS (2001) Climate Change 2001: Impacts, Adaptation, and Vulnerability. Contribution of Working Group II to the Third Assessment Report of the Intergovernmental Panel on Climate Change. Cambridge University Press: Cambridge, UK.

Meena H, Meena RS, Lal R, Yadav GS, Mitran T, Layek J, Somanagouda B. Patil, Kumar S and Verma T (2017) Response of sowing dates and bio regulators on yield of clusterbean under current climate in alley cropping system in eastern U.P., India. *Legume Research*, **41(4):** 563–571.

Meena RS, Bohra JS, Singh SP, Meena VS, Verma JP, Verma SK and Sihag SK (2016) Towards the prime response of manure to enhance nutrient use efficiency and soil sustainability a current need: a book review. *Journal of Cleaner Production*. Elsevier Publication, **112(1):** 1258–1260.

Meena RS, Gogoi N and Kumar S (2017) Alarming issues on agricultural crop production and environmental stresses. *Journal of Cleaner Production*. Elsevier Publication, **142:** 3357–3359.

Milder JC, Majanen T and Scherr SJ (2011) Performance and Potential of Conservation Agriculture for Climate Change Adaptation and Mitigation in Sub-Saharan Africa. Eco agriculture partners, CARE, WWF and World Agroforestry Centre.

Milestad R and Darnhofer I (2003) Building farm resilience: The prospect and challenges of organic farming. *Journal of Sustainable Agriculture* **22(3):** 81–97.

Minguez MI, Ruiz-Ramos M, Díaz-Ambrona CH, Quemada M and Sau F (2006) First-order impacts on winter and summer crops assessed with various high-resolution climate models in the Iberian peninsula. Climate Change (in press).

Mutegi J K, Mugendi DN, Verchot LV and Kung'u JB (2008) Combining Napier grass with leguminous shrubs in contour hedgerows controls soil erosion without competing with crops. Agro-forestry Systems, DOI 10.1007/s10457-008-9152-3.

NAAS (2013) Climate Resilient Agriculture in India. Policy Paper No. 65, National Academy of Agricultural Sciences, New Delhi: 20 p.

NarayanmoorthyA (no date) Drip and Sprinkler Irrigation in India: Benefits, Potential and Future Directions. Available: http://www.iwmi.cgiar.org/Publications/Other/PDF/Paper%2015% 20of%20NRLP%20series%201.pdf.

Neil Savage (2013) Predictive yield. *Nature* **501:** 10–11.

Nelson GC, Rosegrant MW, Koo J, Robertson R, Sulser T, Zhu T, Ringler C, Msangi S, Palazzo A and Batka M *et al* (2010) The costs of agricultural adaptation to climate change. In: Development and Climate Change Discussion Paper 4. The World Bank Washington DC USA.

Neumann K, Verburg PH, Stehfest E and Muller C (2010) The yield gap of global grain production: a spatial analysis. *Agric Syst* **103:** 316–326.

Ngigi SN, Rockström J and Savenije HHG (2006) Assessment of rainwater retention in agricultural land and crop yield increase due to conservation tillage in Ewaso Ngiro river basin, Kenya. *Physics and Chemistry of the Earth* **31:** 910–18.

Olesen JE and Bindi M (2002) Consequences of climate change for European agricultural productivity, land use and policy. *Eur J Agron* **16:** 239–262.

Olesen JE, Carter TR, Diaz-Ambrona CH, Fronzek S, Heidmann T, Hickler T, Holt T, Minguez MI, Morales P, Palutikov J, Quemada M, Ruiz-Ramos M, Rubæk G, Sau F, Smith B and Sykes M (2006) Uncertainties in projected impacts of climate change on European agriculture and ecosystems based on scenarios from regional climate models. Climate Change.

Olive Heffernan (2013) The Dry Facts.Zuma Press/Alamy/Sean Sexton Collection/Corbis/Corbis/Alain Nogues/Sygma/Corbis. Nature 501 Published 26 September 2013.

Ortiz R, Iwanaga M, Reynolds MP, Wu H and Crouch J (2007) Overview on crop genetic engineering for drought prone environments. Journal of Semi-Arid Tropical Agricultural Research 4 http://www.icrisat.org/jornal/SpecialProject/sp3.pdf (9 Sept 2008).

Ortiz R, Sayre KD, Govaerts B, Gupta R, Subbarao GV, Ban T, Hodson D, Dixon JM, Ortiz-Monasterio JI and Reynolds M (2008) Climate change: Can wheat beat the heat? *Agriculture, Ecosystems and Environment* **126:** 46–58.

Passenier A, Lak M (2009) Scarcity and Transition - Research Questions for Future Policy. Ministry of Housing, Spatial Planning and the Environment and Ministry of Foreign Affairs, The Hague.

Peng S, Huang J, Sheehy JE, Laza RC, Visperas RM, Zhong X, Centeno GS, Khush GS and Cassman KG (2004) Rice yields decline with higher night temperature from global warming, *Proc Natl Acad Science* **101:** 9971–9975.

Price Waterhouse Coopers (2011). Making climate finance work for smallholder farmers in sub-saharan Africa. Available at: http://www.pwc.co.uk/assets/ pdf/making-climate-finance-work-for-smallholder-farmers-in-sub-saharanafricapdf.

Ravindran KC, Venkatesan K, Balakrishnan V, Chellappan KP and Balasubramanian T (2007) Restoration of saline land by halophytes for Indian soils. *Soil Biology and Biochemistry* **39(10):** 2661–2664.

Ray DK, Jemes, S, Grahm K, MacDonald and Paul CW (2015) Climate variation explains a third of global crop yield variability. Nature Communication 6:5989 doi: 10.1038/ncomms6989.

Recha J, Kapukha M, Wekesa A, Shames S and Heiner K (2014) Sustainable Agriculture Land Management Practices for Climate Change Mitigation: A training guide for smallholder farmers. Washington DC Eco Agriculture Partners.

Rivero RM, Kojima M, Gepstein A, Sakakibara AH, Mittler R, Gepstein S and Blumwald E (2007) Decayed leaf senescence induces extreme drought tolerance in a flowering plant. *Proceedings of the National Academy of Sciences (USA)* **104:** 19631–19636.

Romanazzi AL, Gentile FL, Polemio M and Trisorio Liuzzi G (no date) Sustainable Agriculture Development under Climate Change in a Karst Coastal Region: The Case of Salento (Southern Italy).

Rosegrant WM, Ximing C and Cline SA (2002) World Water and Food to 2020: Dealing with Scarcity.

Rosenzweig C, Iglesias A, Yang XB, Epstein PR and Chivian E (2001) Climate change and extreme weather events: implications for food production, plant diseases and pests. *Global Change and Human Health* **2:** 90–104.

Rosenzweig C and Parry ML (1994) Potential impact of climate change on World Food Supply. *Nature* **367:**133–138.

Roshetko JM, Lasco RD and de Los Angeles MS (2007) Small holder agroforestry systems for carbon storage. *Mitigation and Adaptation Strategies for Global Change* **12(2):** 219–42.

SCAR (2011) Sustainable food consumption and production in a resource-constrained world 3rd SCAR Foresight Exercise. Report 2011 European Commission Directorate-General for Research and Innovation Directorate E - Biotechnologies, Agriculture and Food Unit E.4 – Agriculture, Forests, Fisheries, Aquaculture.

Seckler D, Amarasinghe U, Molden D, de Silva R and Barker R (1998) World water demand and supply, 1990 to 2025: Scenarios and issues. Research Report 19. Colombo, Sri Lanka: International Water Management Institute.

Serraj R and Siddique KHM (2012) Conservation agriculture in dry areas. *Field Crops Res* **132:** 1–6.

Shetty PK, Ayyappan S and Swaminathan MS (2013) Climate change and sustainable food security. National Institute of Advanced Studies Indian Institute of Science Campus Bangalore - 560 012. ISBN: 978-81-87663-76-8.

Sinclair TR and Muchow RC (2001) Systems analysis of plant traits to increase grain yield on limited water supplies. *Agron J* **93:** 263–270.

Singh S, Kattarkandi B, Deka S and Choudhary R (2010) Impact of climatic variability and climate change on maize productivity in north India. *Current Advances in Agricultural Sciences* **2(1):** 5–9.

Slingo JM, Challinor AJ, Hoskins BJ and Wheeler TR (2005) Introduction: food crops in a changing climate. *Phil Trans R Soc B* **360:** 1983–1989.

Smit B, Burton I, Klein R and Wandel J (2000) An anatomy of adaptation to climate change and variability. *Climatic Change* **45:** 223–251.

Smit B and Wandel J (2006) Adaptation, adaptive capacity, and vulnerability. *Global Environ Change* **16:** 282–292.

Smith P (2005) Agriculture and Climate Change: An agenda for negotiation in Copenhagen, IFPRI Focus 16.

Snyder RL and Melo-Abreu JP (2005) Frost protection: fundamentals, practice, and economics – Volume 1. FAO, Rome.

Soora NK, Singh AK, Agarwal PK, Rao VUM and Venkateswarlu B (2012) Climate Change and Indian Agriculture: Impact, Adaptation and Vulnerability – Salient Achievements from ICAR Network Project IARI Publication 32 p.

Sorrenson WJ, Duarte C and López-Portillo J (1998) Economics of non-till compared to conventional cultivation Systems on small farms in Paraguay, policy and investment implications, Report Soil Conservation Project MAG-GTZ.

Soussana JF (2010) Quelles recherché pour adapter l' agriculture et les ecosystem esanthropisés au change mentclimatique? INRA-ANR, L'atelier de réflexion prospective ADAGE.12 p.

SSAPRI (Stimulating Sustainable Agricultural Production through Research & Innovation) (no date). A Research Agenda to guide public investment in primary agricultural research in Ireland. 2http://www.agriculture.gov.ie/media/migration/agrifoodindustry/foodharvest2020/2020FoodHarvestEng240810.pdf.

Tait J (2001) Science, governance and multi functionality of European agriculture. *Outlook Agric* **30:** 91–95.

Tao F, Yokozawa M, Zhang Z, Hayashi Y, Grassl H and Fu C (2004) Variability in climatology and agricultural production in China in association with the East Asia summer monsoon and El Niño South Oscillation. *Climate Res.* **28:** 23–30.

Thomas D, Osbahr H, Twyman C, Adger N and Hewitson B (2005) Adaptive: Adaptations to climate change amongst natural resource-dependent societies in the developing world: Across the Southern African climate gradient. Tyndall Centre for Climate Change Research Technical Report 35, November.

Tilman D, Cassman Kenneth G, Matson Pamela A, Naylor R and Polasky S (2002) Agricultural sustainability and intensive production practices. Nature Publishing Group **418:** 671–677.

Tubiello FN and Rosenzweig C (2008) Developing climate change impact metrics for agriculture. *Integrated Assessment Journal* **8(1):** 165–184.

Tubiello FN and van der Velde M (no date) Land and water use options for climate change adaptation and mitigation in agriculture. SOLAW Background Thematic Report - TR04A. GET-Carbon, New York, USA.

Tuck G, Glendining MJ, Smith P, House JI and Wattenbach M (2006) The potential distribution of bioenergy crops in Europe under present and future climate. *Biomass Bioener* **30:** 83–197.

UNDP (United Nations Development Programme), Bureau for Crisis Prevention and Recovery (BCPR) (2013) Climate Risk Management for Sustainable Crop Production in Uganda: Rakai and Kapchorwa Districts. New York NY: UNDP BCPR.

UNEP and SEI (Stockholm Environment Institute) (2009) Rainwater harvesting: a lifeline for human wellbeing, United Nations Environment Programme and Stockholm Environment Institute in 2009

Varshney RK, Bansal KC, Aggarwal PK, Datta S and Craufurd PQ (2011) Agricultural biotechnology for crop improvement in a variable climate: hope or hype? *Trends in Plant Science* **16(7):** 363–371.

Venkateswarlu B and Shanker AK (2012) Dryland agriculture: bringing resilience to crop production under changing climate. pp 19–44. In: Crop Stress and its Management: Perspectives and Strategies. Springer Netherlands.

Wall E and Smit B (2005) Climate Change Adaptation in Light of Sustainable Agriculture. *Journal of Sustainable Agriculture* 27(1) by The Haworth Press, Inc. doi:10.1300/J064v27n01_07.

Wang Q (2015) China's scientists must engage the public on GM. Nature Macmillan Publishers Limited 519(7).

Weber DJ, Ansari R, Gul B and Khan MA (2007) Potential of halophytes as source of edible oil. Short communication. *Journal of Arid Environments* **68:** 315–321.

Wilson A and Tyrchniewicz A (1995) Agriculture and Sustainable Development: Policy Analysis on the Great Plains. International Institute for Sustainable Development Winnipeg Manitoba. p. 108.

World Bank (2007) World Development Report 2008: Agriculture for Development. The World Bank Washington DC USA.

World Bank (2009) Agricultural Development under a Changing Climate: Opportunities and Challenges for Adaptation. Internet www.worldbank.org/rural.

World Bank (2009a) Monitoring and Reporting on Financial Flows Related to Climate Change. The World Bank Washington DC USA.

World Bank (2010) Improving wastewater use in agriculture: An emerging priority. Energy Transport and Water Department, Water Anchor (ETWWA). 30 June, 2010.

Xu K, Xu X, Fukao T, Canlas P, Rodriguez RM, Heuer S, Ismail AM, Serres JB, Ronald PC and Mackill DJ (2006) Sub1A is an ethylene response- factor-like gene that confers submergence tolerance to rice. *Nature* **442:** 705–708

Zhang XC and Nearing MA (2005) Impact of climate change on soil erosion, runoff, and wheat productivity in central Oklahoma. *Catena* **61(2/3):** 185–195.

Zhang XJ and Liu W (2005) Simulating potential response of hydrology, soil erosion, and crop productivity to climate change in Changwu Table land on southern Loess Plateau of China. *Agricultural and Forest Meteorology* **131:** 127–142.

Ziervogel (2001) Global science, local problems: Seasonal climate forecast use in a Basotho village, southern Africa, Gina Ziervogel, Environmental Change Institute. Paper prepared for presentation at the Open Meeting of the Global Environmental Change Research Community, Rio de Janeiro October 6–8 2001.

■■■■

12 Present Status and Future Prospects of Organic Farming in India

*Raghuveer Singh[1]**, N.K. Jat[2*], N. Ravisankar[1], Sudhir Kumar[3*], T. Ram[3*] and R.S. Yadav[4*]*

[1]ICAR-Indian Institute of Farming Systems Research, Modipuram, Meerut (UP) – 250 110
[2]ICAR-Central Arid Zone Research Institute, Jodhpur (Rajasthan)- 342 003
[3]ICAR- Indian Agriculture Research Institute, New Delhi- 110 012
[4]ICAR- IISWC RC, Datia (Madhya Pradesh)- 475 661
**Formerly associated with ICAR-Indian Institute of Farming Systems Research, Modipuram, Meerut (UP)*
***Corresponding Author's E mail- rsbicar@gmail.com*

ABSTRACT

Every technology or interference in the natural ecosystem has its own side effect; from last half a century we are using the pesticides and herbicides indiscriminately to raise the crop production, but now its ill effect visible for us. Story of Malwa region of Punjab (cancer belt) and Kerala's endosulfan tragedy are epitome example of it. Now we achieve the self-sufficiency in food grain production our next motto to serve healthy food to people. We are looking for alternative nature based chemical free farming and that is organic farming. It also helps to get the premium price by exporting the chemical free agriculture commodity to the developed countries, day by day demand for healthy and residue free food increase all over the world as well as country itself. People are now ready to pay extra amount if they are ensured that they are purchasing geniunely raised food through natural system. Thus strengthening the certification process in country is must.

***Keywords*:** Conventional farming, Future prospects, India, Organic farming, Present status

1. INTRODUCTION

Organic farming based on "Nature can provides for everyone's need but not for greed"...........Mahatma Gandhi

It is definitely true that India had witnessed a tremendous growth in agricultural production in the era of green revolution. Food grain production, which stood at a mere 50 million tons at the time of independence, had increased almost five and half times to 273.38 million tons by the end of 2016–17 (Press Information Bureau, GOI, 2017) from 159.59 million hectares of cultivated area in country (Agriculture Census, 2010–11). The technologies involved during the inception of green revolution supported by policies and further propelled by agrochemicals, machinery and irrigation were the main driving forces for the enhanced agricultural production and productivity (Roychowdhury *et al.*, 2013). Despite the fact that the food security of India was definitely addressed by these technologies (Charyulu and Biswas, 2010), an important setback was that the farmers using these technologies were still had to depend upon the purchased inputs. With manufacturing of fertilizers and pesticides as the two major inputs of Green Revolution (GR) technologies, an important point of consideration was the need for fossil fuels and/or expensive energy which are associated with serious environmental and health problems. In last 50 years we are using heavy amount of fertilizers and pesticides and we already reach on plateau and diminishing low of return start to work (Venkateswarlu *et al.*, 2008), so we need to apply more input (fertilizer and pesticides) to get small raise in production which cause second generation problem and few of such epitome examples are some regions of Punjab (cancer belt of country) and endosulfan story of cashew plantations area in Kerala (proving finding of Rachel Carson's Silent Spring was published in 1962). Insecticides and herbicides in ideal condition lethal for target group only, for non-target group and human it is safe but this principle is not followed strictly and indiscriminate use of these chemicals put human life and ecosystem health on verge (Aktar *et al.*, 2009). All these thing and un-sustainability issue associated with modern agriculture force us to look back (Balachandran, 2004) in history to know either we are not doing any mistake by depending on off farm inputs because crop production is a recycle system of nature by putting too much off farm input we are making it fragile day by day. One of such natural, recyclable and sustainable approach of farming is Organic farming. It is the effective and cost efficient way to achieve sustainable development in the agriculture sector (IFOAM, 2010). Organic source of nutrient also helps to combat with the problem of multi nutrient deficiency and low organic content in our soil which is affecting productivity of major food crops at farmer field (Singh *et al.*, 2017).

Organic farming is a method of farming system which primarily aimed at cultivating the land and raising crops in such a way, as to keep the soil alive and in good health by use of organic wastes (crop, animal and farm wastes, aquatic wastes) and other biological materials along with beneficial microbes (biofertilizers) to release nutrients to crops for increased sustainable production in an eco-friendly pollution free environment (Narayanan, 2005; Guruswamy and Gurunathan, 2010; Makadia and Patel, 2015).

As per the definition of the United States Department of Agriculture (USDA), study team on organic farming "organic farming is a system which avoids or largely excludes the use of synthetic inputs (such as fertilizers, pesticides, hormones, feed additives etc.) and to the maximum extent feasible rely upon crop rotations, crop residues, animal manures, off-farm organic waste, mineral grade rock additives and biological system of nutrient mobilization and plant protection".

FAO suggested that "Organic agriculture is a unique production management system which promotes and enhances agro-ecosystem health, including biodiversity, biological cycles and soil biological activity, and this is accomplished by using on-farm agronomic, biological and mechanical methods in exclusion of all synthetic off-farm inputs".

2. MAIN PRINCIPLES OF ORGANIC FARMING

The main principles of organic farming by (Chandrashekar, 2010) are as follows:

- To work as much as possible within a closed system, and draw upon local resources.
- To maintain the long-term fertility of soils.
- To avoid all forms of pollution that may result from agricultural techniques.
- To produce foodstuffs of high nutritional quality and in sufficient quantity.
- To reduce the use of fossil energy in agricultural practice to a minimum.
- To give livestock conditions of life that confirm to their physiological need.
- To make it possible for agricultural producers to earn a living through their work and develop their potentialities as human being.

All the above principles based on the four ethical principles (Principle of Health, Principle of Care, Principle of Fairness and Principle of Ecology) IFOAM, 2005.

3. THE MAIN PILLARS OF ORGANIC FARMING

The main pillars of organic farming (Roychowdhury *et al.*, 2013) are as below –

- Organic threshold standards.
- Reliable mechanisms regarding certification and regulatory affairs.

- Technology packages.
- Efficient and feasible market network.

4. ORGANIC FARMING IN INDIA

4.1. Historical background

The concept of organic agriculture is not alien to India. In fact, the first scientific approach to organic farming dates back to the *Vedas* of the later *Vedic* period, the essence of which is to live in harmony with, rather than exploit, Mother Nature. There is brief mention of several organic inputs in our ancient literatures like Rigveda, Ramayana, Mahabharata, Kautilya Arthasashthra etc. In fact, organic agriculture has its roots in traditional agricultural practices that evolved in countless village's and farming communities over the millennium. Major milestones in the area of organic farming are presented in Table 1.

Table 1. Historical perspective of organic farming in India

Ancient period	Oldest practice 10000 years old, dating back to Neolithic age, practiced by ancient civilization like Mesopotamia, Hwang-Ho basin etc.
Ramayana	All dead things - rotting corpse or stinking garbage returned to earth are transformed into wholesome things that nourish life. Such is the alchemy of mother earth - as interpreted by C. Rajagopalachari
Mahabharata (5500 BC)	Mention of Kamadhenu, the celestial cow and its role on human life and soil fertility.
Kautilya Artha-shastra (300 BC)	Mentioned several manures like oil cake, excreta of animals.
Brihad-Sanhita (by Varahmihir)	Described how to choose manures for different crops and the methods of manuring.
Rig Veda (2500–1500 BC)	Mention of organic manure in Ria Veda 1, 161, 10, 2500–1500 BC, is Green Manure in Atharva Veda II 8.3, (1000 BC). In Sukra (IV, V, 94, 107–112) it is stated that to cause healthy growth, the plant should be nourished by dungs of goat, sheep, cow, water as well as meat. A reference of manure is also made in Vrksayurveda by Surpala (manuscript, oxford, No 324 B, Six, 107-164)
Holy u;-an (590 AD)	At least one third of what you take out from soils must be returned to it implying recycling or post-harvest residue.

(*Source*: Bhattacharyva and Chakraborty, 2005)

More recently, Mahatma Gandhi pioneered organic farming through his constructive programmes in several locations in India. It was the aggressive promotion of the green revolution, which led to a reverse in the spread of the

Gandhian movement's organic farming programme. There is evidence to suggest that small farmers, especially in the Third World, are more likely to apply agricultural practices, such as crop rotations and mixed cropping, which are the essential building blocks of organic farming. They also usually combine agriculture with livestock rearing and utilize the manure to replenish the soil fertility.

Empirical evidence further suggests that while conventional agriculture goes better with large holdings, organic farming functions better in small farms. A study by Gupta and Verma (1997), comparing grain production in organic *vis-à-vis* conventional methods, observed that as farm size increases, the advantages of organic rotation become less visible. Further, the study reported that on a smaller scale, organic farming was more profitable and productive than conventional farming.

4.2 Government initiative

The lucrative market of the developed world has so far acted as the primary driving force behind the development of the 'certified organic' sector, which is still in a nascent stage in India. According to one estimate, in 1999 merely 0.001 per cent of the total agricultural land in India was under certified organic cultivation. It is predominantly the NGOs and people's organizations that have been spearheading organic agriculture movement in different parts of the country during the last two decades.

The Government of India has set up a special cell under the Agricultural and Processed Food Export Development Authority (APEDA) of the Ministry of Commerce and Industries (MOCI). The MOCI has come out with the 'National Programme of Organic Products' (NPOP) in 2000, and the 'India Organic' logo in 2002. While these initiatives have been undertaken to promote exports of Indian organic products, the Department of Agriculture and Cooperation has formulated a 'National Project on Organic Farming' to promote organic agriculture as part of an exercise to curb the use of chemical pesticides and make agricultural activity more eco-friendly. Shifting to organic farming, even if it promises higher returns in terms of better prices and international acceptability in the long run, may not be preferred by the majority of farmers as they are dependent on the farm for livelihood and any departure would affect them immediately. To make majority of small farmers shift to organic farming, several subsidies have to be given on organic inputs. Such initiatives are a step in the right direction, but they should be weighed against the huge subsidies that the Central Government has been providing for the production and import of chemical fertilizers and pesticides. There is a great potential for organic farming to flourish in this country and given an appropriate institutional and policy framework, it will not be very difficult to promote the existing *'de-facto organic'* farms to the category of certified organic farms. This would enable the small farmers to take advantage of the lucrative market for certified organic products in the developed world, which could directly contribute towards the improvement of their economic well-being.

4.2.1 *Status of organic farming in India*

Extent of Organic Area and Production

At present in the world Australia at number one position with total 12,29,290 ha area under organic agriculture, in terms of per centage of total land Italy take first position with 9 per cent of total agriculture land cover under organic and with 17,557 number of organic farms Germany recorded highest organic farms in the world. India stand on 14th rank in the world with 528,171 ha area comes under organic agriculture and it covers 0.3 per cent area under organic agriculture of total agriculture land and 44,926 total number of organic farms in the country. The Asian countries together currently account only 7 per cent of the total global organic land, China and India being major contributors (Musa *et al.,* 2015).

Table 2. Rank (on basis of total area) under organic agriculture

Rank (on basis of total area)	Country	Area under organic agriculture (ha)	Percentage of total agriculture land	Number of organic farms
1	Australia	12,294,290	2.8	1550
2	China	2,300,000	0.4	1600
3	Argentina	2,220,489	1.7	1486
4	USA (2005)	1,620,351	0.5	8493
5	Italy	1,148,162	9.0	45,115
6	Uruguay	930,965	6.1	630
7	Spain	926,390	3.7	17,214
8	Brazil	880,000	0.3	15,000
9	Germany	825,539	4.8	17,557
10	UK	604,571	3.8	4485
11	Canada	604,404	0.9	3571
12	France	552,824	2.0	11,640
13	**India**	**528,171**	**0.3**	**44,926**
	World	30,418,261	0.65	718,744

(*Source*: Ramesh *et al.,* 2010)

According to APEDA 2013–14, India ranks 10th in the world in terms of cultivable land under organic certification. The certified area includes 15 per cent cultivable area with 0.72 million Hectare and rest 85 per cent (3.99 million

Hectare) is forest and wild area for collection of minor forest produces. The total area under organic certification is 4.72 million Hectare. India produced around 1.24 million MT of certified organic products which includes all varieties of food products namely Sugarcane, Cotton, Oil Seeds, Basmati rice, Pulses, Spices, Tea, Fruits, Dry fruits, Vegetables, Coffee and their value added products. The production is not limited to the edible sector but also produces organic cotton fibre, functional food products etc. Among all the states, Madhya Pradesh has covered largest area under organic certification followed by Himachal Pradesh and Rajasthan.

Table 3. The status of organic production in India (2013–2014)

Total area under certified organic cultivation	0.72 M ha
Forest and wild area for collection of minor forest produces	3.99 M ha
The total area under organic certification	4.7 M ha
Total production	1.24 million MT
Total quantity exported	177766 MT
Value of total export	1328.61 crores

(Source: APEDA. 2014)

Commodities produced under organic farming

Commodities produced under organic farming in India mainly include tea, basamati rice, cardamom, black pepper, ginger turmeric, mango cashew and herbal extracts.

Table 4. Major products produced in India by organic farming

Type	**Products**
Commodity	Tea, coffee, rice, wheat
Spices	Cardamom, black pepper, white pepper, ginger, turmeric, vanilla, mustard, tamarind, clove, cinnamon, nutmeg, mace, chili
Pulses	Red gram, black gram
Fruits	Mango, banana, pineapple, passion fruit, sugarcane, orange, cashew nut, walnut
Vegetables	Okra, brinjal, garlic, onion, tomato, potato
Oil seeds	Sesame, castor, sunflower
Others	Cotton, herbal extracts

(*Source*: Salvador and Katke, 2003)

Details of herbal extracts

Table 5. Details of herbal extract commonly exported

Sr. No.	Comman name	Scientific name
1	SafedMusali, Dry roots	*Chlorophytum borivillianum*
2	Ashwagandha, Dry roots	*Withania somnifera* (Whole Powder)
3	Ginger root whole & Powder	*Zingiber officinale*
4	Kalmegh, whole plant (leaf)	*Andrographis paniculata*
5	Shatavari, yellow dry roots	*Asparagus racemosa* (Powder)
6	Kapikacchu	*Mucuna pruriens* (Seed)
7	Amalaki	*Emblica officinalis* (Fruit no stone)
8	Haritaki	*Terminalia chebula* (Fruit no stone)
9	Bibhitaki	*Terminalia belerica* (Friut no stone)
10	Guduchi	*Tinosporia cordifolia*
11	Pushkaramool	*Inula racemosa* (Root)
12	Vacha, dry root	*Acorus calamus*
13	Shankapushpi	*Canscora decussata*
14	Gurmar	*Gymnema sylvestre*

(*Source*: Salvador and Katke, 2003)

Products for which Indian production has a comparative advantage

India being a country with different agro-climatic zones, each state produces its own specialty products (Shetty *et al.,* 2014). Based on soil and climate, India is divided into 21 agro-ecological zones. Products for which production in India has a comparative advantage are given in Table 6.

Table 6. Products for which Indian production has a comparative advantage

Product	Season	States	Major Locations
Tea	Throughout the year	Assam, West Bengal, Uttranchal	Darjeeling, Guwahati, Dehradun
Spices	Throughout the year	Kerala, Tamil Nadu, Karnataka	Cochin, Coimbatore, Idduki, Coorg
Coffee	Throughout the year	Kerala, Tamil Nadu, Karnataka	Coimbatore, Coorg, Wayanadu, Peeremade

Product	Season	States	Major Locations
Rice	*Kharif & Rabi*	Punjab, Haryana, Assam, Maharashtra, Tamil Nadu	Amritsar, Jalandhar, Darrang, Ratnagiri, Kanchipuram, Thiruvallur
Wheat	*Kharif & Rabi*	Punjab, Haryana, Uttar Pradesh	Ambala, Patiala, Bhatinda, Faridkot
Vegetables	Throughout the year	All India	Various locations
Fruits	Throughout the year	All India	Various location
Cotton	*Kharif*	Maharashtra, Gujarat Madhya Pradesh	Akola, Amravati, Amreli, Kheda, Indore

**Kharif* and Rabi are two growing seasons in India. *Kharif*is essentially from May to September and Rabi is from November to March (*Source*: Salvador and Katke 2003).

Besides the broad range of products and the seasonal advantages mentioned above, India has other comparative advantages for organic production:

i. India is strong in high quality production of certain crops like tea, some spices, rice specialties, ayurvedic herbs etc.

ii. India has a rich heritage of agricultural traditions that are suitable for designing organic production systems. Sophisticated crop rotation or mixed cropping patterns, for example the famous agro-forestry systems of the Western Ghats, facilitate the management of pests, diseases and nutrient recycling. Botanical preparations, some of which originate from the ancient Veda scripts, provide a rich source for locally adapted pest and disease management techniques. The widespread cultivation of legume crops facilitates the supply of biologically fixed nitrogen.

iii. In several regions of India agriculture is not very intensive as regards the use of agro-chemicals. Especially in mountain areas and tribal areas, use of agro-chemicals is rather low, which facilitates conversion to organic production. On these marginal soils, organic production techniques have proved to achieve comparable or in some cases (especially in the humid tropics) even higher yields than conventional farming.

iv. Compared to input costs, labour is relatively cheap in India, thus favouring the conversion to less input-dependent, but more labour-intensive production systems, provided they achieve sufficient yields.

v. The NGO sector in India is very strong and has established close linkages to a large numbers of marginal farmers. Many NGOs are engaged in promotion of organic farming and provide training, extension services information and marketing services to farming communities.

vi. The Indian Government has realized the potential significance of organic agriculture for the country and has recently started to support organic agriculture on a large scale and on various levels. A national regulatory framework (standards, accreditation regulations) has already been passed in 2000. There are various schemes and events to support and facilitate exports of organic products (*e.g.* the large conference 'Indian Organic Products – Global Markets' held in Delhi in December 2002, mainly sponsored by the Indian Government). The Ministry of Agriculture announced that various forms of support for organic producers, processors and traders were to be included within the latest five-year plan.

4.2.2. *Organic certification and marketing in India*

4.2.2.1. *Accreditation boards in India*

Country has recently started to support organic agriculture on a large scale and on various levels. A national regulatory framework (standards, accreditation regulations) has already been passed in 2000. There are various schemes and events to support and facilitate exports of organic products (*e.g.,* the large conference 'Indian Organic Products – Global Markets' held in Delhi in December 2002, mainly sponsored by the Indian Government). The Ministry of Agriculture announced that various forms of support for organic producers, processors and traders were to be included within the latest five-year plan. The Ministry of Commerce has identified six organisations as accreditation agencies of organic products, they are (1) Agricultural and Processed Food Products Export Development Authority (APEDA), (2) Tea Board, (3) Spices Board, (4) Coconut Development Board, (5) Directorate of Cashew and Cocoa, and (6) Coffee Board. These accreditation boards give permission to certifying agencies for certifying organic products, following the prescribed norms. Certification through these boards and agencies has been made compulsory, particularly for export market, as 'the Government of India has issued a public notice according to which no organic products may be exported unless they are certified by an inspection and certifying agency duly accredited by one 22 of the accreditation agencies designated by the Government of India' (Salvador and Katke, 2003). Several certifying agencies are functioning in India.

4.2.2.2. *List of Certifying agency in India who providing certificates*

Name of the Certification Agency With Certification Mark	Name of the Certification Agency With Certification Mark
Bureau Veritas Certification India (BVCI) Pvt. Ltd.	Indian Organic Certification Agency (INDOCERT)
ECOCERT India Pvt. Ltd.	Lacon Quality Certification Pvt. Ltd.
IMO Control Pvt. Ltd.	OneCert Asia Agri Certification (P) Ltd.
SGS India Pvt. Ltd.	APOF Organic Certification Agency (AOCA)
Control Union Certification	Rajasthan Organic Certification Agency (ROCA)

Name of the Certification Agency With Certification Mark	**Name of the Certification Agency With Certification Mark**
Uttarakhand State Organic Certification Agency (USOCA)	Vedic Organic Certification Agency
ISCOP (Indian Society for Certification of Organic Products)	Chhattisgarh Certification Society, India (CGCERT)
Food Cert India Pvt. Ltd	Tamil Nadu Organic Certification Department (TNOCD)
Aditi Organic Certifications Pvt. Ltd	Intertek India Pvt. Ltd.
Madhya Pradesh State Organic Certification Agency	Natural Organic Certification Agro Pvt. Ltd.

Name of the Certification Agency With Certification Mark	**Name of the Certification Agency With Certification Mark**
Biocert India Pvt. Ltd	Fair Cert Certification Services Pvt. Ltd.
Odisha State Organic Certification Agency (OSOCA)	Gujarat Organic Products Certification Agency (GOPCA)
Uttar Pradesh State Organic Certification Agency	

(*Source*: APEDA, 2014)

4.3 Big Consumers of Indian Organic Products

India exported 135 products last year (2013–14) with the total volume of 177766 MT, with total coast 1328.61 crores. In last three year trend it is found that European Union, USA. and Canada contributes more than 91 per cent of total export from India if Switzerland, Japan and Australia included in this list than it cross over 96 per cent total volume exported from India are consumed by these countries.

4.4. Export Trend

A general trend observed that U.S.A. and Canada nowadays giving fight to European Union breaking monopoly as organic product importer from India. If we observe last three year tend, in 2011–12 total share of European Union is 60 per cent which reduced to 42 per cent in 2013–14 but total remain same above 90 per cent for European Union, USA and Canada.

Table 7. Export volume and value from country in last three year

Countries	2011–12		2012–13		2013–14	
	Export volume (MT)	Value (Rs. crores)	Export volume (MT)	Value (Rs. crores)	Export volume (MT)	Value (Rs. crores)
European Union	51138.8	505.29	82835.37	678.51	56946.72	553.85
U.S.A.	37630.23	197.94	34292.35	228.72	74942.72	498.83
Canada	19848.91	66.66	33645.80	146.05	38545.57	182.41
Switzerland	2161.51	21.04	3455.27	27.57	4306.56	33.88
Japan	232.77	8.79	199.22	11.11	309.07	16.12
Australia	349.14	5.15	468.26	6.60	749.95	14.58
Rest	4055.78	34.47	5380.68	57.25	1964.66	28.94
Total	115417.2	839.3	160276.9	1155.8	177765.2	1328.6

(*Source*: Ministry of Commerce and Industries, GOI, 2014)

5. FUTURE PROSPECTUS OF ORGANIC FARMING

In this India divided under two groups –

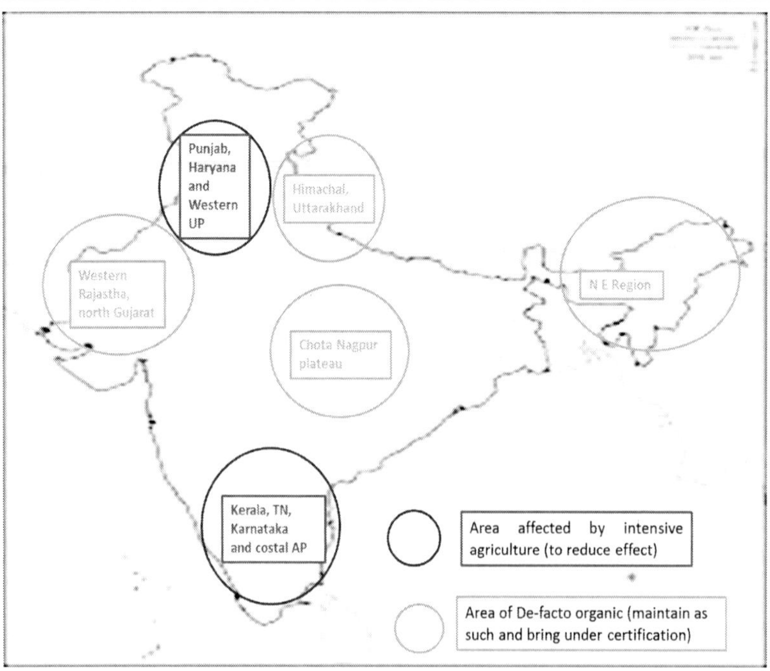

5.1 Area affected by intensive agriculture (to reduce ill effect)

In this we cover North India (Punjab, Haryana and Western UP) and South India (Kerala, Tamil Nadu, Karnataka and costal Andhra Pradesh) which are affected by intensive agriculture, indiscriminate use of off farm inputs (pesticides and fertilizer). Main target by promoting organic farming we reduce ill effect of intensive agriculture

5.1.1. *No more cancer belt and cancer train*

The Malwa region of Punjab, India, is facing an unprecedented crisis of environmental health linked to indiscriminate, excessive, and unsafe use of pesticides, fertilizers, and poor groundwater quality. The region has been described as India's "cancer capital" due to abnormally high number of cancer cases, which have increased 3-fold in the last 10 years. Studies of this region have also highlighted a sharp increase in many other pesticide-related diseases, such as mental retardation and reproductive disorders. The most affected individuals are the agricultural workers who are directly exposed to pesticides. The Malwa region of Punjab, India, is less than 15 per cent of the total area of Punjab (only 0.5 per cent of the total geographical area of India), but it consumes nearly 75 per cent of the total pesticides used in Punjab. As per the 2007 State of Environment Report prepared by the Punjab State Council for Science and Technology. The state, in turn, uses 17 per cent of the total pesticides applied in India. The high use of pesticides, along with environmental and social factors, is responsible for the high concentration of pesticide residues in the food chain of this region. Moreover, many banned and restricted pesticides are still in use in this region, warranting strict periodical health checkups and other interventions. Though several studies in the past decade had showed that Punjab has higher incidence of cancer than the rest of the country, this is for the first time the state government conducted a comprehensive door-to-door survey to quantify the problem. The study underscores that the Malwa region, already infamed as the cancer belt, has the highest number of cancer patients - 107 in 100,000 population. Four districts that topped the cancer incidence list are from this region. With 136 cancer cases per 100,000 populations, Muktsar district fares the worst. It is closely followed by Mansa, Bathinda and Ferozepur districts. Tarn Taran district in the Majha region has the least number of cancer patients - 41 per 100,000 people (Singh J., 2013).

'Cancer train' a lifeline for stricken patients Cancer patients from various towns of Punjab travel on this Lalgarh-Abohar-Jodhpur Train No. 339 to a hospital in Bikaner for cheap treatment. This train famous in farming community and locally known as cancer train. So this is the time to rethink about the our old age practices for farming in place of depending on off farm chemical input which will in long term collapse our health we need to aware our farmer the importance of organic farming so they not look towards market for chemicals and also get themselves to purchase there bad self-health and soil health by waiting in queue.

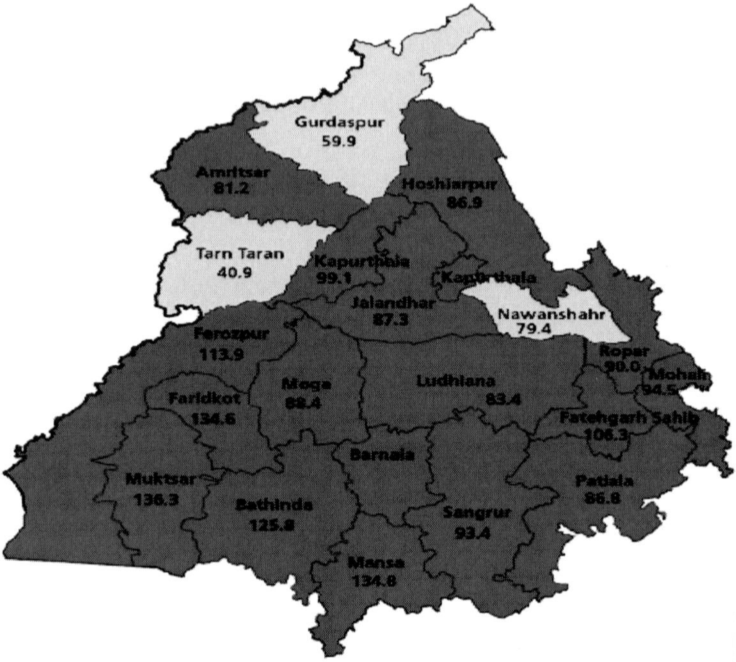

Cancer prevalence per (100,000) peoples in Punjab (Source: Singh J. 2013)

5.1.2. *No more Kerala's Endosulfan Tragedy*

What is Endosulfan?

Endosulfan is a pesticide belonging to the organochlorine group of pesticides, under the Cyclodiene subgroup. It was introduced in the 1950's and it emerged as a leading chemical used against a broad spectrum of insects and mites in agriculture and allied sectors. It is used in vegetables, fruits, paddy, cotton, cashew, tea, coffee, tobacco and timber crops. It is also used as a wood preservative and to control tse-tse flies and termites. The chemical is out of patent and is marketed by many different companies under a variety of names like Agrosulfan, Aginarosulfan, Banagesulfan, Cyclodan, Endocel, Endoson, Endonit, Endomil, Endosol, Endostar, Endodaf, Endosulfer, E-sulfan, Endorifan, Hildan, Redsun, Seosulfan, and Thiodan. Endosulfan is a highly toxic, ubiquitous environmental pollutant that causes long-term harm to humans and wildlife. It is widely considered to be a Persistent Organic Pollutant (POP) and is readily absorbed by stomach, lungs and through the skin. It has been banned in over 68 countries. Some areas where it is banned are Belize, Singapore, Tonga, Syria, Germany, the USA, the Brazilian state Rondonia, the UK, Sweden, Netherlands, Colombia, and the Indian state Kerala. It is severely restricted in Australia, Bangladesh, Indonesia, Cambodia, Japan, Korea, Kazakhstan, Kuwait, Philippines, Lithuania,

Sri Lanka, Taiwan, Thailand, Denmark, Yugoslavia, Norway, Finland, Russia, Venezuela, Dominica and Canada. This pesticide is classified as a Highly Hazardous chemical by the US Environmental Protection Agency (EPA) and the European Union, as a Persistent Toxic Substance by the United Nations Environmental Programme (UNEP), as a Category II - Moderately Hazardous chemical by the World Health Organization (WHO), and as Extremely Hazardous chemical by the Industrial Toxicological Research Centre (ITRC) in India. However, India is the largest producer, consumer and exporter of Endosulfan.

How did the disaster happen?

The endosulfan tragedy happened in Kasaragod district of Kerala in India. It is not a well-known tragedy. It is considered by many experts in the field of pesticide toxicity as one of the world's worst pesticide disasters. The Plantation Corporation of Kerala (PCK), a public sector undertaking under the State Government, which owns three cashew plantations covering 4600 hectares in Kasargod, sprayed endosulfan aerially in these plantations for 24 years (1976 to 2000), three times a year. These uninterrupted spraying in spite of several warnings about its disastrous impact on health and environment has resulted in several chronic, critical and life-threatening ailments in the areas surrounding the plantations.

What are its impacts?

Eleven Gram Panchayats covering twelve villages (group and sub-villages) have been severely affected. These villages are Enmakaje, Belur, Kumbadaje, Badiadka, Muliyar, Karadukka, Periya, Pullur, Ajanur, Kallar, Panathady, Kayyur, Cheemeni. In these villages there is a very high concentration of chronic health complications like Mental, Physical and Behavioural disorders, Sensory loss, Neurological ailments, Cardiovascular diseases, Congenital Anomalies, Dermatological and Musculoskeletal disorders etc. *This tragedy occurred because a toxic chemical was used continuously in a populated area with several (open) natural drinking water sources.*

5.2. Area of de-facto organic (maintains as such and brings under certification)

In this we cover Western India (Western Rajasthan and North Gujarat) and NE Region Main target maintains as such and brings under certification so small farmers to take advantage of the lucrative market for certified organic products in the developed world In India, approximately 70 per cent of arable land, where rainfed or dry farming is practiced, agro-chemicals are not generally used. According to an estimate made by the Institute of Integrated Rural Development in 2001, only one fifth of the dry land farmers in India use chemical inputs, the rest rely on manure and green compost for maintaining soil fertility. The majority of these people use traditional methods, often in highly heterogeneous and risk-prone marginal environments. In fact, given their situation, these farmers have little choice, but to rely upon locally available natural resources to maintain soil fertility and to combat pests and diseases. Whatever may be the reasons, the fact is that the diverse farming systems managed by such small farmers could be considered as organic, as they do not rely on synthetic chemical pesticides or fertilizers and use technologies that optimize nutrient flows and use local resources such as native seeds and traditional knowledge. In technical terms such farms are called *'de-facto organic'* (as distinct from certified organic) farms, *i.e.*, farms that rely exclusively on natural methods of building soil fertility and combating pests and diseases, but are not inspected and verified by any organic certification agency.

Table 8. State wise consumption of plant nutrients per unit of gross cropped area in N E State as compared to national average

	Consumption				**Difference with national average**			
	N	**P_2O_5**	**K_2O**	**Total**	**N**	**P_2O_5**	**K_2O**	**Total**
Arunachal Pradesh	1.9	0.8	0.3	3.0	84.3	41.1	18	143.3
Assam	34.8	14.5	18.2	67.6	51.4	27.4	0.1	78.7
Manipur	21.9	4.7	1.3	27.9	64.3	37.2	17	118.4
Meghalaya	9.0	4.5	1.4	15.0	77.2	37.4	16.9	131.3
Mizoram	16.7	19.8	9.1	45.5	69.5	22.1	9.2	100.8
Nagaland	1.6	1.0	0.3	2.9	84.6	40.9	18	143.4
Tripura	26.0	14.4	11.2	51.6	60.2	27.5	7.1	94.7
All-India	86.2	41.9	18.3	146.3				

(*Source*: FAI, 2011-12)

5.2.1. *N E region, Western India*

There is a great potential for organic farming to flourish in this country and given an appropriate institutional and policy framework, it will not be very difficult to promote the existing *'de-facto organic'* farms N E region, Western India to the category of certified organic farms (Shetty *et al.*, 2013). This would enable the small farmers to take advantage of the lucrative market for certified organic products in the developed world, which could directly contribute towards the improvement of their economic well-being. In India same region like as N E Region, Western Rajasthan Himalayan region and Chota Nagpur plateau are naturally growing crops without fertilizer and pesticides, due to climate reason even same farm, a single particle of urea never used in history so government should cover these areas and reserve for organic region. Government should concentrate these areas and try to give them organic certificate at policy level so they can get benefit of premium price of organic certification. In this direction Sikkim declared first organic state in the country after 13 years effort from 2003 to 2016 (Indian Express, 2016; The Hindu 2016). And now government focusing on Himalayan region adjoining to Sikkim, Ministry of Agriculture and Farmer Welfare has launched a Central Sector Scheme entitled "Mission Organic Value Chain Development for North Eastern Region" for implementation in the states of Assam, Arunachal Pradesh, Meghalaya, Manipur, Mizoram, Tripura, Sikkim and Nagaland, during the 12th plan period (Ministry of Agriculture and Farmer Welfare 2016).

6. CHALLENGES OF INDIAN ORGANIC FARMING

Organic farming in the Indian context has to resolve several issues at both micro and macro level.

6.1. Micro level issues

The micro level issues confronting organic farming include economic viability, particularly for small and marginal farmers, marketing, etc. For example, one of the greatest barriers for organic farming is the so-called conversion period due to the direct and indirect costs. The conversion of a conventional farm to an organic farm requires strictly adhering the rules and standards of production, processing and labelling at prescribed international levels. During the conversion period all the standards required for certifying a product as 'organic' must be fulfilled and verified by a certifying agency. Costs due to things, such as information, marketing charges, inspection, and certification expenses also increase the cost of organic farming. For instance, fees for the inspection and certification can be prohibitively high at Rs. 5000, since this equals the returns from agriculture for many small farmers (Brook and Bhagat, 2004). The often reduced yields of organic farming, as compared to conventional farming particularly during the conversion period before soil nutrients and organic matter are replenished with bio-fertilizers, are an additional liability to the farmer. Particularly during the conversion period when the products are not certified as organic, and thus, they cannot be sold at the organic market price. Farmers often incur expenditures for things, such as farm machinery, bunding and purchase of bioinputs to augment soil fertility and yield. In addition, various barriers like transaction costs (lack of access to relevant knowledge on cultivation practices, market), mandatory documentation required for inspection and certification, lack of demand in domestic market and constraints to enter international market and institutional factors restrict the spread of organic farming (Das, 2004).

6.2. Macro level issues

The macro challenges include impacts mainly on food security, employment, and environment. The question of food security assumes significance considering potential yield reductions of organic farming vis-à-vis conventional farming (Pandey and Singh, 2012), particularly in the two to three year conversion period. Given India's history of inadequate food production, it is necessary to examine food security related issues, taking into account the large number of marginal and small farmers, before organic farming is promoted en masse. Another macro dimension of promoting organic farming is its impacts on rural employment. Organic farming is expected to increase employment opportunities owing to requirement of producing various agricultural inputs, like bio-fertilisers and bio-pesticides, using locally available materials. The scope for increased employment opportunities need to be assessed at the regional and national level. From the environmental point of view, apprehensions have been raised that organic farming might also lead to unsustainable problems, due to increased land and water use to offset decrease in yield. Considering the Indian case, even in organic practices water conservation must ultimately remain the paramount concern. Considering the various challenges to the adoption of organic farming the Working Group on Organic and Bio-dynamic Farming of the Planning Commission (GOI, 2001b) suggested examination of some important issues for effective promotion and practice of organic farming and sustainable agriculture. These include economics of organic crop production, economic and environmental externalities associated with conversion to organic farming, comparative study of chemical based and organic farming covering social, environmental and economic costs.

7. RESEARCH FINDING

Organic farming, in spite of the reduction in crop productivity by 14.6 per cent, provided higher net profit to farmers by 21.5 per cent compared to conventional farming. This was mainly due to the availability of premium price (20–40 per cent) for the certified organic produce and reduction in the cost of cultivation by 15.9 per cent. In cases, where such premium prices were not available and the cost of cultivation was higher primarily due to purchased off-farm inputs, organic farming was not found economically feasible (Dadhwal *et al.*, 2011). However, there was an overall improvement in soil quality in terms of various parameters, *viz.* physical, chemical, biological properties, availability of macro- and micronutrients, indicating an enhanced soil health and sustainability of crop production in organic farming systems

According to Ramesh *et al.* (2010), A survey was made on certified organic farms in the country to ascertain the real benefits and feasibility of organic farming in terms of the production potential, economics and soil health in comparison to the conventional farms, the study revealed that organic farming, in spite of the reduction in crop productivity by 9.2 per cent, provided higher net profit to farmers by 22.0 per cent compared to conventional farming. This was mainly due to the availability of premium price (20–40 per cent) for the certified organic produce and reduction in the cost of cultivation by 11.7 per cent. In cases, where such premium prices were not available and the cost of cultivation was higher primarily due to purchased off-farm inputs, organic farming was not found economically feasible. However, there was an overall improvement in soil quality in terms of various parameters, *viz.* physical, chemical, biological properties, availability of macro- and micronutrients, indicating an enhanced soil health and sustainability of crop production in organic farming systems.

Table 9. Productivity of crops (t/ha) in organic *versus* conventional farming

State	Crop	Organic farming	Conventional farming	Per cent increase (+)/ decrease (-) in organic farming
Maharashtra	Vegetables	11.0	13.0	-15.3
	Fruit crops	11.4	13.6	-16.1
	Rice	2.0	2.5	-20.0
	Wheat	1.2	1.5	-20.0
Karnataka	Soybean	0.9	1.1	-18.2
	Chickpea	0.8	0.8	0.0
	Fruit crops	8.0	9.0	-11.1
	Groundnut	1.2	1.4	-14.2
	Sugarcane	120	140	-14.3
Tamil Nadu and Pondicherry	Cotton	0.6	0.8	-25.0
	Cashew	1.3	1.0	+30.0
	Banana	25.0	30.0	-16.6
	Mango	8.0	6.0	+33.3
	Guava	20.0	23.0	-13.0
	Coconut	28,250 nuts	28,750 nuts	-1.7
Kerala	Rice	5.0	6.0	-16.6
	Pepper	1.38	1.40	-1.4
	Banana	23.6	27.2	-13.2
	Coconut	31,000 nuts	30,500 nuts	+1.6
	Coffee	1.23	1.31	-6.1
	Turmeric	22.5	25.0	-10.0
Uttarakhand	Rice	3.77	3.82	-1.3
	Wheat	3.12	3.92	-20.4
	Potato	12.0	15.0	-20.0
Mean				-9.2

(*Source*: Ramesh *et al.*, 2010)

According to Sahu *et al.* (2010), it may be concluded that majority farmers were found in the range of high level of knowledge of organic farming practices. The wide knowledge gapes are in the areas of organic farming practices like use of Ha NPV, use of trichocards, use of bio-pesticides and use of NADEP compost. The farmers need to be made well aware about the use of such practices so that the basic concept of organic farming and its application part could be familiar to the farmers.

Table 10. Knowledge gap of farmers on the basis of the organic farming practices

S.No.	Organic farming practices	Maximum knowledge (score)	Total obtained Knowledge (score)	Knowledge gap percentage	Rank
1.	Knowledge about concept of organic farming	90	76	15.56	IX
2.	Use of bio-pesticides	90	55	38.88	III
3.	Use of organic manure and crop residues	90	66	26.66	VII
4.	Use of mechanical cultivation	90	61	32.22	VI
5.	Use of vermicompost	90	74	17.77	VIII
6.	Use of bio-fertilizers	90	58	35.55	V
7.	Use of HaNPV	90	51	43.34	I
8.	Use of NADEP compost	90	56	37.77	IV
9.	Use of trichocards	90	53	41.12	II
	Over all knowledge gap	810	550	32.10	

(*Source*: Sahu *et al.*, 2010)

8. CONCLUSION

According to Ramesh P. *et al.* (2005) the following conclusions can be drawn on important issues regarding organic farming:

1. Large-scale conversion to organic agriculture would result in food shortage with the present state of knowledge and technology, as the yield reductions of organic systems relative to conventional agriculture average 10–15 per cent, especially in intensive farming systems. However, in traditional rainfed agriculture, organic farming has the potential to increase the yield, since 70 per cent of total cultivable land falls in this category. Mere 5–10 per cent increase in farm production would definitely help to achieve the targeted growth rate of 4–5 per cent in agricultural production in the Tenth Plan period.
2. Organic manure is an alternative renewable source of nutrient supply. A large gap exists between the available potential and utilization of

organic wastes. However, it is not possible to meet the nutrient requirements of crops entirely from organic sources, if 100 per cent cultivable land is converted to organic farming.

3. Organic farming systems can deliver agronomic and environmental benefits both through structural changes and tactical management of farming systems. The benefits of organic farming are relevant both to developed nations (environmental protection, biodiversity enhancement, reduced energy use and CO_2 emission) and to developing countries like India (sustainable resource use, increased crop yields without over-reliance on costly external inputs, environment and biodiversity protection, etc.).
4. Organic foods are proved superior in terms of health and safety, but there is no scientific evidence to prove their superiority in terms of taste and nutrition, as most of the studies are often inconclusive.
5. Combination of lower input costs and favourable price premiums can offset reduced yields and make organic farms equally and often more profitable than conventional farms. However, studies that did not include organic price premiums have given mixed results on profitability. Thus it is the premium price on the organic food which decides the economic feasibility of organic farming, at least at the current rate of development in organic agriculture.
6. In organic farming systems, pest and disease management strategies are largely preventive rather than reactive. In general, pest and disease incidence is less severe in organic farms compared to conventional farms.

In nut shell the performance of organic agriculture on production depends on the previous agricultural management system. An over simplification of the impact of conversion to organic agriculture on yields indicates that:

- In industrial countries, organic systems decrease yields; the range depends on the intensity of external input use before conversion;
- In the so-called Green Revolution areas (irrigated lands), conversion to organic agriculture usually leads to almost identical yields;
- In traditional rain-fed agriculture (with low-input external inputs), organic agriculture has the potential to increase yields.

REFERENCES

Agriculture Census (2010–11) All India Report on Number and Area of Operational Holdings Agriculture census division Department of Agriculture & Co-operation Ministry of Agriculture, Government of India.

Aktar W, Sengupta D and Chowdhury (2009) Impact of pesticides use in agriculture: their benefits and hazards. *Interdisc Toxicol.* **2(1):** 1–12. doi: 10.2478/v10102-009-0001-7.

APEDA (2014) Agricultural and Processed food product Export Development Authority, Ministry of Commerce and Industries, GOI, www.apeda.gov.in.

Balachandran V (2004) Future in the Past: A study on the status of organic farming in Kerala. Discussion Paper No. 82.

Bhattacharyva P and Chakraborty G (2005) Current status of organic farming in India and other Countries, *Indian Journal of Fertilisers* **1(9):** 111–123.

Brook, Bhagat G (2004) Organic by default: The irony of organic farming in India, available at: http://www.ecoworld.org/Home/ Articles2.cfm? TID=349.

Chandrashekar HM (2010) Changing scenario of organic farming in India: an overview. *Int. NGO J.,* **5:** 34–39.

Charyulu K and Biswas S (2010) Organic input production and marketing in India – Efficiency, issues and policies. CMA Publication No. 239.

Dadhwal KS, Sharma NK and Ghosh BN (2011) Organic farming for resource conservation and soil health improvement in the Himalayan region, India, *Indian Journal of Soil Conservation* **39(3):** 243–250.

Das K (2004) Transitioning to organic, available at: http://www.indiatogether.org/ 2004/jun/ agr-transorg.htm.

Fertilizer Statistics (2011-12) The Fertilizer Association of India, New Delhi.

GOI (2014) Ministry of Commerce and Industries, Lok Sabha Question no. 2393, answered on 25th July, 2014.

Government of India (2001b) Report of the Working Group on Organic and Biodynamic Farming for the Tenth Five Year Plan, TFYP Working Group Sr. No. 46/2001, Planning Commission, New Delhi.

Gupta A and Verma J *et al.* (1997) Vanishing breeds, *Down to Earth,* **11(15):** 27–37.

Guruswamy K and Gurunathan KB (2010) A Need for Organic Farming in India. Journal of contemporary research in management January-march.

IFOAM (2005) Principles of Organic Agriculture Bonn: International Federation of Organic Agriculture Movements.

IFOAM (2010) Organic food and farming: A system approach to meet the sustainability challenge. Kölling, A. (Ed.), Belgium, pp: 05.

Indian Express (2016) Sikkim became India's first fully organic state in 2016. December 31, 2016 10:26 am, Accessed on September 14, 2017. http://indianexpress.com/ article/ india/sikkim-became-indias-first-fully-organic-state-in-2016-4452765/.

Makadia JJ and Patel KS (2015) Prospects, status and marketing of organic products in India - A Review. *Agri. Review* **36(1):** 73–76. DOI: 10.5958/0976-0741.2015. 00009.4.

Ministry of Agriculture and Farmer Welfare (2016) Mission organic value chain development for North Eastern region A Sub-Mission under National Mission for Sustainable Agriculture (NMSA). Operational Guidelines. Accessed on September 14, 2017. http://agricoop.nic.in/sites/default/files/OPG1922016.pdf

Musa M, Bokhtiar SM and Gurung TR (Eds) (2015) Status and future prospect of organic agriculture for safe food security in SAARC countries.

NAAS (2005) Organic farming: Approaches and possibilities in the context of Indian agriculture, Policy paper 30, National Academy of Agricultural Sciences.

Narayanan S (2005) Organic farming in India: Relevance, problems and constraints. Deportment of Economic Analysis and Research, Notional Bonk for Agriculture and Rural Development. Mumbai.

Pandey J and Singh A (2012) Opportunities and constraints in organic farming: an Indian perspective. *Journal of Scientific Research*. **56:** 47–72.

Press Information Bureau, Ministry of Agriculture, Government of India. 09-May-2017 16:49 IST. Accessed on 13 September 2017. http://pib.nic.in/newsite/Print Release.aspx? relid=161670.

Ramesh P, Panwar NR, Singh AB, Ramana S, Yadav SK, Shirvastava R and Rao AS (2010) Status of organic farming in India, *Current Science* **98(6):** 1190–1194.

Ramesh P, Singh M and Rao SA (2005) Organic farming: Its relevance to the Indian context, *Current Science* **88(4):** 561–568.

Roychowdhury R, Gawwad MRA, Banerjee U, Bishnu S and Tah J (2013) Status, Trends and Prospects of Organic Farming in India: *A Review. Journal of Plant Biology Research* **2(2):** 38–48.

Roychowdhury R, Banerjee U, Sofkova S and Tah J (2013) Organic farming for crop improvement and sustainable agriculture in the era of climate change. *OnLine J. Biol. Sci.* **13(2):** 50–65. doi: 10.3844/ojbsci.2013.50.65

Sahu RP, Singh RJ and Singh K (2010) Knowledge Gap about Organic Farming Practices of Farmers of Bageshwar District of Uttarakhand, *Indian Journal of Extension Education* **45(1&2):** 135–136.

Salvador VG and Katke J (2003) Market opportunities and challenges for Indian Organic Products Research Institute of Organic Agriculture (FIBL) and ACNielsen ORG-MARG Swiss State Secretariat of Economic Affairs (SECO) http: // orgprints.org / 000026841.

Shetty PK, Hiremath MB and Murugan M (2013) Status of organic farming in agro ecosystems in India, *Indian Journal of Science and Technology* **6(8):** 5083–5088.

Shetty PK, Alvares C and Yadav AK (eds). Organic Farming and Sustainability, ISBN: 978–93–83566–03–7, National Institute of Advanced Studies, Bangalore. 2014.

Singh J (2013) Punjab, cancer capital of India. Down to Earth, Feb 2, 2013, http://www.downtoearth.org.in/content/punjab-cancer-capital-india.

Singh R, Ravisankar N and Prasad K (2017) Improvement in productivity and economics of major food production systems of India through balanced dose of nutrients. *Current Science* **112 (12):** 2470–2474. doi: 10.18520/cs/v112/i12/2470-2474.

The Hindu 2016. Sikkim becomes India's first organic state. September 23, 2016, Accessed on September 14, 2017. http://www.thehindu.com/news/national/Sikkim-becomes-India%E2%80%99s-first-organic-state/article13999445.ece.

Venkateswarlu B, Balloli SS and Ramakrishna YS (2008) Organic farming in rainfed agriculture: Opportunities and constraints, Central Research Institute for Dryland Agriculture, Hyderabad. pp. 185.

■■■■

13 Organic Farming of Vegetable Crops for Sustainability

S.P. Kanaujia

Department of Horticulture, SASRD,
Nagaland University, Medziphema – 797 106, Nagaland

ABSTRACT

Vegetables play a pivotal role in food and nutritional security and maintenance of human health. Vegetable is the cheapest natural protective food as their consumption prevents several diseases. They also provide valuable roughages in term of dietary fibre which helps in digestion and prevent constipation. Vegetable farming is an important source of income leading to improved livelihood. Indian farmers were basically organic farmers before the advent of inorganic fertilizers and chemical pesticides. To enhance the production and to meet the needs of growing population, intensive production system using chemical fertilizers, weedicides and pesticides were adopted. Indiscriminate use of chemical fertilizers, weedicides and pesticides in modern vegetables production has resulted degradation of soil fertility, declining water table, increase nitrate content in ground water, environment and health hazards problems. Pollution of soil, water and environment caused threat to sustainability of vegetable production. Due to increase in health consciousness, worldwide and in India too, there is an increase in the demand of safe vegetables grown without the use of any chemicals referred as organic vegetables. Therefore, it is need of hour to think over for adopting organic farming of vegetables not only for health sake of the consumers but export also. It has been scientifically proven that organic farming systems of vegetables are the most environment friendly, socially and economically sound production of vegetables and thus sustainable.

Keywords: Vegetables, Organic farming, Sustainability, Organic manure, Biofertilizers, Crop rotation, Biopesticides

1. INTRODUCTION

Vegetables are important components of Indian agriculture due to their nutritional value, medicinal value, industrial value and export potential. It is important in balanced diet. All essential nutrients are available in vegetables especially minerals, vitamins which are required for normal functioning of human metabolic activity. Vegetable farming is profitable venture because vegetables give more economic return to the growers due to their high productivity and more net return/unit area than any other crops. In a span of six decades, India has made quantum jump in vegetables production and emerged as the second largest producer of vegetables next to China and share 18 per cent of world production. It has been possible because of development of high yielding, biotic and abiotic resistant varieties, adopting improved production and protection technology, integrated weed management, integrated nutrient management, use of growth regulator, post-harvest technologies and by bringing more area under irrigation. Presently vegetable occupies 9.396 million hectare with production of 162.89 million tones with productivity of 17.3 tones/ha. The area under vegetable production in the country was increased from 5.593 million hectare in 1991–92 to 9.39 million hectare in the year 2013–14. Likewise, the production was also increased from 58.532 million tones in 1991–92 to 162.89 million tones in the year 2013–14 (IHD 2014). In general nutrient requirement of vegetable crops are high and to meet the nutrient requirement of vegetable crops for getting optimum yield, large amount of chemical fertilizers are used. Most of the vegetables are succulent and attract several insects and diseases, in order to control them, heavy pesticides are used in vegetable crops especially okra, tomato, brinjal, cole crops etc. These pesticides remain chemically active for long period and produce hazardous effect on the environment and human health. Indiscriminate use of chemical fertilizers, weedicides and pesticides in modern vegetables production has resulted pollution of soil, water and environment. Overtime, the use of synthetic inputs has come to the level of causing a concern to the environment and human health. To overcome these challenges, a need is felt today to reduce the use of these harmful chemicals slowly and slowly by use of organic manures, green manuring, biofertilizers, bio-pesticides, insect, pest and disease resistant varieties, crop rotation, trap crop etc. for regeneration of the soil, reduction of environment pollution and producing the quality (residual and disease free) vegetables to make it a sustainable and profitable system. Worldwide and in India too, there is an increase in the demand for healthy, organically produced vegetables which is tasty, rich in nutrition and safe. It has been scientifically proven that organic farming systems of vegetables are the most environment friendly and thus sustainable.

Organic farming is an old practice of natural farming, which avoids the use of synthetic compounds including inorganic fertilizers, pesticides, herbicides, growth regulators and other chemicals. Hence, question of environment pollution does not arise. Organic farming is a method of farming system which primarily aimed at cultivating vegetables in such a way, as to keep the soil alive and in good health by use of organic wastes (crop, animal and farm wastes) and other biological materials along with biofertilizers and crop rotation to release nutrients to crops for increased sustainable production in an eco-friendly pollution free environment. Use of synthetic fertilizers, pesticides, growth hormons etc. is prohibited under organic farming. Organic farming is eco-friendly and sustainable which promotes and enhances agro-ecosystem including biodiversity, biological cycles and soil biological activity. There is ever growing export market for organic vegetable products in the domestic and international market. The demand of organic products has been steadily increasing due to the growth of consciousness of health hazards due to contamination from chemicals. The consumers are now willing to pay a premium for the safe vegetables that result from organic farming. The major vegetable produced in India by organic farming is potato, tomato, brinjal, chilli, okra, onion, garlic, leafy vegetables etc. Nursery raising is a very important operation in successful production of organic vegetable crops. Weak and unhealthy seedlings result into low yield. Nursery should also be grown organically for organic vegetable production. For that it is necessary to raise virus free healthy seedlings of these crops either in the greenhouse or by covering the nursery beds with insect proof net.

2. ORGANIC NURSERY RAISING

It is almost impossible to do direct sowing of small seeds properly in the main field. Raising the seedlings in nursery beds is an economical, easier and convenient way to nourish and take special care of young, tender seedlings against pest, diseases, weeds, hails, heavy rains, hot sun and other natural hazards. The ideal soils for nursery raising is well drained, fairly deep and high in organic matter. Healthy and productive soil helps seedling to develop good root system. Nursery soil pH should be 5.5 to 7.0. About 3 kg well rotten FYM/Compost or 1.0 kg vermicompost per square meter area should be mixed in nursery bed. Neem cake @100g/m^2 is also be added to the nursery bed. If beds are dry, irrigate the beds for proper decomposition of manures and fertilizers. The preparation of nursery beds is an important operation of seedling raising. Germination and growth of seedlings are highly influenced by bed preparation. Dig the nursery area to a depth of 25–30 cm. Remove all the clods, stones and weeds from the beds. The size of nursery bed should be kept 3 × 1 × 0.15m and 30 cm space between beds. Length of the bed may be kept as per the availability of space, but width is restricted to 1 meter only which facilitates intercultural operations. The space between two beds helps in weeding, application of pesticides, for drain out the excess water from the nursery beds. Nursery beds should be prepared thoroughly. The soil of beds should be reduced to the finest tilth before sowing. The surface of nursery beds should be smooth and slightly raised in centre compared to the margin so that water may not be stagnated and drain out easily from the beds. During rainy season, raised bed should be prepared. Flat bed is generally practiced during summer and winter season. The main drawback in this system is water stagnate during rainy season. For getting healthy nursery, soil must be treated to kill the harmful soil borne pathogens and harmful pests. Soil may be treated with soil solarization or with bioagent. Soil solarization is a cheapest and eco-friendly approach for soil disinfection of nursery beds. For soil solarization, dig the nursery soil to a depth of 25–30 cm, moist it and cover the beds with transparent polythene sheet of 100–200 gauge thickness for 10–15 days during April–June. All the margins of the polythene are buried in the soil. In this way, the soil temperature inside the polythene reaches up to 48–56ºC. It reduces the soil borne pathogens, harmful pests and weeds to a substantial level. After 10–15 days, polythene sheet is removed and seeds are sown. Biotreatments are another eco-friendly method of soil treatment to manage many soil borne pathogens. Soil should be treated with *Trichoderma harzianum* or *Trichoderma viride* @ 10g/m^2 area. Soil drenching should be done with *Trichoderma* solution @ 20g/litre of water. The seed should be sown 2 days after bioagent application. The soil of bed should be moist at the time of bio-agent application.

Organic produced good seed is a pre-requisite in organic farming of vegetables. Seeds should be purchased from reliable source preferably from public sector/research station/authentic sources. Seeds should be free from insects, diseases, weed seeds and other crop seeds. Preferably use quality, certified or foundation seed for raising the seedlings. Seed also carries pathogens of certain diseases. If not treated, seed should also be treated to avoid soil borne disease. Seed may be treated with *Psuedomanas fluorescens* or *Trichoderma viride/Trichoderma harzianum* @ 5g/kg of seeds. Seed treatment with *Trichoderma* is an eco-friendly way to manage fungal diseases. The seeds may be tied in muslin cloth and dip into hot water (45°C) for 20 minutes to manage seed borne bacterial diseases like black rot in cole crops. It is an important and eco-friendly seed treatment method to manage bacterial diseases. Dense sowing results thin, spindly seedlings with small root which do not good for transplant. Therefore, seed should be sown in line about 1–2 cm apart and 0.5–1cm depth. Line to line distance should be kept 5–8 cm. Line sowing facilitates less incidence of damping off disease as compared to broad cast method. It also helps in raising healthy, bold and uniform seedlings. Less seeds are also required as compared to the broadcasting method. Nursery bed should be checked regularly for appearance of weeds, diseases, pests and any other physical problem. Over crowded, weak, diseased and insect damaged seedling should be thinned out at the time of weeding. Thinning facilitates balance light and air to each and every seedling. It is good to protect nursery beds with 40–50 mesh nylon net cover to protect insect vector. Prior to transplanting, seedlings should be hardened by gradually reducing or withholding of watering to the nursery bed for 4–5 days. Hardening of seedlings is necessary to prepare them for withstanding transplanting shock. Hardened seedlings bear the transplanting shock better and give less mortality. Hardened plants produce new roots faster than non-hardened plants. During the hardening process, prune some older leaves or a portion of leaves in order to minimize transpiration, otherwise need not to remove any leaves/part. The nursery beds must be watered 6 hours before uprooting the seedlings to avoid root damage and facilitate smooth uprooting. Do not uproot too old and too young seedlings that are unsuitable for transplanting in the main field. Seedlings of most of the vegetable crops become ready for transplanting in 3–4 weeks depending upon the crop and season. About 3–4 weeks old seedlings having 4–6 true leaves or 10–15 cm tall are ideal for transplanting except onion.

Onion seedlings are ready for transplanting after 6–8 weeks of seed sowing. Uproot only required number of seedlings preferably in the after noon. BT (*Bacillus thuringiensis*) formulation 1 ml/litre should be sprayed to the seedlings one day before transplanting the seedlings. Root dip treatment with *Trichoderma viride* solution at 20 g/litre of water for 15–20 minutes is also good for managing the soil borne fungal diseases. Similarly, biofertilizers such as *Azatobactor*, *Azospirillum* and PSB should also be inoculated @ 2 kg/ha before transplanting the seedlings.

3. TRANSPLANTING

The field in which transplanting has to be done is prepared thoroughly by repeated harrowing and planking. It is divided into small beds depending upon the source of irrigation. About 3–4 weeks old seedlings having 4–6 true leaves or 10–15 cm tall should be transplanted in the main field. Healthy and uniform seedlings free from insects, pests and diseases have good root system should be selected for transplanting. Always transplant the seedlings in the main field at proper spacing. Planting is done either in flat beds or on ridges. Planting on ridges is advisable during rainy season. Transplanting should be done preferably in the evening to avoid mortality. This provides ample moisture for the plants to survive because evaporation and transpiration are lesser at this time. In cloudy weather, transplanting may be done whole day. The interval between uprooting and transplanting the seedlings should be minimum to protect them from excessive drying. Gap filling must be done to replace dead seedlings within a week after transplanting.

4. IRRIGATION

Light irrigation is done immediately just after planting in order to bring the soil in close contact with roots and repeat irrigation till the seedlings are well established. Frequent irrigation is essential in vegetables for optimum plant growth and yield. Maintenance of optimum moisture supply in vegetable crops is also important to avoid disease incidence in organic vegetable production. Drip irrigation should be installed in organic vegetable production. Drip irrigation helps to maintain better soil moisture regimes and also reduces pest and disease incidence in addition to saving of irrigation water through reduced evaporation losses as well as preventing weed growth. Mulches may be used to conserve soil moisture.

5. WEED MANAGEMENT

Weeds compete for moisture, nutrients, sunlight and space with the crop thus reducing the yield and quality of the vegetables. Weed intensity is high in *kharif* season than in *rabi* season. Vegetable crops are mostly widely spaced and heavy fertilized hence, scope of weeds is more in initial stages. It is essential to keep weeds under control from the initial growth itself. The most critical period of crop for weed competition is between 30–50 days after transplanting in vegetable crops. Plastic mulch is a widely used to cover the soil in organic vegetable production to control weeds. Mulches are also used to reduce or increase soil temperature and conserve soil moisture. The decomposition of organic mulches like paddy straw or grasses in soil increase the organic matter. The mulch provides a physical barrier between soil and plant and reduces the diseases. Mulching should be done 15 to 20 days after planting the crop. The weed biomass should be recycled in organic vegetable production.

6. NUTRIENT MANAGEMENT IN ORGANIC VEGETABLE FARMING

Being short duration and high yielding, nutrient requirement of vegetable crops are high. Hence, nutrient management in organic vegetable farming is the biggest challenge. Organic manures coupled with biofertilizers may be applied to meet the nutrient requirement of vegetable crops in organic vegetable production. In general the application of 10–20 tonnes/ha of any locally available organic manures coupled with the inoculation of biofertilizers can take care of the nutrient requirement for most of the vegetable crops. Crop rotation is also important for soil fertility management. Following materials can be used in organic vegetable production.

(i) Organic manures

Organic manures are organic materials derived from animal and plant residues that contain plant nutrients in complex organic forms. In vegetables production, role of organic manure is much more as compared to other crops. It has been proved experimentally that vegetable production with the use of organic manures possess good firmness and having long life. Organic manures which are being used in vegetable cultivation are FYM, pig manure, sheep manure, poultry manure, compost, vermicompost, crop residues, biogas slurry, agro industrial waste (pressmud, etc.) and oil cakes (mahua cake, neem cake, safflower cake etc). Organic manures increase organic matter in the soil. As soil organic matter increases, the ability to supply nutrients to the crops also increases. The increase of soil organic matter to optimum level is a key aspect of any organic vegetable production. Organic matter improves physico-chemical and biological properties of soil especially improves cation exchange capacity, soil aeration and water holding capacity. Organic manures are slow releasing, hence nutrients are less prone to loss than inorganic fertilizers. Hence, soil, water and environment pollution can be reduced. Organic manures are also source of many minor elements which is most important for vegetable production. Organic manures must be applied at least 20–30 days before planting the vegetable crops to avoid immobilization of nitrogen. The application of FYM (20 t ha^{-1}) was significantly superior in increasing the head size and head yield as compared to recommended dose of NPK in cabbage (Chaurasia *et al.* 2004). On the other hand, FYM (60 t ha^{-1}) was found superior among the different manures in cabbage in terms of yield, quality and economics (Zango *et al.* 2009). Vermicompost is rich in plant nutrient than FYM and is an important component of organic farming (Table 1).

The nutrient content of vermicompost varies with the inputs. Using residues of leguminous crops for composting would increase the nitrogen contents. Besides the major nutrients, it is also a rich source of other nutrients *viz* calcium, magnesium, sulphur, zinc, copper, iron etc. (Sudha and Chandani, 2003). These nutrients are easily available to the vegetable crops. It also contains growth promoting substances *viz,* growth hormones, enzymes and vitamins and beneficial microorganisms *viz*, nitrogen fixing, phosphorus solubilizing and cellular decomposing organisms which help in improving soil productivity. Application of vermicompost increases physico-chemical properties of soil particularly increase water holding capacity of soils and reduces water requirement of vegetable crops. Its application also reduces pest and diseases attacks in vegetable crops. Vermicompost is prepared by converting organic wastes into compost through certain species of earthworms. Earthworms are used as the natural bioreactor for decomposition of waste materials. Hence, species of earth worms to be used for vermicomposting should have good survival rate in dense organic bed. *Eisenia fetida*, *Eudrilus eugeniae*, *Perionyx excavatus*, *Lampito mauritii* and *Octochatona serrata* are important species which are being used in vermicomposting in India (Singh and Thomas 2010). The red earth worm or tiger worm (*Eisenia fetida*) is the world's most widely used species for the process of vermicomposting because they need very little care. It can be used at wide range of temperature varying from 0–40ºC. These are hardy and live on wide variety of degradable organic wastes. *Eudrilus eugeniae* is fastest growing, second most widely earthworm. It is popularly known as night crawler. They moderately tolerate disturbance and show some preference in selection of feed substrate. These species can be cultured on animal dung, poultry dropping and other types of biodegradable waste. Earthworms consume all kind of organic matters and excreted it in digested form. Organic matter undergoes complex biochemical changes in intestine and excreted out as granular casts of earthy odour. Earthworms excreta, undigested organic matter and their cocoons make up vermicompost. Nowadays vermicompost is being popularized among vegetable growers. Role of vermicompost as organic component in vegetable production is well established.

Table 1. Approximate NPK values of various organic manures

Organic manures	N (%)	P (%)	K (%)
FYM	0.5	0.2	0.5
Farm compost	0.5	0.15	0.5
Town compost	1.4	1.0	1.4
Pig manure	0.75	0.2	0.5
Vermicompost	1.0	0.5	1.5
Poultry manure	3.0	2.6	1.4
Neem cake	5.2	1.0	1.5

Source: Bora *et al.* (2003)

Nonedible oil cakes are not suitable for feeding the cattle like mahua cake, neem cake, safflower cake etc. can be applied to the soil. They supply nutrients to the crops in about 7–10 days after application. The non-commercial portion of the crops commonly known as straw is used as crop residue. Crop residues can be recycled by composting or used as mulch or incorporated in the soil. On an average, cereal straw and residues on maturity contain about (0.5%) N, 0.6% P and 1.5% K (Katiyar 2000). The quantity of nutrients in legume residues is much high than in cereal straw. Pressmud (by product of sugarcane industry) is also potential source for organic farming of vegetable crops. It contains N (1–2.5%), P_2O_5 (0.5%) and K_2O (0.4–0.8%). Green manuring consists of turning over fast growing legumes. It is one of the most effective and environmentally sound methods of organic farming of vegetable crops. Green manures not only provide organic matter, nitrogen and other nutrients but also improve physico-chemical and biological properties of soil. A large number of fast growing nitrogen fixing legumes crops like dhaincha (*Sesbania acculeata*), sunhemp (*Crotolaria juncia*) and cowpea (*Vigna* sp.) during *kharif* season and berseem (*Trifolium alexandrinum*) during *rabi* season may be used as green manures in organic vegetable production. These crops are ploughed in the soil about 6–8 weeks of sowing when adequate growth is attained at least 3 weeks before transplanting the vegetables. Generally, green manure crops add 100–150 q/ha green matters and 60–100 kg nitrogen/ha (Meisheri *et al.* 2001).

(ii) Biofertilizers

Biofertilizers are very important component in organic vegetable production. In general vegetable crops give good response to biofertilizers. Biofertilizers are the product of living cell of different micro-organism that benefit the plants by providing nitrogen or phosphorus or bring rapid mineralization. Biofertilizers are important component of organic farming of vegetables. Biofertilizers are eco-friendly, economically viable, social acceptable and less expansive. Biofertilizers increase the sustainability of soil and make it more productive. Uses of biofertilizers in combination with organic manures offer a good opportunity to increase the vegetable production and achieve sustainability. They improve growth and yield of vegetables by producing plant growth substances. They act as antagonist for soil borne pathogens. The fertility of soil depends not only on its chemical composition but also on the qualitative and quantative nature of micro-organism inhabiting it. The population of soil microorganism is greater in the rhizosphere than the soil away from the plant roots. They play a role in the carbon, nitrogen, phosphorus and sulphur cycles and availability of certain trace elements like Mn, Cu, and Fe in the soil. Hence, biofertilizers play a significant role in nutrient supply to vegetable crops. However, their performance is not stable due to their biological nature. Neutral pH is generally considered optimum for bioferotilizers. In vegetable crops, biofertilizers are applied through soil inoculation, seed inoculation and seedling inoculation. Seed of vegetable crops should be inoculated @ 500 g/ha. In seedling inoculation method, biofertilizers should be applied @ 2kg/ha. (2 kg of culture in 20 liter of water) and in soil inoculation, biofertilizers should be applied @ 5 kg/ha. The biofertilizers that can be used in vegetable production are nitrogen fixer, phosphate solubilizer and phosphorus mobilizers.

Nitrogenous biofertilizers

Rhizobium is used for leguminous vegetable crops like pea, cowpea and beans and *Azotobacter* and *Azospirillum* are used for non-leguminous vegetables. *Rhizobium* is symbiotic nitrogen fixer bacteria, which fix nitrogen in association with root of leguminous vegetables. *Rhizobium* can fix up to 120 kg N/ha in vegetable crops under optimum conditions. *Rhizobium* inoculation increase nodulation and nitrogen capacity in garden pea and increase pod length, pod girth, number of pods/plant and pod yield in garden pea. *Rhizobium* inoculation increased 15.85 per cent higher pod yield over uninoculated (Kanaujia *et al.* 1998). *Azotobacter* and *Azospirillum* fix up to 30 kg N/ha (Pandey and Kumar 2002). Besides, they also produce growth promoting substances like IAA, GA_3 etc. along with chemicals, which are inhibitory to certain root pathogen. The inoculation of *Azotobacter* and *Azospirillum* increased root length, diameter and yield in carrot and radish as compared to uninoculated. They convert atmospheric nitrogen into ammonical form which is made available to plants. Hence it is very important that biofertilizers should be used right type. Nitrogenous biofertilizers can save 25–50 per cent nitrogen and increase the yield of vegetable crops 15–30 per cent (Kanaujia and Narayan 2003). Chattoo *et al.* (2003) reported that inoculation with *Azotobacter* and *Azospirillum* enhanced 6.67 per cent and 11.09 per cent higher capsicum yield over uninoculated. However, *Azospirillum* proved superior to *Azotobacter* in capsicum. Pandey and Kumar (2002) also noted positive effect of *Azotobacter* on the yield of different vegetable crops. *Azotobacter* and *Azospirillum* biofertilizers when applied with organic manures increase higher yield in vegetable crops. Avitoli *et al.* (2012) reported that *Azotobacter* gave the highest bulb yield (203.33 q ha^{-1}), protein conten (6.30%) and economics in onion when applied with poultry manure (20 t/ha) as compared to FYM (40 t/ha), pigmanure (30 t/ha), vermicompost (10 t/ha).

Phosphatic biofertilizers

Next to nitrogen, phosphorus is the other major nutrient. Phosphatic biofertilizers solubilize fixed phosphorus already present in the soil and make it available to the plant. The phosphatic fertilizers which are applied to crop only 15–20 per cent is consumed by crop, rest get fixed in the soil which is unavailable to the plants. Another factor which effects the utilization of phosphorus is its low mobility in the soil. Phosphate solubilizing microorganism like *Pseudomonas* and *Bacillus* possess the ability to bring insoluble phosphate in soil into soluble phosphate by secreting organic acids. Phosphatic biofertilizers (Phosphotica) can be used with all the nitrogenous biofertilizers *i.e. Rhizobium*, *Azotobacter* and *Azospirillum* biofertilizers. They make more nutrients available to the crops and improve soil properties (soil texture, WHC. etc). The Vesicular Arbuscular mycorrhzal (VAM) are symbiotic association with root of higher plant. VAM produce vesicle and arbuscules inside the root system. Phosphatic biofertilizers can fix 40–100 kg P/ha in vegetable crops under optimum conditions. Phosphatic biofertilizers save 25–50 per cent phosphatic fertilizer and increased yield upto 30 per cent. Carrot and chilli yield are influenced by VAM application. Carrot and chilli yield increased 42.0 per cent and 5.7 per cent respectively over check. (Kanaujia and Narayan 2003).

(iii) Organic fertilizers

A number of approved organic fertilizers are available for organic vegetable cultivation. They are sodium nitrate (mined), potassium sulphate (mined), Rock phosphate, green sand (for K), gypsum, lime, wood ash etc. Certain by product of the meat processing industry such as blood and bone meal have recently come under scrutiny because of food safety concern.

(iv) Crop rotation

Crop rotation is the important way to maintaine soil fertility. It also controls weeds, insect, pest and diseases in organic farming of vegetable crops. Rotation consists in moving the crop to a different plot each year because the same crop takes the same nutrient from the plot year after year. So that finally the soil becomes lacking in the particular nutrients. Deep rooted vegetable (beet, etc.) should follow shallow rooted ones (leafy vegetable, cole crops etc.) as both take their nutrients from different depth of soil. Legumes are essential in all cropping systems because legumes are known to fix atmospheric nitrogen and improve the nitrogen status of soil. Amount of residual nitrogen vary from 30–60 kg/ha.

7. DISEASES AND PESTS MANAGEMENT IN ORGANIC VEGETABLE FARMING

The wide use of synthetic chemicals for control of insect, pest and diseases proved to be effective and convenient but their use is causing serious ecological and social problems. Crop protection is limiting factor in organic farming of vegetable crops. Organic farmers have observed consistently that the potential of diseases under organic condition appears to be lower than under conventional conditions. Some organic disease management practices should be carried out before the planting of vegetable crops, such as soil solarization, collection of debris and burning them. Seed and seedling used in organic farming should be clean and disease free. Remove and dispose of diseased plants to reduce the source of inoculums. The mulch provides a physical barrier between soil and plant and reduces the diseases. Insect, pest and diseases in organic vegetable farming can be controlled by solarization, biopesticides, botanical pesticides, resistant varieties, crop rotation, trap crops, predators, parasites and protected cultivation.

(i) Bio-pesticides

Biopesticides are economically and ecologically sound and do not harm the environment, but effectively reduces the pest population. Recently, encouraging results have been obtained with the application of microbial pathogen, such as necleo polyhedrosis virus (NPV), *Bacillus thuringiensis* and *Trichogramma brassiliensis* suppressed fruit borer in tomato, shoot and fruit borer in brinjal and okra, diamond back moth in cabbage *etc. Trichoderma harzianum* and *Pseudomonas fluorescens* @ 10 g/litre are used against soil borne diseases like damping off, root rot, wilt, collar rot *etc* (Murthy and Nigam 2001).

(ii) Botanical pesticides

Botanical pesticides are having a prominent position in pest management in organic farming of vegetable crops, as they do not leave harmful residues. Certain pesticides or chemicals that are derived from plants or minerals such as Tulsi, Lantana, tobacco decoction (0.05%), sabadilla, rotenone, pyrethrins, ryania, pongamia oil (1%), neem soap (1%), neem oil (1%), neem seed kernel extract (4%), onion extract (5%), neem cake (250 kg/ha) etc., can be used in organic vegetable farming to control pest. Neem based pesticides are getting importance because of their versatile ability to control many pests in vegetable crops. Azadirachtin is the active principle having insecticidal property present in neem tree. At IIHR, Banglore it was observed that the application of neem cake @ 250 kg/ha at planting was very effective in reducing fruit borer population in tomato (Anon., 2003). Singh (2006) reported that biopesticides *Bacillus thuringiensis* and Neem based pesticides were found equally effective in controlling Diamondback moth of cabbage, leaf caterpillar of radish and shoot borer of brinjal as compared to chemical endosulfan. Foliar spray of onion extract (5%) is found effective in reducing podery mildew in vegetable crops.

(iii) Crop rotation

Crop rotations are very effective in breaking insect build up, especially for insects of limited host range. It also enhances the biodiversity which inturn is beneficial to the natural enemies. Vegetable crops belonging to the same family do not grow in the same plot year after year because diseases will cling to the soil and repeatedly attack the crops.

(iv) Disease and pest resistant varieties

Vegetable crops are highly susceptible to number of bacterial, fungal, phytoplasma and viral diseases, insect, pests and nematodes. Indiscriminate use of chemicals to control the diseases, insect, pests and nematodes have several disadvantages, particularly increase in the cost of production and increase environment pollution and residual toxicity of the chemicals (fungicide, insecticides/ pesticides etc.) in edible produce, it causes health hazards. Hence resistance varieties of vegetable crops are important options. Thus disease and insect resistance variety must be considered during selecting variety. Diseases and pests resistant varieties are given in Table 2 and 3 (Singh *et al.*, 2009)

Table 2. Disease resistant varieties of vegetable crops

Crops	Resistant varieties
Tomato	
Fusarium wilt	Columbia, Walter Floraded, HS-110, Meenakshi, Menka, Maitri, Manmohano, Homestead, Rasika, Ratna, Pant Bahar, Rishi
Bacterial wilt	BT-1, BT-10, LE-79, Suraksha, Arka Alok, Arka Abha, Arka Vardan, Arka Abhijit, Arka Shreshta
Early blight	Floraded, Abhiman, Sungro-830, Kalyanpur No.1, S-12
Leaf curl virus	Hisar Anmol, H-36, H-86, H-88, Avinash-II, Sungro-501, Arka Ananya
Brinjal	
Bacterial wilt	BB-7, BWR-12, SM-67, Pant Rituraj, Arka. Keshav, Arka. Nidhi, Arka Neelkant, PPC , Pant Brinjal Hyb-1, Arka Sheel
Little leef	PPC, Hisar, Shymal
Phomopsis blight	Pusa Bhairav, Pusa Anupam
Chilli	
Leaf curl virus	Jowahar-218, Punjab Lal, Pusa Sadabahar, Pusa Jwala, No-5, No-6, No-9, Hyb No-1, Pant C_1, Pant C_2, Jwala Sakshi, Bijli, Agni
Anthracnose	No-5, No-6, No-9, Hyb No-1, Punjab Surkha, Punjab Lal, Jwawar-218, Vardan
Capsicum	
Bacterial wilt	Arka Gaurav
TMV	Green Gold, Indira
Potato	
Early blight	Kufri Sinduri, Kufri Badshah, Kufri Pukhraj
Late blight	Kufri Badshah, Kufri Swarna, Kufri Sutlej, Kufri Chipsona, Kufri Giriraj
Cabbage	
Black rot	Pusa Mukta, Bahar, Pragati, Geetanjali, Harirani Gol, Kranti
Black leg	Pusa Drum Head

Crops	Resistant varieties
Fusarium	Swati, Sarita, Bajarang, Shanti, BSS 44, Swarna, Sonali, Quisto, Vishwas
Cauliflower	
Black rot	PSBK-1, Pusa Shubra, Pawas, Early Himlata, Kranti, Harirani Gole
Downy mildew	Pusa hyb-2, Pawas
Radish	
White rust	Arka Nishant
Pea	
Powdry mildew	JP-4, PP-83, PRS-4, NDVP-4 Arka Ajit (FC-1), Pusa Pragati
Fusarium wilt	JM–1, JM –2, Kala Nagin
Cowpea	
Bacterial wilt	Pusa Komal
Rust	Arka Suman, Arka Garima
French bean	
Powdery mildew	Cnten Pusa Parvati
Common mosaic virus	Pant Anupama, Pusa Parvati, Contender, Pant bean – 2
Rust	Pant Anupama, Arka Komal, Arka Suvidha, Arka Anoop
Angular leaf spot	SVM – 1
Okra	
YVMV	P-7, P-8, Punjab Padmini, Parbhani Kranti, Varsha, Vijay, Vihal, Uphar, Arka Anamika, Arka Abhay, Supriya, Tara, Panchalic, Andhunik, Hisar Unnat, Pusa A-4, Varsha Uphar, No. 7, No. 8, No. 10
Cucumber	
Powdery mildew	Poinsette
Downy mildew	Poinsette
Anthracnore	Poinsette
Angular leaf spot	Poinsette
Musk melon	
Powdery mildew	Arka Rajhans, Punjab Rasila
Downy mildew	Punjab Rasila

Crops	Resistant varieties
Water melon	
Powdery mildew	Arka manik
Downy mildew	Arka manik
Arthacnise	Arka manik
Onion	
Purple blotch	Arka Kalyan, Arka Niketan

Table 3. Insect resistant varieties of vegetable crops

Crops	Resistant varieties
Tomato	
Nematode	Hisar Lalit, Sel - 120 , NT - 3 , NT - 8, NT – 12
Brinjal	
Shoot and fruit borer	Punjab Jamuni, Gole, Punjab Sadabahar, Punjab Barsati, Punjab Neelam, Punjab Chamkila, Anamalai, PPC,
Jassids	Black beauty, Vijay hybrid, Punjab Chamkila, Manjari Gota
Nematode	PPC, Pant Rituraj, Arka Nidhi, Arka Kesav, Arka Neelkanth
Chilli	
Thrips	K-1, Pusa Jwala, Chamtkar
Aphid	Pusa Jwala
Potato	
Nematode	Kufri Swarna
Cabbage	
Aphid	All season, Red Drum Head, Sure Head, Express Mail
Cauliflower	
Stem borer	Early Patna, Kathmandu local
Okra	
Jassid	Punjab Padmini, New selection, Pusa A-4
Shoot & fruit borer	Pusa A-4, Arka Abhay, P-8
Round gourd	
Fruit fly	Arka Tinde

Crops	Resistant varieties
Pumpkin	
Fruit fly	Arka Suryamukhi
Bitter gourd	
Fruit fly	Hisar-II

(v) Trap crops

Trap crops are species of plants which are planted to attract and retain a pest species. Trap crop must be definitely and distinctly attractive to the pest than main crop. Use of mustard and marigold as trap crops has become highly popular in management of pests affecting vegetable crops. Diamond back moth of cabbage can be controlled by mixed crop with mustard and tomato fruit borer can be controlled if grow tomato with marigold.

(vi) Protected cultivation of vegetables

Protected cultivation is another important tool to overcome many pests and diseases in organic production of vegetables. Foliar disease of tomato can be controlled by installing rain shelter under heavy rainfall area. Insect proof net house is used to produce safe vegetables. Vegetable crops like tomato, chilli, sweet pepper, okra etc. under open fields affected by leaf curl virus and yellow vein mosaic. These viruses are mainly spread by insects (vectors) like whitefly or aphids. The population of white fly after on start of monsoons is very high and it remains in the environment up to middle of November depending upon the temperature. It is very difficult to control these vectors and tomato, brinjal, chilli, sweet pepper or okra crops are highly infected with viruses. The second most common and most severe problem in tomato, brinjal and okra is the fruit borer against which also the growers are unable to control this insect perfectly. The only way to control the virus and fruit borer is to put a mechanical barrier between the crops and open environment, and this is possible with the use of insect proof nylon nets of 40 to 50 mesh densities in form of net houses or insect proof net covered walk in tunnels. By this way the growers can grow virus and borer free crops of tomato, sweet pepper, chilli brinjal and okra during rainy season.

8. CONSTRAINTS IN ADOPTING ORGANIC VEGETABLE FARMING

Though organic vegetable farming is one of the best way of sustainable vegetable production. There are some constraints to adopt organic cultivation of vegetables in India. They are

1. Lack of quality organic manures is major difficulty in organic cultivation of vegetables

2. Lack of awareness to among farmers about scientific methods of the making compost and vermicompost.
3. Lack of awareness to nutrients managements.
4. Biofertilizers are not available easily
5. Lack of awareness in biological control of pest.
6. Lack of financial support
7. Farmers do not want to take risk in vegetable production during initial stage.
8. Certified organic seeds or planting materials are not available easily.

9. CERTIFICATION OF ORGANIC VEGETABLE PRODUCE

Certified organic vegetable produce refers to vegetable produce that have grown and processed according to strict uniform standards. Certification includes inspection of farm fields and processing facilities. Farm practices inspected include long term soil management, buffering between organic farms and any neighbouring conventional farms, product leveling and record keeping. Processing inspection includes review of the facility's cleaning and pest control methods, ingredient transportation and storage and record keeping. With passage of the US organic Food Product Act of 1990, all organic food will be required to be certified to meet the National and International standards. The International Federation of Organic Agriculture Movement (IFOAM) has developed basic standards for organic agriculture including organic vegetables. The IFOAM India has prepared and adopted the basic standard as applicable to Indian certification. India has implemented the national standards for organic products under the National Programme for Organic Production (NPOP). A trademark of "Indian Organic" will also be granted on the basis of compliance with national standards for organic production. Agricultural and Processed Food Products Exports Development Authority (APEDA) is responsible for export of organic vegetable products. APEDA has been assisting the Ministry of Commerce in preparing various documents under the National Programme for Organic Production (NPOP) and has also developed the National Organic Logo. In order to vegetables certified as organic, crops must be grown on land which has been free from prohibited substances for three years prior to harvest. A transition period of 3 year is required during which no inorganic fertilizer and agro-chemicals should be used. Crops grown on land in transition to organic from conventional farming can not be labeled as organic. Presently, eleven certification agencies are accredited by APEDA. These are listed below with their branch.

1. Indian Organic Certification Agency (Indocert), Cochin
2. IMO, Control Pvt. Ltd., Bangalore
3. SKAL International, Bangaluru
4. ECOCERT International, Aurangabad

5. SGS India Pvt. Ltd., Gurgaon
6. BCVQI (India) Pvt. Ltd., Mumbai
7. International Resources for Fairer Trade, Mumbai
8. Lacon Quality Certification Pvt. Ltd., Theepany, Kerala
9. Natural Organic Certification Association, Pune
10. One Cert Asia Agri Certification Pvt. Ltd., Jaipur
11. Uttranchal State Organic Certification Agency, Uttranchal

CONCLUSION

Vegetables are highly prospective for organic cultivation as they are consumed fresh. Organic vegetable cultivation can supply safe vegetables for human consumption and export. India is land of small holdings. Organic farming on such small holdings can be promoted. Most of the North Eastern states of India are having virgin land without using any chemicals. These states are very much suitable for organic cultivation of vegetable crops. There have been significant advancements in the developments of organic vegetable cultivation. Now-a-days scientific methods are available for making organic manures like compost, vermicompost etc. Organic manures coupled with biofertilizers may be applied to meet the nutrient requirement of vegetable crops in organic vegetable production without any ecological damage. Crop rotation is the important way to maintaining soil fertility. It also controls weeds, insect, pest and diseases in organic farming of vegetable crops. Insect, pest and diseases in organic vegetable farming can be controlled by solarization, biopesticides, botanical pesticides, resistant varieties, crop rotation, trap crops, predators, parasites and protected cultivation. Organic standards, certification, labeling and inspection are most important for organically grown vegetable crops.

REFERENCES

Anonymous (2003) Use of neem cake in pest management of vegetables. IIHR website.

Avitoli K, Singh AK, Kanaujia SP and Singh VB (2012) Quality production of kharif onion (*Allium cepa*) in response to biofertilizers inoculated organic manures. *Indian Journals of Agricultural Sciences* **82(3):** 236–240.

Bora SS, Bora TS and Rajput PR (2003) Organic farming: a way to sustainability. *Indian Farmers Digest*, March pp 7–10.

Chattoo, M , Zargar, MY, Mir, SA, Wani, MS and Kanaujia SP (2003) Effect of *Azotobacter* and *Azospirillum* inoculation on growth and yield of capsicum. *Applied Biological Research* **5:** 10–14.

Chaurasia, SNS, De Nirmal and Singh KP (2004) Effect of organic manures on the growth and yield attributing characters of cabbage. In: Proceeding of National Symposium on Organic Farming in Horticulture (Eds Pathak RK, Kishan R and Khan RM), CISH, Lucknow.

IHD (2014) Indian Horticulture Database, National Horticulture Board, Gurgoan, India.

Katiyar AK (2000) Organic wastes also preserve soil fertility. *Indian Farmers Digest*, January, pp 29–31.

Kanaujia SP and Narayan R (2003) Biofertilizers for sustainable vegetable production. *Intensive Agriculture*, May-June, 15–17.

Kanaujia SP, Sharma, SK and Rastogi, KB (1998) Effect of phosphorus, potassium and *Rhizobium* inoculation on growth and yield of pea. *Ann. Agric. Res.* **19(2):** 219–221.

Meisheri TG, Usadadia VP, Vaidya AC and Oatel JB (2001) Green manures. *Indian Farming*, Nov. pp. 11–12.

Murthy KS and Nigam GL (2001) Organic farming and pest management. *Indian Farming*, **51(3):** 27–28.

Pandey V and Kumar D (2002) Biofertilizers for sustainable Agriculture. *Agriculture Today*, September. Pp. 44–47.

Singh, SR (2006) Response of biofertilizers and pesticides on yield and quality of cabbage, radish and brinjal in vegetable based rotation system. *Applied Biological Research* **8 (1/2):** 33–36.

Singh, HP and Thomas GV (2010) Organic Horticulture Principles, Practices and Technologies. Westville Publishing House, New Delhi.

Singh HP, Rai M, Pandey S and Kumar, S (2009) Vegetable varieties in India. Studium Press (India) Pvt . Ltd. New Delhi.

Sudha B and Chandani S (2003) Vermicompost – A potential organic manure for rice. *Indian Farming*, Nov. pp. 21–22.

Zango K, Kanaujia SP, Singh VB and Singh PK (2009) Effect of manures and biofertilizers on growth, yield, and quality of cabbage (*Brassica oleracea* var. *capitata*) under foot hill condition of Nagaland. *Environment and Ecology* **27(3):** 1127–1129.

14 Integrated Crop Management Practices for Enhancing Productivity, Resource use Efficiency, Soil Health and Livelihood Management

Pradip Tripura

Department of Agricultural Chemistry and Soil Science, College of Agriculture, JAU, Junagadh, Gujarat, India
Email: prdpt001@gmail.com

ABSTRACT

Integrated crop management (ICM) is a holistic approach to sustainable agriculture. It considers the situation across the whole farm, including socio-economic and environmental factors, to deliver the most suitable and safe approach for long-term benefit. This means carefully considering site selection, soil management, seed and planting material, crop rotation, crop nutrition, pest management, water management and landscape management that fit the local conditions and climate. ICM is not rigidly-defined. It is a dynamic system that adapts to changing conditions by combining local knowledge with new research and technologies. We often work where resources for production are limited, and in these situations the best solution is often to focus on optimising existing resources. In such a way, ICM delivers sustainable agricultural production that safeguards a farm's natural assets and surrounding community, now and in the future.

1. INTRODUCTION

The ICM team uses a multi-stakeholder approach to work with local partners at all levels, from governments to farming households. Together, we promote the every-day use of ICM as a long-term strategy to improve agricultural production, farmer livelihoods and food security. We work through steps of participatory development, validation, implementation and finally dissemination of economically-viable ICM solutions. Through continuous exchange of knowledge with our local partners and experts from other CABI centres around the world, we ensure that ICM conceptual frameworks are developed according to local needs. Building capacity of local partners to monitor and evaluate their work empowers them to adapt their approaches when needs change (Fig. 1).

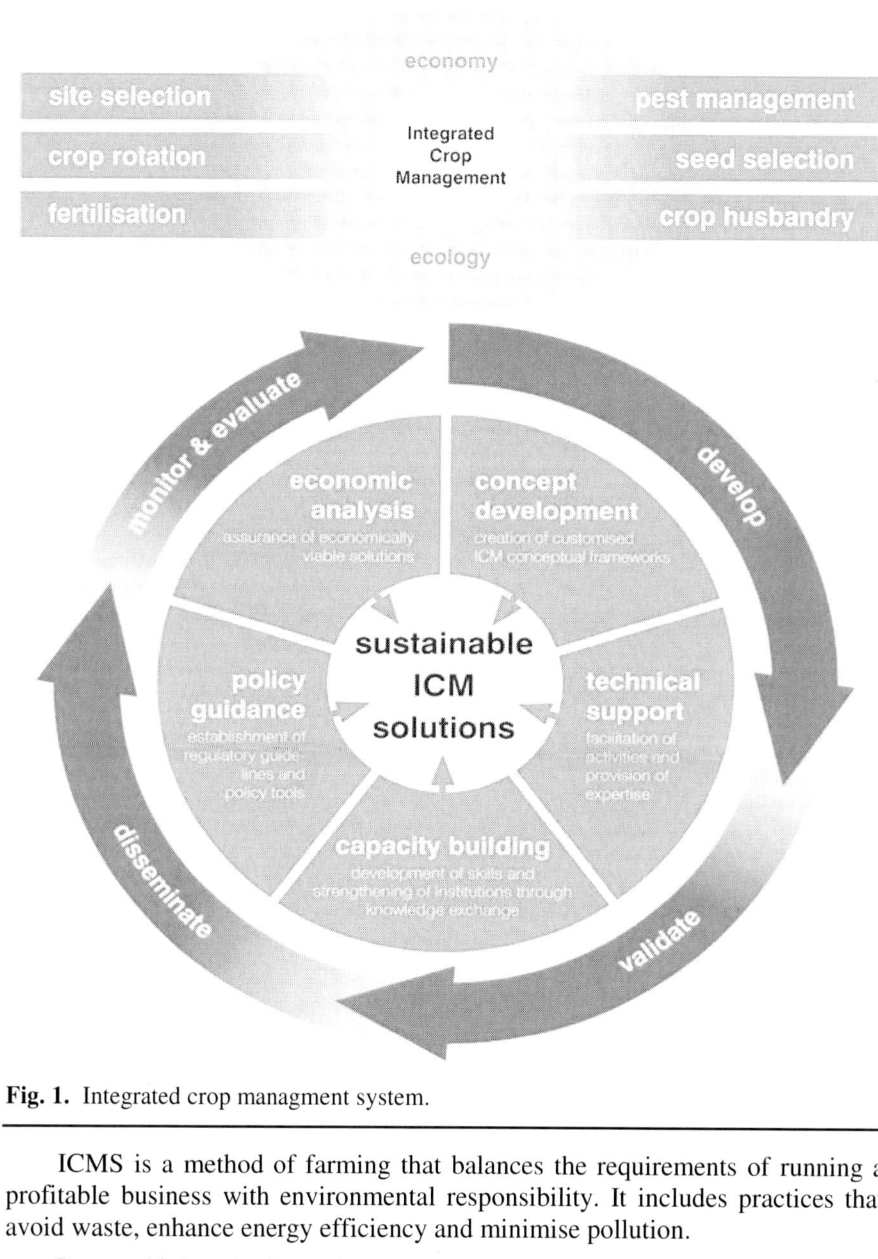

Fig. 1. Integrated crop managment system.

ICMS is a method of farming that balances the requirements of running a profitable business with environmental responsibility. It includes practices that avoid waste, enhance energy efficiency and minimise pollution.

By combining the best of modern technology with good farming practice, we provide a whole-farm, long-term strategy. It is not a quick fix that can be applied to one crop, or one field or one season. Although primarily concerned with crop production, livestock management is equally important on mixed farms because livestock are consumers of crops and providers of organic nutrients (Kumar and Jain, 2005) (Fig. 2).

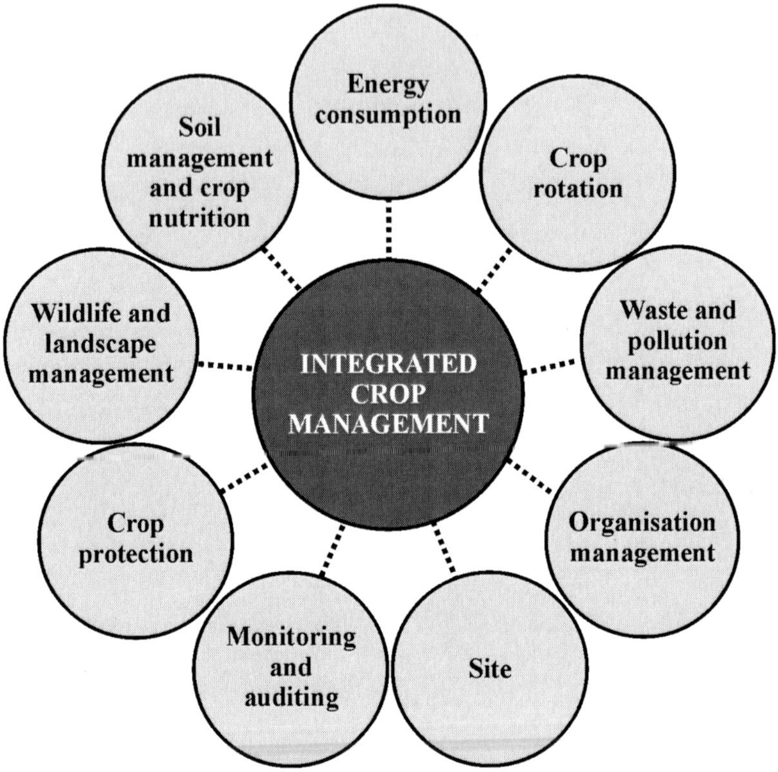

Fig. 2. Integrated crop management

2. THE ROLE OF ECM'S AGRONOMIST

Our agronomists have vast experience of working in partnership with farmers to assess, monitor and plan all of their activities in an integrated way that both respects biodiversity and ensures profitability. Since our system involves the whole farm and is site-specific, there are no hard and fast rules about how we achieve this. Individual farms differ in many ways: location, climate, soil type and cropping pattern, to name a few. Even so, we have established general guidelines that can help all farmers and growers take practical steps to improve their management practices. These cover everything from use of fertilisers and crop protection products to the maintenance and enhancement of landscape features and wildlife habitats. Whatever the farm, wherever it is located, planning, taking action and monitoring outcomes are the vital processes in our system.

3. WHY WE NEED INTEGRATED CROP MANAGEMENT

Providing expanding population with enough to eat is a huge challenge. How we meet this challenge and at the same time respect biodiversity and take into account climate change are matters of the utmost importance. Our Integrated Crop Management System allows farming to be practised in a way that safeguards the environment. At the same time it recognises that the quality, quantity and price of produce, the profitability of the farm and the adoption of new technologies are all essential if the overall stability of agriculture is to be preserved (Alam *et al.* 2013). The ICM is a method of farming that balances the requirements of running a profitable business with responsibility and sensitivity to the environment. It includes practices that avoid waste, enhance energy efficiency and minimize pollution. ICM combines the best of modern technology with some basic principles of good farming practice and is a whole farm, long term strategy.

ICM is a 'whole farm approach' which is site specific and includes:

- The use of crop rotations.
- Appropriate cultivation techniques.
- Careful choice of seed varieties.
- Minimum reliance on artificial inputs such as inorganic fertilisers, pesticides and fossil fuels.
- Maintenance of the landscape.
- The enhancement of wildlife habitats.

One of the main objectives of ICM is the reduction or replacement of external farm inputs, such as inorganic fertilizers, pesticides and fuel, by means of farm produced substitutes and better management of inputs. Total replacement is not possible without significant loss of yields, but partial substitution of inputs can be achieved by the use of natural resources, the avoidance of waste and efficient management of external inputs. This would then lead to reduced production cost and less environmental degradation. The principals and practices of ICM are outlined below:

3.1. Crop protection

- Increase diversity of crop species to prevent disease and pest carry over from crop to crop.
- Ensure effective nutrient uptake by scheduling crops with different nitrogen demands in the correct sequences.
- Preserve soil fertility, structure and minimize erosion by ensuring adequate crop cover, good rooting depth and reduction of compaction.
- Use of disease resistant cultivars to minimize the need for agro chemical inputs.

3.2. Soil protection

Minimal cultivations to:

- Reduce energy usage (*i.e.* fuel).
- Reduce soil erosion.
- Reduce adverse effects on soil invertebrates such as earthworms and predatory beetles and spiders.
- However, there should be effective seedbed preparation and crop establishment.
- Cultivations dependent on soil type, climate and topography of individual farms.

3.3. Crop nutrition

Nutrient inputs should be carefully balanced in respect of:

- Individual crop requirements.
- Crop off takes.
- Soil residues and residues from previous crop.
- Regular soil analysis is recommended.
- Use of cover crops/green manures before spring sown crops to minimize leaching and erosion.

3.4. Integrated pest management

- Minimal use of well selected pesticides, *i.e.* ones that have minimal off target effects.
- Alternative husbandry techniques such as mechanical weeding.
- In crop monitoring systems (such as traps) to assess pest levels to scale pesticide use to the level of the problem.
- Improve habitat for predators to increase natural level of biological control.

3.5. Wildlife and landscape

Planning a programme for the whole farm (cropped and non-cropped areas) to enhance biodiversity and landscape features:

- Hedges, ditches, field margins, beetle banks and conservation headlands allowing wild species to establish and migrate, and to provide recreational areas for people.
- A greater diversity of broad leaved weeds may be left within crops to provide food sources for birds and insects, providing the aggressive crop damaging weeds are contained.

3.6. Energy efficiency (Fig. 3)

- Detailed analysis of energy use, especially fossil fuels.
- Remedial action to minimize waste.
- Consider alternative energy sources.
- Change in cultivation practice, *i.e.* less passes.
- Replacement of high fuel consumption machinery, with more efficient alternatives.
- Rationalization of vehicle movements.

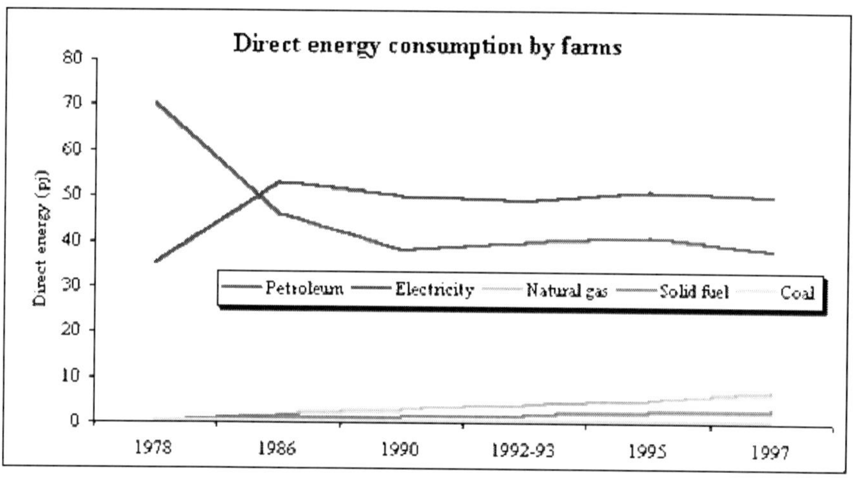

Fig. 3. Direct energy consumption by farms

3.7. Pollution and waste

- Strict following codes of practice.
- Recycling of crop residues.
- Recycling or safe disposal of non-organic wastes.

4. ICM IN PRACTICE

The environmental benefits of ICM are difficult to quantify and are related to longer term processes. On the long term projects, biodiversity has increased, there have been improved bird numbers and reduced nitrate leaching and soil erosion. Data from the experimentation, trial farms and various projects has indicated (Roling and Wagemakers, 1998):

- Generally a 5–15 per cent yield reduction, but indications that this is reducing as experience grows.

- Quality of produce is generally maintained.
- Variable costs reduced by 20–30%.
- Gross margins maintained or slightly increased.
- Pesticide inputs reduced by 30–70%.
- Nitrogen inputs reduced by 16–25%.

5. ENHANCING PRODUCTIVITY AND RESOURCE USE EFFICIENCY BY ICM

The term 'resource use efficiency in agriculture' may be broadly defined to include the concepts of technical efficiency, allocative efficiency and environmental efficiency. An efficient farmer allocates his land, labour, water and other resources in an optimal manner, so as to maximise his income, at least cost, on sustainable basis (Haque, 2000). However, there are countless studies showing that farmers often use their resources sub-optimally. While some farmers may attain maximum physical yield per unit of land at a high cost, some others achieve maximum profit per unit of inputs used (Chaudhary and Harrington, 1993). Also in the process of achieving maximum yield and returns, some farmers may ignore the environmentally adverse consequences, if any, of their resource use intensity.

A more intensive use of natural resources, causes increasing pressure on biological systems and on production, and consumption systems in agriculture. Due to increasing global urbanisation, the distribution of the flow of energy, materials and waste is shifting on a large scale, and the quality of ecosystems (for example soil and water) is under pressure (Jayanthi *et al.* 2000). The flow of nutrients, residue and waste is becoming concentrated in prosperous, densely populated urban areas, while elsewhere the soil is becoming depleted.

We are investing in the transition to a more sustainable and efficient system of production and consumption. We are developing new knowledge and technology to deal more efficiently with the available raw materials. We are changing primary production streams through new combinations of various sectors, product groups and raw materials. We are working on the intelligent closure of previously separate cycles of energy, materials and nutrients through the optimal use of plant and animal sources (Singh *et al.* 1998). In addition, we are contributing to a high-efficiency revolution in the use of raw biological materials through the efficient conversion and distribution of streams, raw materials and products, as well as the prevention of waste and disposal of nutrients (Peng *et al.*, 2015). This transition is also accompanied by new business opportunities and risks, which were previously unknown. For the effective support of this transition, governance is crucial; new networks must be formed between the most important actors in production and consumption, innovations in institutions must be required and new practices must be developed. Socio-economic analyses show how this transition can be shaped.

Integrated crop management (ICM) is a combination of the traditional methods with appropriate modern technology, balancing the economic production of crops with positive environmental management (Mishra, 2013). Basic components of ICM are crop management, nutrient management and pest management. Through the process of ICM, farmers make better use of on-farm resources. One of the main objectives of ICM is the reduction of external farm inputs, such as chemical fertilizers and chemical insecticides or pesticides (Kumar and Singh, 2014). Total replacement is not possible without significant loss of yields, but partial substitution of inputs can be achieved by the use of organic inputs (Chu *et al.* 2016). This would then leads to reduced production cost and less food contamination and environmental degradation (Ashby, 2001). The major components of ICM strategies and approaches are seedling establishment, integrated nutrient management (INM), integrated pests and disease management (IDM/IPM) and integrated weed management (IWM) (Kumar and Shivay, 2008). There is increasing pressure on vegetable producers in the country to intensify their production. For the local farmers, information on sustainable and profitable permanent vegetable production systems is of high value. A major limitation in application of INM and IPM is the lack of trained personnel. Many farmers are not trained adequately in use of organic fertilizer and augmentative biological control, leading to misunderstanding of its potential efficacy. Farmers often believe that organic fertilizer and natural enemies do not work well. The use of bio-pesticides is limited due to moderate toxicity and slow action. Many farmers are not yet aware of the proper usage of bio-control agents and bio-pesticides. A number of microbial consortium and botanicals such as neem, karanj, pongmia, mahua, garlic and tobacco, have been found to be effective against insect pests and diseases (Mamun and Ahmed, 2011). INM and IPM adoption is influenced by the cost versus efficacy of products, ability to integrate new products and techniques into existing farm management practices and managerial skills. In this way ICM enhance the resource use efficiency, the livelihood security, increasing the productivity of oilseed and pulses crops (Gautam *et al.*, 2007).

Balanced application of appropriate fertilizers is a major component of INM. Over application of fertilizers induces neither substantially greater crop nutrient uptake nor significantly higher yields (Qin *et al.*, 2013). Rather, excessive nutrient applications are economically wasteful and can damage the environment. Under application, on the other hand, can retard crop growth and lower yields in the short term, and in the long term jeopardize sustainability through soil mining. Balanced fertilization should also include secondary nutrients and micronutrients, both of which are often most readily available from organic fertilizers. Improved application and targeting of inorganic and organic fertilizer not only conserves nutrients in the soil, but makes nutrient uptake more efficient (Ghosh *et al.*, 2015). This paper contributes to the improvement of vegetable production using tools of ICM techniques like INM and IPM, leading to more sustainable production methods and higher farmer's incomes.

Obtained results also showed that not only vegetable yield, but also the efficiency of fertilizer application in vegetable production were improved with the application of the INM system. The increase in fertiliser efficiency due to the application of the INM technology not only reduced production costs but also the negative effects of nutrient losses on the environment (Haque, 1998). Microbial consortium contains N fixing, P and Zn solubilizing and plant growth promoting microbes as a single formulation and helped in reducing N and P fertilizer requirement by 25–30 per cent and also increases yield of 13–20 per cent in vegetables. Foliar application of secondary and micronutrient formulation supplied essential nutrients other than NPK and increases resistance to diseases.

These nutrients are also necessary for enhancement of fruit appearance, fruit keeping quality and taste of vegetables. The soap spray of Neem/pongmia reduces the insect-pests problem to a great extent. Seed treatment with *Trichoderma harzianum* manages fungal pathogens, such as *Fusarium* species and *Phytophtora* species as well as nematodes. The IVCM system provides recommendations that are the 'best management practices' for vegetable growing, based on knowledge from this on-farm trial and farming experiences. The output recommendation serves as a criterion for the evaluation of farmers' success in managing profitable vegetable crop establishment. Adopting improved practices like vegetable integrated crop management (IVCM) enhanced the yield more than 100 per cent over farmer's practice. IVCM system with improved crop variety, INM and IPM paid more dividends with BC ratio more than 7.0 in all the three vegetable crops (Table 1).

Table 1. Effect of conventional, INM, IPM and ICM system on vegetable

Crop Management System	Tomato	Green Chilli	Brinjal
Yield under conventional method (t/ha)	60.40	32.30	35.30
Yield under INM method (t/ha)	69.90	34.60	35.90
Yield under IPM method (t/ha)	71.35	35.15	38.70
Yield under ICM method (t/ha)	72.50	36.57	42.00
Yield under farmer's practice (FC) (t/ha)	25.50	16.00	20.40
% change of ICM over FC	+190	+128	+106
Net profit of ICM (Rs. in Lakh)	5.39	4.64	3.12
Cost of cultivation of ICM (Rs. in Lakh)	0.65	0.48	0.41
B:C (ICM)	8.29	9.67	7.61

(*Source*: Singh and Gohain, 2017)

Information on maximum grain yield and its attributes are limited for double-season rice crop grown under the subtropical environment. This study was conducted to examine key characteristics associated with high yielding double-season rice crop through a comparison between an integrated crop management (ICM) and farmers' practice (FP). Field experiments were conducted in the early and late seasons in the subtropical environment of Wuxue County, Hubei Province, China in 2013 and 2014. On average, grain yield in ICM was 13.5 per cent higher than that in FP. A maximum grain yield of 9.40 and 10.53 t ha^{-1} was achieved under ICM in the early- and late-season rice, respectively. Yield improvement of double-season rice with ICM was achieved with the combined effects of increased plant density and optimized nutrient management. Yield gain of ICM resulted from a combination of increase in sink size due to more panicle number per unit area and biomass production, further supported by the increased leaf area index, leaf area duration, radiation use efficiency, crop growth rate, and total nitrogen uptake compared with FP. Further enhancement in the yield potential of double-season rice should focus on increasing crop growth rate and biomass production through improved and integrated crop management practices (Table 2).

Table 2. Integrated crop management practices for maximizing grain yield of double-season rice crop

Year	season	treatment	Daily grain yield (kg ha^{-1} d^{-1})	Total dry weight (g m^{-2})	N uptake (kg ha^{-1})	1000-grain weight (g)
2013	Early	N0^{a}	69.0^{c}	1136.0^{c}	109.4^{c}	23.3^{a}
		FP	89.9^{b}	1376.1^{b}	172.2^{b}	22.2^{c}
		ICM	101.5^{a}	1911.5^{a}	237.9^{a}	22.7^{b}
		Mean	86.8	1474.5	173.2	22.7
	Late	N0^{a}	73.2^{c}	1354.4^{c}	114.9^{c}	22.2^{a}
		FP	84.6^{b}	1537.0^{b}	145.6^{b}	22.4^{a}
		ICM	100.3^{a}	1976.3^{a}	212.2^{a}	22.4^{a}
		Mean	86.0	1622.6	157.6	22.3
2014	Rearly	N0^{a}	60.7^{c}	865.4^{c}	89.6^{c}	25.2^{a}
		FP	98.1^{b}	1494.8^{b}	176.4^{b}	24.1^{b}
		ICM	110.6^{a}	1702.2^{a}	205.7^{a}	24.1^{b}
		Mean	89.8	1354.1	157.3	24.5
	Late	N0^{a}	78.8^{c}	1185.0^{c}	128.4^{c}	22.8^{b}
		FP	99.9^{b}	1488.8^{b}	162.8^{b}	22.9^{b}
		ICM	109.8^{a}	1860.7^{a}	208.3^{a}	23.2^{a}
		Mean	96.2	1511.5	166.5	22.9

(*Source*: Wang *et al.*, 2017)

Rice crops were grown in their paddies using their normal (farmer) practices and ICM in the 2007 season. In this experiment the input costs, labour and yield were recorded. ICM practices increased yields and gross margins and reduced labour and costs compared to farmer practices (Table 3).

Table 3. Yield, labour requirement, cost and gross margin between Farmer Practice (FP) and Integrated Crop Management (ICM) in Timor-Leste.

Parameter	**FP**	**ICM**
Yield (ton/ha)	2.5	3.5
Labour (person day/ha)	113	91
Cost ($/ha)	462	440
Gross margin ($/ha)	-$9.90	$103.70

(*Source*: Ogoshi *et al.*, 2008)

The increased yield was not as high as expected due to an in-season drought and severe locust infestation. On-farm tests conducted by the SM CRSP in 2005 produced yields as high as 5 tons/ha.

6. SOIL HEALTH BENEFITS FOR SUSTAINING CROP PRODUCTION

The terms 'soil health' and 'soil quality' are becoming increasingly familiar worldwide. A modern consensus definition of soil health is "the continued capacity of the soil to function as a vital living ecosystem that sustains plants, animals and humans". In general, soil health and soil quality are considered synonymous and can be used interchangeably, with one key distinction conceptualized by scientists and practitioners over the last decades: soil quality includes both inherent and dynamic quality.

The benefits of healthy soil in sustaining crop production are most evident when growing conditions are less than ideal (Craswell, 1998). Healthy soils increase the capacity of crops to withstand weather variability, including short term extreme precipitation events and intra-seasonal drought. The extreme drought in 2012 resulted in variable yield reduction to corn and soybean production in Iowa with the worst impact on fields with conventional tillage systems (i.e., chisel plow, deep ripping, etc.). Increasingly highly variable weather conditions present increased risks to crops and require more careful attention to conservation planning to mitigate impacts on soil health and crop productivity (Selvarajan *et al.,* 1997).

Soil health is defined by the level to which it is able to continually provide multiple functions to sustain plants, animals, and humans' lives. The complex biological, physical, and chemical interlink of a healthy soil can influence plant water availability under dry conditions, off-field nutrient losses to nearby streams during rain events, and the availability of nutrients through nutrient cycling for food and fibre production (Regmi *et al.,* 2009). Furthermore, healthy soils maintain or enhance water and air quality through the improvement of soil C storage and water infiltration, and support human health and wildlife habitat.

6.1. What influence soil health?

Soil management practices, cropping systems, and weather conditions influence soil health. Therefore, a healthy soil that is well managed can increase soil water infiltration and storage, storage and supply of nutrients to plants, microbial diversity, and soil carbon storage. Soil organic matter (SOM) is a central soil property that is heavily affected by management practices, which in turn influences soil physical, biological, and chemical functions.

The relationships between soil organic matter and management inputs such as tillage and cropping systems can be documented through the evaluation of soil health indicators (Fig. 4). Those indicators reflect the level of response of the soil system to different management inputs. Field and laboratory evaluation of these different indictors can aid in fine tuning management practices to optimize soil biological, physical, and chemical functions.

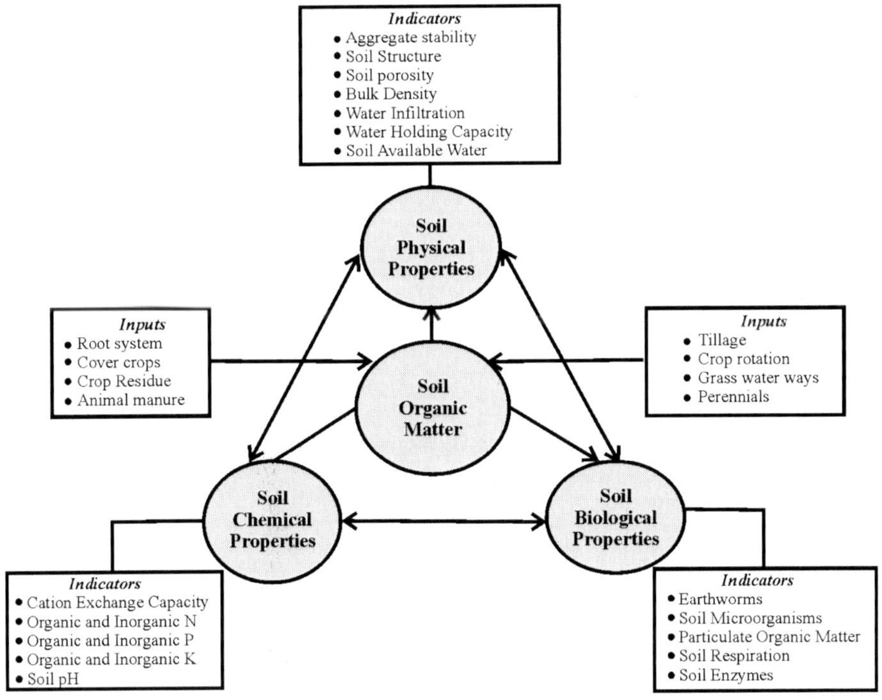

Fig. 4. Soil health indicators and systems inputs

The central soil property that influences soil functions is organic matter. The organic matter component of the soil system is only a small fraction of the topsoil horizon (ranging from 1–5% or greater by dry weight depending on the soil type and other formation factors), but essential for the soil physical, biological, and chemical functions and general soil ecosystem services. The key services for agriculture production are: nutrient provision and cycling, pest and pathogen protection, production of growth factors, water availability, and formation of stable aggregates to reduce the risk of soil erosion (Solanes and Villarreal 1999). However, these functions are sequentially influenced by each other starting with organic matter as the building block for the well linked functions.

6.2. Tillage effects on soil health

The increased use of intensive tillage and other management practices in row crop production systems can increase soil erosion, reduce soil health and water quality, and the capacity to achieve sustainable agricultural production systems. Soil erosion is always associated with tillage intensity, especially during the spring season when soils are most vulnerable to water erosion due to lack of vegetation or residue cover to protect the soil surface from high rain intensity.

Many factors contribute to this problem, but tillage is the prime contributing factor. Soils under modern production agriculture have lost significant amount of their carbon pool because of erosion, decomposition, and leaching. The magnitude of soil organic carbon (SOC) loss from cultivated soils in the Midwest region of the United States is estimated to be in the range of 30 per cent to 60 per cent of the amount present under virgin soil conditions since the conversion from prairie system in late 1800s. This loss in soil organic matter by cultivation is in part caused by the oxidation of organic matter and CO_2 release in addition to losses through surface runoff and soil erosion.

6.3. Conservation practices improve soil health

Soil management and conservation practices that protect soil health are not only economically and environmentally necessary, but the right approach to sustain and increase soil resiliency. This can be achieved by adopting conservation plans that are practical, site specific and an integral component of the overall agriculture production system to achieve intended objectives.

These conservation plans would include no-tillage and reduced tillage (i.e., strip-tillage), which leave post-harvest crop residue to cover the soil surface (Fig. 5). In addition, many soil conservation plans include practices, such as cover crops, the construction of grass waterways, terraces, buffer strips, and pasture erosion control systems with manure application and soil testing.

Summary

- The benefits of healthy soils in sustaining crop production are most evident when growing conditions are less than ideal. Healthy soils increase the capacity of crops to withstand weather variability and short term extreme precipitation events and intra-seasonal drought.

- Soil health functionality is highly influenced by soil organic matter, a central soil property that influences soil physical, biological, and chemical functions.
- The interrelationships between soil organic matter and management inputs such as tillage and cropping systems can be documented through the evaluation of soil health indicators of biological, physical, and chemical properties.
- Best management practices that build soil health and sustain productivity are many, and can lead to better ecosystem and societal services. The implementation of such practices should be considered on regional and site specific basis. Site specific adoption of different tillage and conservation practices integrated within the overall production system can effectively increase crop productivity and soil ecosystem services. These conservation practices include, but are not limited to, no-tillage, strip-tillage, cover crops, perennials, grass waterways, terraces, buffer strips, and other measures for erosion control.
- A systems approach to conservation management of row cropping systems is important to enhance soil health and improve water quality. As an effective solution to building soil health and improving water quality, conservation practices should be an integral and essential component of nutrient reduction loss, sediment and nutrient loading plans.

Fig. 5. Corn grown in No-till system

7. LIVELIHOOD MANAGEMENT

A person's **livelihood** refers to their "means of securing the basic necessities - food, water, shelter and clothing of life". Livelihood is defined as a set of activities, involving securing water, food, fodder, medicine, shelter, clothing and the capacity to acquire above necessities working either individually or as a group by using endowments (both human and material) for meeting the requirements of the self and his/her household on a sustainable basis with dignity (Shivaji *et al.,* 2013). The activities are usually carried out repeatedly. For instance, a fisherman's livelihood depends on the availability and accessibility of fish.

The concept of **Sustainable Livelihood (SL)** is an attempt to go beyond the conventional definitions and approaches to poverty eradication.

These had been found to be too narrow because they focused only on certain aspects or manifestations of poverty, such as low income, or did not consider other vital aspects of poverty such as vulnerability and social exclusion. It is now recognized that more attention must be paid to the various factors and processes which either constrain or enhance poor people's ability to make a living in an economically, ecologically, and socially sustainable manner.

The SL concept offers a more coherent and integrated approach to poverty. The sustainable livelihoods idea was first introduced by the Brundt and Commission on Environment and Development, and the 1992 United Nations Conference on Environment and Development expanded the concept, advocating for the achievement of sustainable livelihoods as a broad goal for poverty eradication. In 1992, Robert Chambers and Gordon Conway proposed the following composite definition of a sustainable rural livelihood, which is applied most commonly at the household level: "*A livelihood comprises the capabilities, assets (stores, resources, claims and access) and activities required for a means of living: a livelihood is sustainable which can cope with and recover from stress and shocks, maintain or enhance its capabilities and assets, and provide sustainable livelihood opportunities for the next generation; and which contributes net benefits to other livelihoods at the local and global levels and in the short and long term.*" People are getting poor due to inflation.

Integrated crop management imperative to develop strategies and agricultural technologies that enable adequate employment and income generation, especially for small and marginal farmers who constitute more than 80 per cent of the farming community (Behera *et al.*, 2004). The integrated crop management approach is considered to be the most powerful tool for enhancing profitability of farming systems. These integrated crop management required to be planned, designed, implemented and analyzed for increasing productivity and profitability (Xue *et al.,* 2012). These systems also need to be socially acceptable, economically viable and eco-friendly. Integration of enterprises lead to greater dividends than single enterprise based farming, especially for small and marginal farmers. It also leads to improvement in nutritional quality of daily diet of farmers.

7.1 Why integrated crop management for livelihood management

7.1.1. *Deteriorating resource base*

During post-green revolution period, our attempt to solve food problem and attain self-sufficiency in food production through excess use of agrochemicals, inevitable dependence on irrigation and high cropping intensity has led to contamination of food with harmful chemicals, pollution of ground water, degradation of soil quality and damage to agriculturally beneficial micro-organisms. In many regions both surface and ground water are already becoming unfit for human and animal consumption due to high concentration of pesticide residues. Available estimates revealed that nearly 120.72 million ha of land in the country is degraded. Intensified agriculture, coupled with indiscriminate use of irrigation water and no judicious fertilizer application, especially in irrigated areas of the country has led to various kinds of physical and chemical degradation of the soil (Prasad, 2003).

7.1.2 *Climate change*

The increasing green-house gases resulted in global warming. The Intergovernmental Panel for Climate Change (IPCC) projections on temperature predicts an increase of 1.8 to 4.0°C by the end of this century. Temperature and sea level changes will affect agriculture through their direct and indirect effects on crops, soils, livestock, fisheries and pests. The brunt of environmental changes is expected to be very high in India due to greater dependence on agriculture, limited natural resources, alarming increase in human and livestock population, changing pattern in land use and socio-economic factors that pose a great threat in meeting the food, fibre, fuel and fodder requirement. Recent studies done at the Indian Agricultural Research Institute indicated the possibility of loss of 4–5 million tonnes in wheat production in future with every rise of 1°C temperature throughout the growing period.

7.1.3 *Narrowed biodiversity*

The narrowing of genetic biodiversity occurs as traditional crop varieties and local animal breeds are being replaced by modern ones. These new varieties/ breeds are certainly better matched to modern intensive agriculture, but rarely any consideration is given to preserving the bio-diversity of an agricultural ecosystem (Jirli *et al.*, 2008). In addition, the increased farming density tends to erode the biodiversity of flora and fauna in the agricultural ecosystems. For example, extensive adoption of rice-wheat monoculture in the Indo-Gangetic Plains has replaced the other traditional crops (Chand and Haque, 1997). Soil micro-flora is also adversely influenced on account of large-scale use of agro-chemicals and lack of recycling of crop residues in the region.

7.1.4 *Multiplicity of integrated farming systems*

Very often, almost all Indian farmers, in pursuit of supplementing their needs of food, fodder, fuel, fibre and finance resort to adopt integrated farming systems, majority of them revolving around the crops + livestock components (Devasenapathy *et al.,* 1995). Livelihood of small and marginal farmers, comprising more than 80 per cent of total farmers, depends mainly on crops and livestock, which is often affected by weather aberrations (Jayanthi *et al.,* 2009). Under present scenario, in the absence of scientifically designed, economically profitable and socially acceptable appropriate integrated farming systems models, they are unable to harness the benefits of integration. An important consequence of this has been that their farming activities remain, by and large, subsistent in nature rather than commercial and many a times uneconomical.

7.1.5 *Low rate of farm resource recycling*

In the absence of adequate knowledge among farmers about techniques and benefits of recycling of farm, industrial and municipal organic wastes in agriculture, these remain unutilized. A vast untapped potential exists to recycle these solid and liquid organic wastes of farm origin. Recycling of crop residues may be a potential organic source to sustain the soil health. Incorporation of crop residues of either rice or wheat increases the yield of rice and nutrient uptake and also improves the physico-chemical properties of the soil, ensuring better soil environment for crop growth (Ladha *et al*., 2009).

7.1.6 *Technology adoption gaps*

In our efforts to develop and improve upon existing technologies, involvement of people in conceptualization and extension of technologies would appear very important. The farm family had never been the focal point of our investigations. This top down approach had given a poor perception of the problems that they tried to solve. Due to poor extension mechanisms at national as well as state levels, many farmers, especially those at lower strata of social structure, remain uninformed about many of the development schemes and the desired impact of such schemes is not obtained. One of the reasons for poor rate of transfer of agricultural technologies is poor linkages between the different clientele groups of agriculture. Practically linkages among farmers, service providers, technological and financial institutions are either weak or nonexistent (NAAS, 2009).

CONCLUSION

India's agricultural economy and food security depend vitally on the small holder farmers. In relation to their aggregate land holding, the holdings smaller than 1.0 ha contribute proportionately more. Despite this contribution, these farmers comprise almost three fifth of the nation's hungy and poor. In that time, holistic approach of Integrated crop management (ICM) is introduce to sustainable agriculture. It considers the situation across the whole farm, including socio-economic and environmental factors, to deliver the most suitable and safe approach for long-term benefit. Integrated Crop Management System is a method of farming that balances the requirements of running a profitable business with environmental responsibility. It includes practices that avoid waste, enhance energy efficiency and minimise pollution. ICM is not rigidly-defined. It is a dynamic system that adapts to changing conditions by combining local knowledge with new research and technologies. We often work where resources for production are limited, and in these situations the best solution is often to focus on optimizing existing resources. In such a way, ICM delivers sustainable agricultural production that safeguards a farm's natural assets and surrounding community, now and in the future.

The anthropologists and sociologists studied age-old practices again as they survived the test of time and stated that these methods have something to offer due to their sustainability. The recently developed systems of integrated crop management which includes both traditional and modern technology are required to be adopted at large-scale after its evaluation in all agro-ecological situations (Marothia, 1997). The future of success of vegetable production lies in the growth of crops with less environmental stress. Thus, emphasis needs to be placed on practical aspects of such simple techniques. Yield improvement of double-season rice with ICM was achieved with the combined effects of increased plant density and optimized nutrient management. In ICM, the problem of malnutrition of food insecurity can be solved by obtaining production and money.

REFERENCE

Alam MM, Karim Md R and Ladha JK (2013) Integrating best management practices for rice with farmers' crop management techniques: A potential option for minimizing rice yield gap. *Field Crops Res.* **144:** 62–68.

Ashby JA (2001) Integrating research on food and the environment: An exit strategy from the rational fool syndrome in agricultural science. *Ecol. Soc.* 5.

Behera UK, Jha KP and Mahapatra IC (2004) Integrated management of available resources of the small and marginal farmers for generation of income and employment in eastern India. *Crop Research* **27(1):** 83–89.

Chand R and Haque T (1997) Sustainability of rice wheat crop system in Indo Gangetic region. *Economic and Political Weekly* **32 (13):** March 29.

Chaudhary MK and Harrington LW (1993) Rice-wheat system in Haryana, Working paper, CIMMYT, Mexico.

Chu G, Wang Z, Zhang H, Yang J and Zhang J (2016) Agronomic and physiological performance of rice under integrative crop management. *Agron. J.* **108:** 117–128.

Craswell ET (1998) Sustainable crop and soil management on sloping lands. Paper presented at the International Symposium on Asian.

Devasenapathy P, Mytswamy V, Christopher Louduraj A and Rabindran R (1995) Integrated farming systems for sustained productivity. *Madras Agricultural Journal* **82:** 306–307.

Gautam US, Paliwal DK and Naberia S (2007) Improvement in livelihood security for small and marginal farmers through front line demonstrations on oilseed and pulse crops in Central India. Indian Research Journal of Extension Education **7(1):** 1–5.

Ghosh BN, Singh RJ and Mishra PK (2015) Soil and input management options for increasing nutrient use efficiency. A. Rakshit *et al.* (Ed.) Nutrient use efficiency: from basics to advances, Springer. pp. 17–27.

Haque T (1998) Interaction and impact of technology, infrastructure and policy variables on agricultural growth, (Unpublished report).

Haque T (2000) Contractual Arrangements in Land and Labour Markets in Rural India.

Jayanthi C, Rangasamy A and Chinnusamy C (2000) Water budgeting for components in lowland integrated farming systems. *Agricultural Journal* **87:** 411–416.

Jayanthi C, Vennila C, Nalini K and Chandrasekaran B (2009) Sustainable integrated management of crop with allied enterprises- Ensuring livelihood security of small and marginal farmers. Special Feature: Sustainable Agriculture.

Jirli B, Bhati DS and De D (2008) Diversifying cropping system with rapeseed mustard – An approach to sustain livelihood. In proceedings of international seminar on strategies for Improving Livelihood Security of Rural Poor, Goa, India, pp: 201–202.

Kumar D and Shivay YS (2008) Integrated crop management. Modern concepts of agriculture. Indian Agricultural Research Institute (IARI), New Delhi, India.

Kumar S and Jain DK (2005) Are linkages between crops and livestock important for the sustainability of the farming system? *Asian Economic Review* **47(1):** 90–101.

Kumar S and Singh A (2014) Biopesticides for integrated crop management: Environmental and regulatory aspects. *J. Biofertil. Biopestici.* **5:** 121. doi:10.4172/2155-6202. 1000e121.

Ladha JK, Tirol-Padre A, Reddy CK, Cassman KG, Verma S, Powlson DS, Kessel C, Daniel de B Richter, Chakraborty D and Pathaket H (2009) Integrating crop and resource management technologies for enhanced productivity, profitability and sustainability of the rice-wheat system in South Asia. In Integrated Crop and Resource Management in the Rice-Wheat System of South Asia (eds Ladha, J. K. *et al.*). pp. 69–108 (International Rice Research Institute, Los Baños, Philippines.

Mamun MSA and Ahmed M (2011) Prospect of indigenous plant extracts in tea pest management. Internat. *J. Agril. Res. Innov. & Tech.* **1(1 & 2):** 16–23.

Marothia DK (1997) Agricultural technology and environmental quality: An institutional prospective. *Indian Journal of Agricultural Economics* **52(3):** 473–487.

Mishra M (2013) Role of eco-friendly agricultural practices in Indian agriculture development. Internat. *Agric. Food Sci. Techn.* **4(2):** 2249–3050.

Ogoshi R, Balasubramanian V and Jones M (2008) Integrated crop management (ICM) implementing ICM Technology in Timor-Leste, United States Agency for International Development (USAID)

Peng X, Khush GS, Virk P, Tang Q and Zou Y (2015) Crop management for increasing rice yield and nitrogen use efficiency in northeast China. *Agron. J.* **107:** 1682–1690.

Prasad R (2003). Efficient fertilizer use: The key to food security and better environment. Fertilizer News, **48(8):** 13–26.

Qin J, Xiong J, Ding C, Wei G, Ding Y and Wang S (2013) Integrated nutrient, water and other agronomic options to enhance rice grain yield and N use efficiency in double season rice crop. *Field Crops Res.* **148:** 15–23.

Regmi AP, Hobbs PR, Acharya GP and Giri GS (2009) Improving food security through integrated crop and resource management in the rice-wheat system in Nepal. In Integrated crop and resource management in the rice-wheat system of South Asia (eds Ladha, J.K. *et al.*). pp. 69–108 (International Rice Research Institute, Los Baños, Philippines).

Roling NG and Wagemakers MAE (1998) http://www.ecifm.rdg.ac.uk/integrated_crop_management.htm

Selvarajan S, Aggarwal PK, Pandey S, Lansigan FP and Bandapadhaya SK (1997) Systems approach for analysing trade-offs between income, risk and water use in rice-wheat production in Northern India. *Field Crops Research* **51:** 147–161.

Shivaji A, Dadabhau and Wadkar SK (2013) Sustainable rural livelihood security through integrated farming systems - A Review, *Agri. Reviews* **34(3):** 207–215.

Singh AK and Gohain I (2017) Integrated crop management to enhance vegetable productivity and farm income through INM and IPM practices, *The Asian Journal of Horticulture* **12(1):** 165–168.

Singh S, Mittal JP and Panu CIS (1998) Frontier energy use for the cultivation of wheat crop in Punjab. *Energy Conservation and Management* **39(576):** 485–491.

Solanes M and Villarreal FG (1999) The dublin principles for water as reflected in a comparative assessment of institutional and legal arrangements for integrated water resource management. Global water partnership technical adviseory committee.

Wang D, Huang J, Nie L, Wang F, Ling X, Cui K, Li Y and Peng P (2017) Integrated crop management practices for maximizing grain yield of double-season rice crop. Science Report, www.nature.com/scientificreports/

Xuc Y, Xue A, Duan H, Liu L, Wang Z, Yang J and Zhang J (2012) An improved crop management increases grain yield and nitrogen and water use efficiency in rice. *Crop Science* **53:** 271–284.

15 Climate Resilient Agro-Technologies for Enhanced Crop and Water Productivity under Water Deficit Agro-Ecologies

Anshul Gupta and Sandeep Kumar[1]

Department of Agronomy, SKN College of Agriculture, SKNAU, Jobner (Rajasthan) -303329
[1]*Department of Agronomy, CCS Haryana Agricultural University, Hisar – 125005 (Haryana), India*
Email: anshulgupta2803@gmail.com

ABSTRACT

Impact of climate change on agriculture is being witnessed all over the world, but countries like India are more vulnerable due to a large portion of population is dependent on agriculture, excessive pressure on natural resources and poor coping mechanisms. Temperature increases of 0.60°C over past 100 years in India. That leads to negative impact on yield of crops and changing the cropping patterns due to increased temperature, water stress and reduction in number of rainy days. Thus it affects the food security of country. Agriculture contributes approx. 15 per cent of India's GDP, a 4.5 to 9.0 per cent negative impact on production implies cost of climate change to be roughly at 1.5 per cent of GDP per year. To feed the growing population, agricultural systems need to be more productive while helping mitigate the climate change and preserving the natural resource base. They need to use inputs more efficiently, have greater stability in their outputs and are more resilient to risks, shocks and long-term climate variability.

1. INTRODUCTION

Climate alteration affects on agriculture in a numeral of ways, *viz.* rainfall, temperatures, and climate extremes (e.g., cold and heat waves); increase in plant pests and diseases; alarming change in biosphere environment; changes in the food quality (Meena *et al.*, 2017). Climate change which is largely a result of burning fossil fuels leads to global warming, imbalance in precipitation and hydrological cycles (Fig. 1). Frequently changing rainfall pattern, heat waves, and other extreme weather events affects agricultural production. Different parts of India have different magnitude of impact due to climate change. Parts of western Rajasthan, Southern Gujarat, Madhya Pradesh, Maharashtra, Northern Karnataka, Northern Andhra Pradesh, and Southern Bihar are likely to be more vulnerable in terms of extreme events (Fig. 1) (Mall *et al.*, 2006). Climate change generally increased pressure on already scarce water resources. Effect of climate change on water resources shows a varied response in Indian river basin depending upon the hydrological model, climate change scenarios and downscaling approaches used (Gosain *et al.*, 2011; Islam *et al.*, 2014).

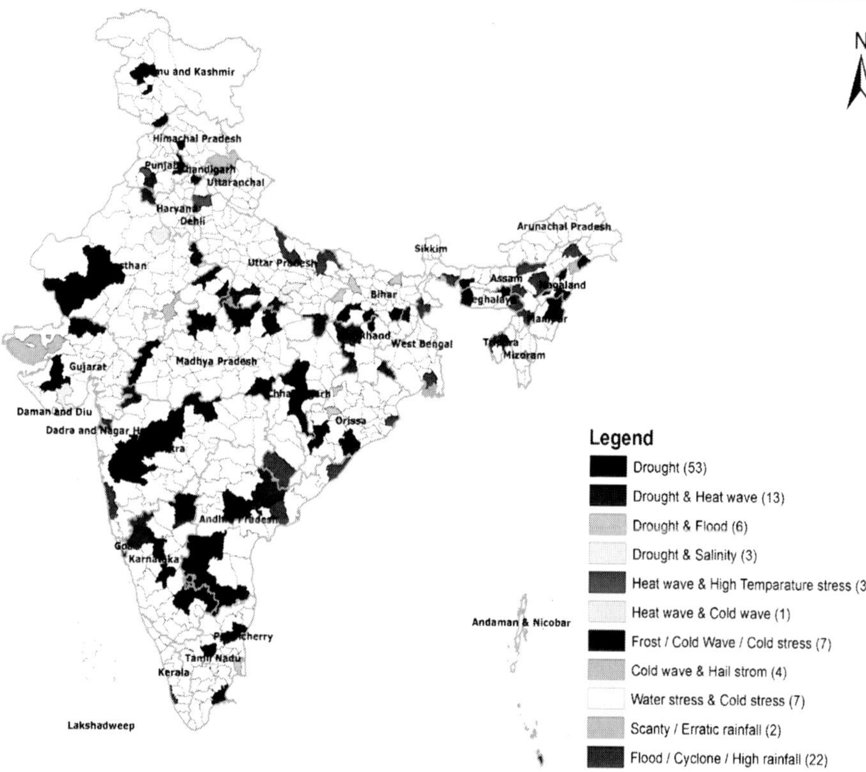

Fig. 1. Climatologically vulnerable districts of India

Rainfed agriculture occupies 67 per cent net sown area, contributing 44 per cent of production. With the increasing population of country, there is an urgent need to increase the productivity of rainfed areas from the current 1 t ha^{-1} to 2 t ha^{-1} in the next two decades. Climate shift is considered most prominent when one considers the climatic spectrum of the dryland and arid regions (Ramakrishna *et al.*, 2000) as these marginal areas provide early signals of the effect of climate variability and change (Sinha *et. al.,* 1998). The arid, semi-arid and dry sub humid regions (rainfall less than 1150 to 1200 mm) are mostly affected by climatic shifting as in these eco-systems, drought is the main part of the natural cycles affecting productivity and leading to desertification. So Resilience is defined as the propensity of a system to retain its organizational structure and productivity following a perturbation (Holling, 1973).

2. PILLARS OF CLIMATE RESILIENCE AGRICULTURE

- **Productivity:** Climate resilience agriculture aims to maintain sustainability and increase agricultural production without hampering climate. This, in turn, will raise food and nutritional security.
- **Adaptation:** It leads to exposure of farmers to short-term risks, while also strengthening their resilience by acquiring adaptation to face of shocks and longer-term stresses. Particular attention is given to protecting the ecosystem services which ecosystems provide to farmers and others. These services are essential for maintaining productivity and our ability to adapt to climate changes.
- **Mitigation:** Climate resilient technologies should help to reduce and lower down greenhouse gas (GHG) emissions. This states reduction of emission of per kilo of food, fibre and fuel production by reducing deforestation from agriculture, and managing soils and trees to improves carbon sinks and absorb CO_2 from the atmosphere.

3. IMPACT OF CLIMATE CHANGE

3.1 Hydrological cycle

The changes in hydrologic cycle cause variation in monsoon rainfall that leads to more frequent and intense droughts and floods in many agricultural regions. Climatic disasters like droughts, extreme precipitation events, cyclones and heat waves caused to greater instability in biomass production.

Following hydro-climatic changes occurs in Indian agriculture:-

- Kharif (monsoon) season crops have risk of facing flood and droughts.
- Rabi and summer crops vulnerable to higher ET losses and require larger, more frequent irrigation.
- Surface water storage structures get water from runoff, but also suffer increased evaporation from large open surfaces of reservoirs and open canal networks as a result of higher mean temperature.
- More frequent droughts require more frequent irrigation through multi-year reservoir storage capacity, of which India has very little as of now.

An early withdrawal of monsoon, heavy rainfall events, late onset of the monsoon, dry spells etc. have been witnessed in recent years, reduces the crop production. In India, during the last 15 years, less than normal rainfall is recorded during 7 years, 6 years recorded more than normal rainfall and 2 years recorded average rainfall (Srinivasa *et al.,* 2015). Analysis of rainfall data of 2001–2015 resulted that the rainfall deficit found more than 10 per cent in 6 years, i.e. 2001, 2002, 2004, 2009, 2014 and 2015, with maximum deficit being 23 per cent in 2009 followed by 19 per cent during 2002. The extreme weather events (Table 1) occur in recent decade's poses difficulty in maintaining sustainable production and this could be attributed to climate change.

Table 1. Monsoon rainfall and extreme weather events during 2001–2015 in India

Year	SWM rainfall departure from normal (%)	Extreme events
2001	-15	Drought
2002	-19	20-day heat wave during May in Andhra Pradesh
2003	+2	Extreme cold winter in 2002–03
2004	-13	Abnormal temperatures during March 2004; drought-like situation in India in July 2004
2005	-1	Flood in 2005
2006	-1	Cold wave during 2005–06; floods in arid Rajasthan and Andhra Pradesh, and drought in north-eastern regions in 2006
2007	+5	Abnormal temperatures during third week of January to first week of February 2007
2008	-2	-
2009	-23	Severe drought all over India
2010	+2	One of the warmest years
2011	+1	Failure of September rains in Andhra Pradesh
2012	-8	Drought in Punjab, Haryana, Gujarat and Karnataka. Cyclone, floods in Andhra Pradesh
2013	+6	Flood in Uttarakhand, cyclone *Phalin*
2014	-12	Flood in Jammu and Kashmir
2015	-14	Extreme heat waves, drought

Source: Srinivasa *et al.* (2015)

3.2. Drought

Drought is one of the short term extreme climatic events. That is a period of abnormally dry weather long enough to cause a serious hydrological imbalance.

Results indicated that, impact of drought on food grain production is more than that of flood (Singh *et al.,* 2011). Drought as a result of climate change severely affects the agricultural production most extremely in arid and semi-arid regions, farmer's livelihood and ultimately the direct impacts on economics (Kumar *et al.*, 2016).

3.3. Heat

Climatic changes is the main reason to increase the earth temperatures due to rise of greenhouse gases levels, such as carbon dioxide (CO_2), methane (CH_4), ozone (O_3), nitrous oxide (N_2O) and chlorofluoro carbons (CFCs). Due to increasing concentrations of these greenhouse gases, there is urgent need to concern about future changes in our climate and direct or indirect effect on agriculture (Garg *et al.*, 2001; IPCC, 2001; Aggarwal, 2003; Krupa, 2003).

3.4. Carbon dioxide

Over the next 30–50 years, CO_2 concentrations will rise to about 450 parts per million by volume (ppmv) (Fig. 2a and 2b). The response to increasing CO_2 level is higher on C_3 species (wheat, rice, and soybeans) species than on C_4 species (maize, pearl millet and sorghum). Uprety *et al.* (2003) observed an increase in rice grain yield due to increase in CO_2 concentration. That helps in increasing the net photosynthetic rate and greater accumulation of sugar that leads to accelerated development of leaves and tillers and finally grain yield.

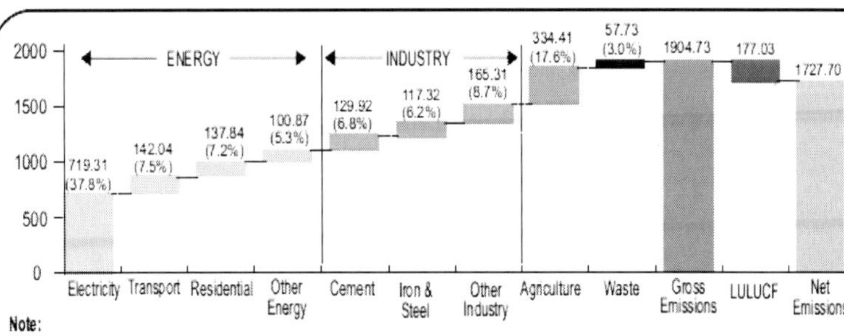

Note:

Other Energy: includes GHG emissions from petroleum refining. manufacturing of solid fuel commercial & institutional sector, agriculture & fisheries and fugitive emissions from mining, transport and storage of coal, oil and natural gas.

Other Industry: includes GHG emissions from production of glass and ceramics, soda ash, ammonia, nitric acid carbides, titanium dioxide, methanol, ethylene oxide, acrylonitrile, carbon black, caprolactam, ferro alloys, aluminum, lead zinc, copper pulp and paper, food processing, textile, leather, mining and quarrying. non specific industries and use of lubricants and paraffin wax.

Agriculture: includes CHG emissions from livestock, rice cultivation, agricultural soils and burning of crop residue.

Waste: includes CHG emissions from municipal solid waste (MSW), industrial and domestic waste water.

LULUCF: Includes CHG emissions and removals from changes in forest land, crop land, glass land, wet land, settlements and combustion of fuel wood in forests.

Fig. 2a. Sectorial GHG emission (Mt CO_2 eq.) in 2007 (*Source*: India: GHG emission-2007 (May 2010), Indian network for climate change assessment, MoEF, Govt. of India)

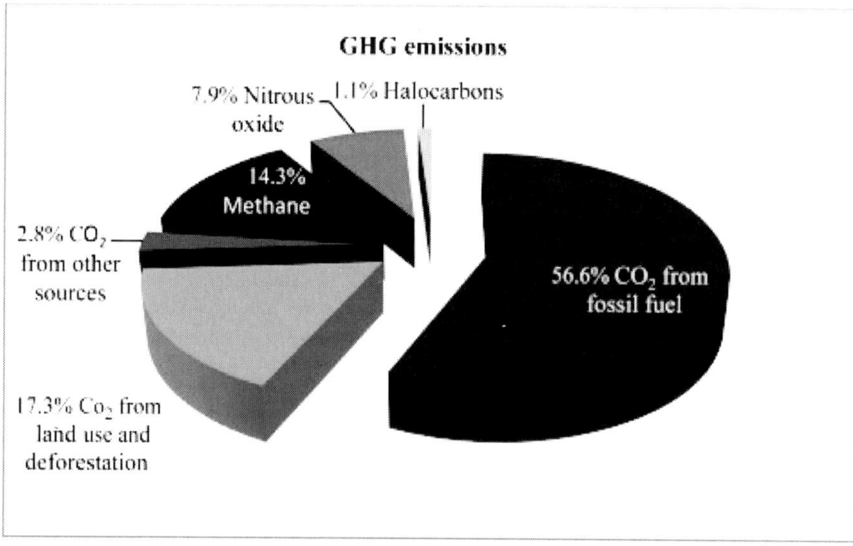

Fig 2b. Greenhouse gas emission system

Higher CO_2 level disturbs the climate in the following ways, (1) changes in air temperature, particularly enhances night temperature and other trace gases and changes in moisture availability and their effect on vegetative versus reproductive growth; (2) need for more fertilizers (3) heavy pest infestation (Krupa, 2003).

3.5. Insect-pest

Gradual warming of climate will lead to changes in the composition of pest fauna in different areas. The high population growth rate of many species will ensure changes in pest distribution. Increasing winter temperature increases the duration of hibernation of pests may decrease, that increasing their activity (Khan *et al.*, 2009).

4. CLIMATE RESILIENT TECHNOLOGIES

4.1. Use of climate ready crops

Breeding of new crop varieties with higher yield potential and tolerant to multiple stresses (drought, flood, salinity) will be the key to maintain yield stability (Lobell *et al.*, 2008). Rice genotypes Shushk Samrat found to perform better under drought conditions. With one irrigation the average yield of Shushk Samrat was found to be 3.12 t/ha, compared to only 1.43 t/ha of Moti-local and 2.75 t/ha of Sarju 52-local (DRR, 2011). Different findings in different wheat growing zones of country *i.e.*, North Western Plain Zone, North Eastern Plain Zone, Central Zone and Peninsular Zone, wheat varieties NI-5439, WH-1021 and

HD-2733 were found having higher stress tolerance and with better yield potential under both normal and restricted irrigation conditions of India.

Table 2. Short duration drought tolerant cultivars: Answer to reduced duration due to climate change

Location	Crop	Variety	Duration
Varanasi	Rice	Vandana	93
	Pigeonpea	T-21	176
Phulbani	Pigeonpea	T-21	160
Arjia	Maize	Surya	70-75
Anantapur	Groundnut	Vemana	105-110
Indore	Soybean	JS-90-41	87-98
Rewa	Rice	Kalinga-3	90
Akola	Cotton	AKH-081	150-160
Bijapur	Sunflower	KBSH-1	90-95
Solapur	Sorghum	Mauli	105-110
Hisar	Pearlmillet	HHB-67	60-62
Bangalore	Fingermillet	GPU-26	90-105

Source: CRIDA, Hyderabad, India

Development of an efficient cropping system is a tool that can help agriculture to adapt according to the climatic situation. For example practice of multiple cropping that have several crops growing at one time can help systems to exhibit greater durability during periods of high moisture or heat stress.

4.2. Adjustment of planting dates

Adjustment of cropping season to reduce the effect of temperature fluctuations that induced spikelet sterility can be used to reduce yield instability, by avoiding flowering during warmest period of cropping season. A strong correlation was found between yield reduction in rice and wheat, and the shortening of crop duration under high temperature, whereas shifting the transplanting date seems to be an effective measure to shortening of crop duration for both crops (Jalota *et al.*, 2013).

To abate the shortening effect of crop duration due to high temperature changed sowing dates and late maturing varieties can be adapted. For kharif crops using early sowing or longer maturing varieties will counteract the detrimental effect of climate changes. Increasing sowing time by 10 days in late-sown varieties of wheat and delaying of sowing by 10 days in normally sown cultivars increase yields under a modified climate (Attri and Rathore, 2003).

4.3. Efficient irrigation management

Climate change has increased the possibility and intensity of drought. So that efficient water management would be able to increase the resilience of agricultural production to climate change. Irrigation water can be managed more efficiently by adopting appropriate ways of water used, by developing system to collect and supply surface water and by judiciously regulating groundwater irrigation. It is necessary to adopt policies ensuring development of efficient water markets. Adoption of efficient irrigation methods such as sprinkler and drip irrigation and rainwater harvesting ensuring "more crops per drop" are steps in the right direction (GoI, 2016). So we can improve the water use efficiency by adopting the following practices:

4.3.1. *Rain water harvesting*

Farms in Rainfed areas have uncertainty in rainfall pattern. These areas have light texture soils that lead to loss of fertile soil during monsoon season due to heavy runoff. In those areas, measures which reduce the infiltration rates are practiced. By diverted the run-off of the harvested water, it is diverted to tanks, underground aquifers, or dedicated surface ponds (though this is the least costly alternative, it results in excessive water loss via evaporation), from where water can be extracted and used for irrigation). Production of *rabi* crops is low due to the unavailability of water for pre-irrigation and sowing and also to provide irrigation at critical crop growth stages.

Rainwater harvesting and recycling done with the following ways:–

***Temporary check dams*:** Temporary check dams constructed by using low cost gunny bags which are filled with sand from stream beds. The sand filled bags were placed one above the other in two or more rows. Clay is used to fill the gap between two rows to check water leakage. Excess water is drained through an outlet. These check dams help in ground water recharge and rising of water table in the area. Stored water is used for life saving irrigation in *rabi* and summer crops.

***Farm ponds*:** A farm pond is a large hole dug out in the earth, usually square or rectangular in shape, which harvests rainwater and stores for supplemental irrigation to overcome the rainfall variability and protects crop from stress due to climate change (Fig. 3). There is an inlet is made up to regulate inflow and an outlet to discharge excess water. Pond edges are surrounded by a small bund to get protection from erosion. The depth and size of the pond is depending on field area, the type of soil and water requirements. Water from pond is distributed to field either by pumping or manually (Uddin *et al.*, 2017).

Fig. 3. Farm pond

4.3.2. *In situ soil moisture conservation*

At the farm area *in situ* moisture conservation helps in coping the moisture stress, occurs due to climatic vagaries. Ridge and furrow method and broad bed furrow system helps to conserve the rainwater and drain out excess water to utilize it for recharging ground water. Broad bed furrow planting improved relative leaf water content and water use efficiency in chickpea as saving of irrigation water up to 33.3 per cent was recorded over that in flat (Kumar *et al.*, 2015). In the bed planting, only furrows are irrigated. The area of furrows per unit area in the wider beds is lower than the narrow beds (Fig. 4a). So, it received lower amount of irrigation water and helps in saving the water over the conventional method (Mollah *et al.*, 2009) (Fig. 4b).

Fig 4a. Planting of crop in furrow

Fig 4b. Ridges and furrows system in cotton irrigated raised bed

Benefits of *in situ* moisture conservation techniques –

1. It helps to retain the moisture in the soil profile which supports plants in mid-season or late-season spells of drought. Double cropping / intercropping can be practiced in water scarcity areas. The large soil water storage capacity supports more crop growth during the subsequent dry post-rainy season.
2. It helps in discharge of surplus surface runoff.

4.3.3. *Efficient irrigation methods*

Considerable water savings can be acquired with sprinkler, drip/micro-sprinkler/drip irrigation systems in less water availability areas which have conditions to operating them efficiently. Drip irrigation as a promising climate adaptation technology with enhances productive and economic efficiency and reduces water expanses (Venot, 2016).

In the drip irrigation technique, water is directly applied to the root zone of the crop in small quantities using a low pressure delivery system with a network of pipes with small emitters (or drippers) built in to them. This method helps to retain the soil moisture at consistent levels against the flood irrigation method where there is a huge variation in soil moisture levels.

Application of irrigation water intermittently with ponding depth 5 ± 2 cm in rice, after 3 days of disappearance of previously ponded water saves water as well and also reduces global warming by 25–30 per cent over continuous flooding (Pathak, 2015; Chaudhari *et al.*, 2015). Micro-irrigation helps in saving water and energy and lower down carbon emissions (Shah, 2009). Drip irrigation helps in lower down the over exploitation of ground water which occurs due to surface irrigation methods like wild flooding and other problems as waterlogging, heavy weed infestation, salinity and high evaporation losses are absent in it (Narayanamoorthy, 1997).

Table 3. Water saving through drip method of irrigation

Crop's Name	Water Consumption (mm/ha)		Water Saving over FMI	Water consumed (mm) / Yield (quintals)	
	Flood	Drip	(%)	Flood	Drip
Ash gourd	840	740	12	77.49	61.51
Beet root	857	177	79	187.53	36.2
Sweet potato	631	252	61	148.82	42.78
Onion	602	451	25	64.73	36.97
Radish	464	108	77	441.9	90.76
Tomato	498	107	79	80.58	12.06
Chillies	1097	417	62	259.34	68.47
Cauliflower	389	255	34	46.67	22
Papaya	2285	734	68	175.77	31.91
Banana	1760	970	45	30.61	11.09
Grapes	532	278	48	20.15	8.55
Pomegranate	1440	785	45	26.18	7.2
Sugarcane	2150	940	65	16.79	5.53
Cotton	856	302	60	329.23	92.64
Coconut	–	–	60	–	–
Groundnut	500	300	40	292.4	105.63

(*Source*: Narayanamoorthy, 2005)

4.3.4. *Irrigation scheduling*

Optimal allocation of available water among the competing crops and optimum timing of application is to be decided under adequate and limited water supply situation, so as to maximize economic returns from available water. Under limited water supply situation, irrigation water must be allocated so that periods of possible water deficits coincide with the least sensitive growth periods. Thus irrigation scheduling should be decided based on the water availability. Under limited water supply, the objective is to achieve maximum WUE which can be improved by adjusting time, method and spacing of planting, better fertilization, irrigation method and timing, and weed control (Dahiya *et al.*, 2017). There are different methods for irrigation scheduling *viz.*, critical crop growth stages, feel and appearance method, soil moisture depletion approach, irrigation water at different cumulative pan evaporation method. Baviskar (2017) conducted a field experiment at Anand, Gujrat to scheduling irrigation in wheat crop. Results revealed that Irrigation scheduled at 0.8 IW:CPE ratio and irrigations at CRI + tillering + jointing + flowering + milking + dough stage gave significantly higher mean grain, and straw yields compared to irrigation with Irrigations at crown root initiation (CRI) + tillering + jointing + flowering and Irrigations at CRI + tillering + jointing + flowering + milking; treatments.

4.4. Crop diversification

Extreme temperature, frequent and intensive flood, cyclone and other natural disasters due to climate change became acute and expecting to be severe in future. In the same time, crop diversification could be an effective adaptation option under this situation as it protects natural biodiversity, strengthening the ability of the agroecosystem to respond to these stresses, minimizing environmental pollution, reducing the risk of total crop failure, reducing incidence of insect pests, diseases and weed problems and secure food supply opportunities, and also providing producers with alternative means of generating income (Lakhran *et al.,* 2017). Crop diversification has proven to be rational and cost effective method responses to climatic variability and change, particularly in countries which are predominantly of rural economics and are at most risk (Bradshaw *et al.*, 2004). Climatic resilience can be achieved by crop diversification, by engendering a greater ability to suppress pest outbreaks and dampen pathogen transmission, which may worsen under future climate scenarios, that also helps in buffering crop production from effect of greater climate variability and extreme events (Lin, 2011). Crop diversification is commonly practiced to fulfil the demand and stabilized the income during adverse weather conditions. Due to changing monsoon patterns and water resources depletion, the traditional cropping pattern is becoming less productive. Sole cropping is predominant in scarce rainfall regions of India, but it resulted low yields or sometimes even in crop failure due to erratic monsoon rainfall and skewed distribution. In such areas crop intensification can be done through mixed cropping, intercropping and post harvest value addition processes for gaining prominence as a climate change adaptation strategy (Mahdi *et al.,* 2015).

4.5. Conservation agriculture (CA)

Conservation agriculture is a practice that is based on three principles promoted as a means for sustainable agricultural intensification. These practices include minimum soil disturbance (no till) (disturbed area must be less than 15 cm wide or 25% of cropped area), mulching with crop residue (ground cover must be more than 30%), and crop rotation (rotation should involve at least three different crops). That practice helps in increased soil fertility, reduced input demand, and reduced risk to yields from rainfall shocks (Brouder and Gomez-Macpherson, 2014). It improves water infiltration and moisture conservation through mulching during scarce rainfall condition (Thierfelder and Wall, 2009) and reduces erosion rates during excess rainfall duration (Schuller *et al.*, 2007). So CA has become a important element to maintain sustainably in agricultural system, increase crop productivity, and a fixture in the promotion of 'climate resilient' agriculture.

CA helps to mitigate climate change through soil carbon sequestration. CA practice estimated to sequester 0.4–0.8g carbon in world agricultural soils per year (Lal, 2004). Soil organic carbon improves near-surface soil from CA cause a better soil structure, bulk density, infiltration and aeration; and expected to contribute to increased sustainability and climate change adaptation. Several practices, such as no-till or reduced tillage, residue retention, conversion to permanent pasture, crop rotation and fertilizer application have sequester Carbon in various countries, including India (West and Marland 2002; Vanden Bygaart *et al.,* 2003; Lal 2004; Patle *et al.*, 2013).

4.6. Agro-forestry system

Water is already scarce resource in agricultural system, and changing climate expected to make this situation worse. There are several mechanisms whereby, agroforestry may increase water use efficiency. To mitigate climate extremes in high temperature areas as well as under intra-annual climatic fluctuations, agroforestry play its role by covering the surface with tree canopies to create a more favourable microclimate for crops and more resilient ecosystems for better agricultural production (Luedeling *et al.*, 2014). Favourable microclimate benefited farmers by maintaining resilience through enhanced productivity and coping with climate change at the farm level. Agroforestry enhances soil fertility with several benefits and services that trees on farm provide, such as fuel wood, timber, fruits or fodder. Trees with crop plants help to in conserve water and maximize use of limited water supply (Ong *et al.,* 2006). Below-ground diversity increases with trees and helps to maintain local ecosystem stability and resilience (Barrios *et al.,* 2012).

In rainfed agriculture, soil moisture is depended only on the rainfall which infiltrate in to deeper soil layers. Particularly annual crop production is often affected by limited water supplies, as residual water in the soil profile following harvest of annual crops and off-season rainfall remained unused (Ong *et al.*, 2006). A tree with crop plants acts as a shelterbelt that helps in reducing the transpiration and evaporation losses, helps to enhance water use efficiency of the system. Trees in agroforestry provide watershed protection as well as biodiversity conservation and carbon storage. A mechanism Hydraulic lift' is found in several tree species, in which tree roots lift water from moist deep soil layers to the upper soil layers, where it is accessible to crops (Ong and Leakey, 1999; Bayala *et al.,* 2008). Belt of the trees in agroforestry system known as riparian buffers, reduces water pollution from agricultural fields and helps to clean runoff water by reducing runoff velocity, thereby improves infiltration, sediment deposition and nutrient retention (Jose, 2009).

Fig. 5a. Neem + cowpea

Fig. 5b. Leucaena + Sunhemp

Agroforestry improves productivity of system by efficient use of available light, water and nutrients by both trees and crop plants to increase dry matter production (Black and Ong, 2000). Trees should have complementary relationship with associated crops (Fig. 5a and 5b). In agroforestry systems, the key to success is choosing the right combination of trees and crops to exploit spatial and temporal complementarities in resource use.

5. CONCLUSION

In agriculture climate, change will be one of the important factors to affect the food security. Climate change mitigation strategies are mainly related to technological interventions, management skills, and sound governmental policies to overcome the adverse effect of climate. With the technological interventions of climate change, supportive government policies should also be undertaken. These are some policies interventions:

- Mainstreaming the climate change vulnerable areas through climate-smart and efficient water interventions, such as different governmental policies: importantly, the *Pradhan Mantri Krishi Sinchayee Yojana* (PMKSY), the National Mission for Sustainable Agriculture (NMSA), the National Water Mission and Mahatma Gandhi National Rural Employment Guarantee Act (MGNREGA), Soil health card scheme and evaluating the effect of increasing resilience to climate change and reducing greenhouse gases potential. This will also helps in acquiring India's Intended Nationally Determined Contribution (INDC) target for 2030 by reducing GHG emissions, increasing carbon sequestration and enhancing resource use efficiency besides ensuring national food security;
- Increasing the promotion of micro-irrigation and solar pumps for proper land and water management.
- Managing the runoff water by maintaining proper storage sites for diverting the flow from rivers and canals to reduce the flood risk and recharge ground water.
- Made up of contingency crop plan at farm level as a real time implementation to combat with drought stress through strong and supportive government policies and well established institutions.
- Improve the precision farming by adopting different precision tools like enhanced weather forecasting, hydrological monitoring, synoptic charts, crop weather advisory services, improved information and communication technologies (ICT) for prevention and early preparedness to cope with climatic risk.
- Village level farming which is dependent on land and water conditions, to improve water availability to improving resilience by maintaining the balance between supply and demand.

REFERENCES

Aggarwal PK (2003) Impact of climate change on Indian agriculture. *J. Plant Biology* **30(2):** 189–198.

Attri SD and Rathore LS (2003) Simulation of impact of projected climate change on wheat in India. *Int. Journal of Climatology* **23:** 693–705.

Barrios E, Sileshi GW, Shepherd K and Sinclair F (2012) Agroforestry and soil health: linking trees, soil biota and ecosystem services.

Baviskar VS, Patel JB, Takale SR, Shitap MS, Patel VJ and Sankar GR Maruthi (2017) Efficient irrigation management for enhancing productivity and profitability of wheat under semi-arid inceptisols in Gujarat. *Indian Journal of Soil Conservation* **45(2):** 183–191.

Bayala J, Heng LK, van Noordwijk M and Ouedraogo SJ (2008) Hydraulic lift study in two native tree species of agroforestry parklands of West African dry savanna. *Acta Oecologia* **34:** 370–378.

Black, Ong (2000) Utilization of light and water in tropical agriculture. *Agricultural and Forest Meterology* **104:** 25–47.

Bradshaw B, Dolan H and Smit B (2004) Farm level adaptation to climatic variability and change: crop diversification in the Canadian prairies. *Clim Change* **67:** 119–141.

Brouder SM and Gomez-Macpherson H (2014) The impact of conservation agriculture on smallholder agricultural yields: A scoping review of the evidence. *Agriculture, Ecosystems & Environment* **187(1):** 11–32.

Chaudhari SK, Islam A, Biswas PP and Sikka AK (2015) Natural resource management: problems and prospects. *Indian Journal of Fertilizer* **11(4):** 16–23.

Dahiya S, Kumar S, Kumar S, Khedwal RS, Harender, Chaudhary C and Ankush (2017) Management practices for improving water use efficiency of crops for boosting crop production. In: Agricultural, Allied Sciences & Biotechnology for Sustainability of Agriculture, Nutrition & Food Security, edited by Ratnesh Kumar Rao, P.K. Sharma, M. Raghuraman and J.K. Singh, Mahima Publications, ISBN: 978-81-926935-8-3, pp. 115–121.

DRR (2011) Directorate of Rice Research: Production oriented survey, All India Coordinated Rice Improvement Programme (ICAR), Directorate of Rice Research, Rajendranagar, Hyderabad-500 030, Andhra Pradesh, India. pp 163.

Garg A, Shukla PR, Bhattacharya S, Dadhwal VK (2001) Sub-region (district) and sector level SO2 and NOx emissions for India: assessment of inventories and mitigation flexibility. Atmospheric Environment 35: 703–713.

GOI (2010) India: GHG emission-2007 (May 2010), Indian network for climate change assessment, Ministry of Environment and Forest, Govt of India.

Gosain AK, Sandhya Rao S and Arora A (2011) Climate change impact assessment of water resources of India. *Current Science* **101:** 356–371.

Government of India (2016) Economic Survey 2015–16. New Delhi: Ministry of Finance.

Holling CS (1973) Resilience and stability of ecological systems. *Annual Review of Ecology and Systematics* **4:** 1–23.

IPCC (Intergovernmental Panel for Climate Change) (2001) Climate Change 2001 – The Scientific Basis, Contribution of Working Group I to the Third Assessment Report of the Intergovernmental Panel on Climate Change [Houghton, J.T., Y. Ding, D.J. Griggs, M. Noguer, P.J. van der Linden, impact of climate change on indian agriculture: a review 475 X. Dai, K. Maskell and C.A. Johnson (eds.)], Cambridge University Press, Cambridge, UK, 881 pp.

Islam A, Shirsath PB, Naresh Kumar S, Subhash N, Sikka AK and Aggarwal PK (2014) Use of models in water management and food security under climate change scenarios in India. In Ahuja LR, Ma L, Lascano RJ (eds). Practical Applications of

Agricultural System Models to Optimize the Use of Limited Water. Advances in Agricultural Systems Modelling 5.

Jalota SK, Kaur H, Kaur S and Vashisht BB (2013) Impact of climate change scenarios on yield, water and nitrogen-balance and use efficiency of rice–wheat cropping system. *Agric Water Manage* **116:** 29–31.

Jose S (2009) Agroforestry for ecosystem services and environmental benefits: An overview. *Agroforestry Systems* **76:** 1–10.

Khan SA, Kumar S, Hussain MZ and Kalra N (2009) Climate Change, Climate Variability and Indian Agriculture: Impacts Vulnerability and Adaptation Strategies. *Climate Change and Crops* 19–38.

Krupa S (2003) Atmosphere and agriculture in the new millennium. *Environmental Pollution* **126:** 293–300.

Kumar N, Singh MK, Praharaj CS, Singh U and Singh SS (2015) Performance of chickpea under different planting method, seed rate and irrigation level in Indo-Gangetic Plains of India. *Journal of Food Legumes* **28(1):** 40–44.

Lakhran H, Kumar S and Bajiya R (2017) Crop diversification: An option for climate change resilience. *Trends in Biosciences* **10(2):** 516–518.

Lal R (2004) Carbon sequestration in soils of Central Asia. *Land Degrad Dev* **15(6):** 563–72.

Lin BB (2011) Resilience in agriculture through crop diversification: Adaptive management for environmental change. *Bio Sci* **61(3):** 183–193.

Lobell DB, Burke MB, Tebaldi C, Mastrandrea MD, Falcon WP and Nylor RL (2008) Prioritizing climate change adaptation needs for food security in 2030. *Science* **319(5863):** 607–610.

Luedeling E, Huth NI, Kindt R and König K (2014) Agroforestry systems in a changing climate - challenges in projecting future performance .*Curr Opin Environ Sustain* **6:** 1–7.

Mahdi SS, Dhekale, Choudhury BS, Roy S, Bangroo, Shabir A, Gupta, Kumar S (2015) On the climate risks in crop production and management in India: A review. *Australian Journal of Crop Science* **9(7):** 585–595.

Mall R Singh, Gupta A, Srinivasan G and Rathore L (2006) Impact of climate change on Indian agriculture: a review. *Climate Change* **78:** 445–478.

Meena RS, Gogoi N and Kumar S (2016) Alarming issues on agricultural crop production and environmental stresses. Journal of Cleaner Production. *Elsevier Publication,* **142:** 3357–3359.

Mollah MIU, Bhuiya MSU and Kabir MH (2009) Bed planting: A new crop establishment method for wheat in rice-wheat cropping system. *J. Agric. Rural Dev.* **7:** 23–31.

Narayanamoorthy A (1997) Drip irrigation: A viable option for future irrigation development. *Productivity* **38(3):** 504-511.

Narayanamoorthy A (2005) Efficiency of Irrigation: A case of drip irrigation (Occasional Paper: 45). Mumbai, India: Department of Economic Analysis and Research, National Bank for Agriculture and Rural Development.

Ong CK, Black CR and Muthuri CW (2006) Modifying forests and agroforestry for improved water productivity in the semi-arid tropics. *CAB Reviews: Perspectives in Agriculture, Veterinary Science, Nutrition and Natural Resources* **65:** 1–19.

Ong CK and Leakey RRB (1999) Why tree-crop interactions in agroforestry appear at odds with tree-grass interactions in tropical savannahs. *Agroforestry Systems* **45:** 109–129. doi:10.1023/A:1006243032538.

Pathak H (2015) Greenhouse gas emission from Indian agriculture: trends, drivers and mitigation strategies. *Proceedings of the Indian National Science Academy* **81(5):** 1133–1149.

Patle GT, Bandyopadhyay KK and Singh DK (2013) Impact of conservation agriculture and resource conservation technologies on carbon sequestration–a review. *Indian J Agr Sci* **83(1):** 3–13.

Ramakrishna YS, Rao AS, Rao GGSN and Kesava Rao AVR (2000) Climatic constraints and their management in the Indian arid zone. Symposium on Impact of Human Activities on the Desertification in the Thar Desert held at Jodhpur on 14th February, 2000.

Schuller P, Walling DE, Sepulveda A, Castillo A and Pino I (2007) Changes in soil erosion associated with the shift from conventional tillage to a no-tillage system, documented using 137Cs measurements. *Soil & Tillage Research* **94(1):** 183–192.

Shah T (2009) Climate change and groundwater: India's opportunities for mitigation and adaptation. *Environmental Research Letters* **4(3):** 1–13.

Singh A, Phadke VS and Patwardhan A (2011) Impact of drought and flood on Indian food grain production. In Challenges and opportunities in Agro-meteorology, (Eds. Attri SD, Rathore LS, Sivakumar MVK and Dash SK. Springer Berlin Heidelberg, pp. 421–433.

Sinha SK, Kulshreshtha SM, Purohit AN and Singh AK (1998) Climate change and perspective for agriculture, Base Paper. National Academy of Agricultural Sciences. 20 p.

Srinivasa Rao Ch, Rao KV, Chary GR, Prasad YG, Subba Rao AVM, Ramana DBV, Prasad JVNS, Rama Rao CA, Pankaj PK, Gopinath KA, Kandpal BK, Maheswari M, Rao VUM and Sikka AK (2015) Compensatory Rabi Production Plan—2015. Technical Bulletin 1/2015. Central Research Institute for Dryland Agriculture and Natural Resource Management Division, Indian Council of Agricultural Research: Hyderabad; 66 pp.

Thierfelder C and Wall PC (2009) Effects of conservation agriculture techniques on infiltration and soil water content in Zambia and Zimbabwe. *Soil & Tillage Research* **105:** 217–227.

Uddin Jalal Md, Sravya V and Abdullah Omer Bin (2017) Rain Water Harvesting (Farm Pond). *International Journal of Emerging Research in Management &Technology* **6:** 2.

Uprety DC, Dwivedi N, Jain V, Mohan R, Saxena Jolly M and Paswan G (2003) Response of rice cultivars to the elevated CO_2. *Biologia Plantarum* **46(1):** 35–39.

Vanen Bygaart AJ, Gregorich EG and Angers DA (2003) Influence of agricultural management on soil organic carbon: a compendium and assessment of Canadian studies. *Can J Soil Sci.* **83:** 363–380.

Venot JP (2016) A success of some sort: Social enterprises and drip irrigation in the developing world. *World Development,* **79:** 69–81.

West TO and Marland G (2002) A synthesis of carbon sequestration, carbon emissions, and net carbon flux in agriculture: comparing tillage practices in the United States. *Agric Ecosyst Environ* **91:** 217–232.

■■■■

16 Significance of Soil Organic Matter to Soil Quality and Evaluation of Sustainability

Chetan Kumar Jangir[*], *Sandeep Kumar*[1] *and Ram Swaroop Meena*[2]

Department of Soil Science, Chaudhary Charan Singh Haryana Agricultural University, Hisar, Haryana – 125004, India
[1]*Department of Agronomy, Chaudhary Charan Singh Haryana Agricultural University, Hisar, Haryana – 125004, India*
[2]*Department of Agronomy, Institute of Agricultural Sciences, Banaras Hindu University, Varanasi, Uttar Pradesh – 221005, India*
[*]Author for correspondence: *chetanjangir710@gmail.com*

1. INTRODUCTION

Soil health is the capacity of soil to function as a vital living system, within ecosystem and land-use boundaries, to sustain plant and animal productivity, maintain or enhance water and air quality, and promote plant and animal health. Anthropogenic reductions in soil health, and of individual components of soil quality, are a pressing ecological concern. A conference entitled 'Soil Health: Managing the Biological Component of Soil Quality' was held in the USA in November 1998 to increase awareness of the importance and utility of soil organisms as indicators of soil quality and determinants of soil health. To evaluate sustainability of agricultural practices, assessment of soil health using various indicators of soil quality is needed. Soil organisms and biotic parameters (e.g. abundance, diversity, food web structure, or community stability) meet most of the five criteria for useful indicators of soil quality. Soil organisms respond sensitively to land management practices and climate. They are well correlated with beneficial soil and ecosystem functions including water storage, decomposition and nutrient cycling, detoxification of toxicants, and suppression of noxious and pathogenic organisms. Soil organisms also illustrate the chain of cause and effect that links land management decisions to ultimate productivity and health of plants and animals. Indicators must be comprehensible and useful to land managers, who are the ultimate stewards of soil quality and soil health. Visible organisms such as earthworms, insects, and moulds have historically met this criterion. Finally, indicators must be easy and inexpensive to measure, but the need for knowledge of taxonomy complicates the measurement of soil organisms. Several farmer-participatory programs for managing soil quality and health have incorporated abiotic and simple biotic indicators. The challenge for the future is to develop sustainable management systems which are the vanguard of soil health; soil quality indicators are merely a means towards this end.

Soil organic matter (SOM) is the organic matter (OM) component of soil, consisting of plant and animal residues at various stages of decomposition, cells and tissues of soil organisms, and substances synthesized by soil organisms. Soil organic matter exerts numerous positive effects on soil physical and chemical properties, as well as the soil's capacity to provide regulatory ecosystem services. Particularly, the presence of SOM is regarded as being critical for soil function and soil quality. The positive impacts of SOM result from a number of complex, interactive edaphic factors; a non-exhaustive list of SOM's effects on soil functioning includes improvements related to soil structure, aggregation, water retention, soil biodiversity, absorption and retention of pollutants, buffering capacity, and the cycling and storage of plant nutrients. SOM increases soil fertility by providing cation exchange sites and acting as reserve of plant nutrients, especially nitrogen (N), phosphorus (P), and sulfur (S), along with micronutrients, which are slowly released upon SOM mineralization. As such, there is a significant correlation between SOM content and soil fertility. Soil organic matter also acts as a major sink and source of soil carbon (C). Although the C content of SOM is known to vary considerably, SOM is typically estimated to contain 58 per cent C, and the terms 'soil organic carbon' (SOC) and SOM are often used interchangeably, with measured SOC content often serving as a proxy for SOM. Soil represents one of the largest C sinks on the planet and plays a major role in the global C cycle. Therefore, SOM/SOC dynamics and the capacity of soils to provide the ecosystem service of C sequestration through SOM management have received considerable attention in recent years. The concentration of SOM in soils generally ranges from 1 to 6 per cent of the total topsoil mass for most upland soils. Soils whose upper horizons consist of less than 1 per cent organic matter are mostly limited to desert areas, while the SOM content of soils in low-lying, wet areas can be as high as 90 per cent. Soils containing 12–18 per cent SOC are generally classified as organic soils.

It can be divided into three general pools: living biomass of microorganisms, fresh and partially decomposed residues, and humus: the well-decomposed organic material. Surface plant litter is generally not included as part of SOM. Organic compounds found in plant residues include:

- **Carbohydrates** are made up of C, hydrogen (H) and oxygen (O), and range in complexity from rather simple *sugars* to the big molecules of *cellulose*.
- **Fats** consist of glycerides of fatty acids, like butyric, stearic, and oleic. They are also made up of C, O and H atoms.

- **Lignins** are complex compounds that form the older parts of wood, and consist also mainly of C, O and H. They are resistant to decomposition.
- **Proteins** contain N in addition to C, H and O, and also small amounts of S, Fe, and P.
- **Charcoal** is elemental C derived from incomplete combustion of OM. Charcoal is resistant to decomposition.

The breakdown of the organic compounds is done at very different rates, depending on their nature. The ranking, from fast to slow rates are as follow.

1. Sugars, starches and simple proteins.
2. Proteins
3. Hemicelluloses
4. Cellulose
5. Lignin and fats

Humus – As vegetal material undergoes decomposition, some microbial resistant compounds are formed. These include modified lignin, oils, fats and waxes. Secondly, some new compounds are synthesized, like polysaccharides and polyuronides. These materials form the basis for humus. New reactions take place between these compounds and some proteins and other N containing products, incorporating thus N and avoiding its mineralization. Other nutrients are also protected in this way from mineralization.

Humic substances classification – There is a classification into three groups, based on solubility in acids and alkalis, and also related to stability.

Fulvic acid is the group which contains the materials that have the lowest molecular weight, and are soluble in acids and alkali, and susceptible to microbial attack.

Humic acid group contains the intermediate materials, with medium molecular weight, soluble in alkali, but insoluble in acid and intermediate resistance to microbial attack.

Humin is the generic name for the materials with highest molecular weight, that are darkest in colour, insoluble in acid and alkali, and with the most resistance to microbial attack.

2. SOIL ORGANIC MATTER: CHEMICAL NATURE AND FRACTIONS

The classic method of fractionating SOM into humic acid, fulvic acid, and humin has been known for hundreds of years, but is no longer considered meaningful because the fractions are artificially defined and do not exist in soils *per se*. Additionally, this method of fractionation does not produce chemically discrete SOM fractions, but, rather, fractions that are heterogeneous and non-reproducible (Fig. 1).

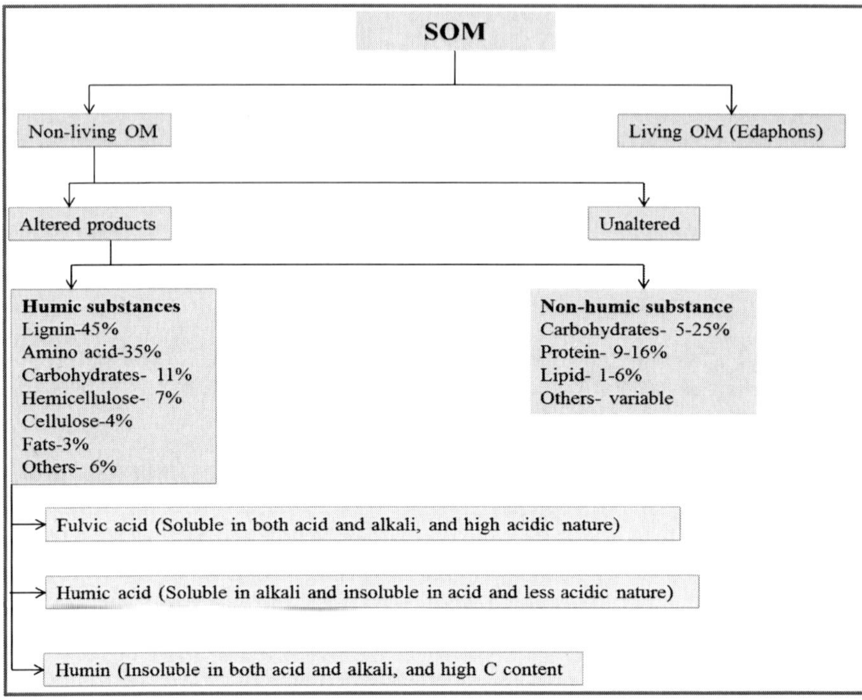

Fig. 1. Fractionation of SOM

A more biologically and agriculturally meaningful method of describing SOM is by dividing it into various "pools" which are sorted by how easily the material is decomposed (*e.g.* active or labile; slow or intermediate; and passive or stable). Pools, which have measurable OM components, are theoretically separate entities and are more concisely designated by fractions. This method of SOM classification is far more commonly used now than the out dated measurements of humic and fulvic acid separation.

2.1 Active/labile fraction

In general, younger organic material, from recently deposited roots and residue, dead organisms, or waste products, is the most biologically "active" fraction of the SOM, meaning that it serves as a food source for the living soil biological community. The younger fraction is also referred to as the "labile" SOM fraction, indicating that it is more readily decomposed than the passive/stable fraction. Generally, this fraction of the SOM is less than five years old. There are many ways to measure the active fraction, but one of the most commonly used methods is to measure the particulate organic matter (POM). Particulate organic matter is defined as the microbial active fraction of SOM. The reason that POM has become so frequently used is that it has been shown to have a strong response to management decisions, such as tillage, residue handling and levels, and crop rotation. There are a number of different ways to measure POM, but they all rely on separation techniques that are based on size and density of the SOM material.

2.2 Passive/stable fraction

Many soil organisms assist in the process of decomposing plant and animal tissues. During the process of decomposition, chemical transformations take place, creating new organic compounds in the soil. After years or decades of these transformations, the original organic materials are converted into chemically complex, nutrient-poor compounds that few microbes can degrade. These compounds are referred to as "passive" or "stabilized" and can make up a third to a half of SOM. Such passive or stabilized materials are what we commonly refer to as "humus" or the "stable fraction". The stable fraction does not contain many nutrients, and so is not directly important for soil fertility. However, the stable humus fraction of soil is chemically very reactive and contributes to the soil's net chemical charge, known as the cation exchange capacity (CEC) and anion exchange capacity. In this way, humus temporarily and reversibly binds plant nutrients in the soil, preventing them from leaching, so that they are available for plant uptake. The stable fraction also modifies and "stabilizes" toxic materials, so that they are less reactive and/or dangerous. Finally, the stable fraction enhances soil aggregation that reduces a soil's vulnerability to erosive forces and thereby reduces soil loss by erosion.

3. ROLE OF SOM

Once a land manager begins working towards enhancing SOM, a series of soil changes and environmental benefits follow. The rate and degree of these changes and the best suite of practices needed to achieve results vary with soil and climate. Initially, managing for greater SOM may require higher pesticide, herbicide, or nutrient applications. In time, productivity and environmental quality will be enhanced.

1. Apply practices that enhance SOM

- Diverse, high biomass crop rotations
- Cover crops
- Reduced tillage
- Rotational or prescribed grazing

2. Organic matter dynamics change

- Increased surface residue forms a physical barrier to wind and water erosion.
- Higher residue rotations and cover crops contribute more OM and nutrients to the soil.

- Less soil disturbance means lower OM losses.

3. Soil properties change

- Surface structure becomes more stable and less prone to crusting and erosion.
- Water infiltration increases and runoff decreases when soil structure improves.
- Soil organic matter holds 10 to 1,000 times more water and nutrients than the same amount of soil minerals.
- Beneficial soil organisms become more numerous and active with diverse crop rotations and higher OM levels.

4. Air quality, water quality, and agricultural productivity improve

- Dust, allergens, and pathogens in the air immediately decline.
- Sediment and nutrient loads decline in surface water as soon as soil aggregation increases and runoff decreases.
- Ground and surface water quality improve because better structure, infiltration, and biological activity make soil a more effective filter.
- Crops are better able to withstand drought when infiltration and water holding capacity increase.
- Organic matter may bind pesticides, making them less active. Soils managed for OM may suppress disease organisms, which could reduce pesticide needs.
- Crop health and vigour increase when soil biological activity and diversity increase.
- Wildlife habitat improves when residue management improves.

Soil organic matter influences on soil quality indicator properties and functions- Most arable soils contain only 2 to 4 per cent OM by weight, yet very little about these soils is not significantly influenced by the OM in them. Organic matter provides much of the soil's capacity to store nutrients and water. It plays a critical role in the formation and stabilization of soil structure, which in turn produces good tilth and drainage, and resistance to erosion. It can not only carry and make available N, S, and P, but also improve the availability of nearly all nutrients, whether applied as fertilizer or weathered from minerals. It promotes the health of the soil ecosystem and stimulates organisms that cycle C and protect plants from disease. Humic fractions of SOM can directly influence plants through enhanced nutrient uptake, effects on intermediary metabolism, and, according to some authors, possibly some hormone like reactions. However, knowledge of the mechanisms involved is quite limited and the importance of these direct effects in field soils is still quite uncertain. This chapter therefore focuses on impacts of SOM on the physical, chemical, and biological properties of soil, many of which indirectly influence plant growth.

4. EFFECT OF SOM ON SOIL PROPERTIES

4.1 Effect on physical properties

To a remarkable degree, increased OM can counteract the ill effects of too much clay or too much sand. Increasing the SOM content usually increases total porosity and therefore decreases bulk density. Friability is a term that describes the tendency of soil clods to easily crumble into their constituent natural aggregates. Highly friable soils are easy to cultivate or to sow with a no-till seeder. Soil management can have pronounced effects on friability. Cropping systems that promote SOM accumulation usually also promote soil aggregation. The soil properties that contribute to the formation and stabilization of macro-aggregates include soil texture, clay mineralogy, exchangeable cations, Fe and Al oxides, and $CaCO_3$, as well as SOM. In three sites from semiarid to sub-humid Spain, micro-aggregate stability positively correlated with clay content, whereas macro aggregates stability correlated mainly with SOM. However, below threshold SOM content, aggregate stability was more closely associated with the soil carbonate content. The water regime in soils is influenced by SOM in several ways. First, OM increases the soil's capacity to hold plant available water, defined as the difference between the water content at field capacity and that held at the permanent wilting point. It does so, both by direct absorption of water and by enhancing the formation and stabilization of aggregates containing an abundance of pores that hold water under moderate tensions.

4.2 Effect on chemical properties

Nutrient storage and release nearly all the N and large proportions of the P and S found in soils occur as constituents of SOM. The SOM serves as both the principal long-term storage medium and as the primary short-term source of these and other nutrients.

4.2.1 *Microbial enhancement of nutrient availability*

The main contribution of soil organisms to make plant nutrition is through cycling nutrients contained in crop residue and other organic amendments. As organisms in the soil food web use stored photosynthetic energy and nutrients in organic residue, mineralization of N, P, S, and other elements occur, rendering them more available to plants. However, there are a number of other ways by which soil organisms can enhance the availability of nutrients to plants:

- Biological N_2 fixation can make important contributions to both the short- and long-term N status of plants. For example, the mututalist Rhizobium (in legume root nodules), free-living *Azotobacter* and associate symbiotic *Azospirillum*, and those that form associations with non-agricultural plants enhance N availability.

- Arbuscular mycorrhizal fungi form mutualist associations with most crop plants, greatly enhancing the effective surface area of the root systems, and thus contact with soil solids. The greater surface area produced by mycorrhizal hyphae promotes uptake of water as well as P and other nutrients.
- Siderophores, low-molecular-weight compounds that chelate Fe and increase its availability to plants, are produced in low-Fe soils by a variety of microorganisms, such as *Agrobacterium, Bacillus, Pseudomonas*, and *Rhizobium*. A wide variety of soil microorganisms are capable of enhancing P solubility in artificial media, and some have been shown to increase P availability in soils.

4.2.2 *Cation Exchange Capacity*

The colloidal fraction, which consists of both clay and humified OM, is recognized as the seat of chemical activity in soils, including the capacity for ion exchange. The relative contribution of clay and SOM to soil CEC is largely determined by the amount of SOM, the amount and mineralogy of clay, and the soil pH. For soils low in clay, SOM can be responsible for essentially all of a soil's CEC. Even for clay soils, SOM can contribute a significant fraction of the CEC. For example, illitic and chloritic clays have CECs *ca.* 20 cmol (+)/kg of clay. Composting is a practical means for storing organic residues and stabilizing their N for later use as a soil fertility amendment. Due to the low lability of its N, large amounts of compost might be needed initially to supply enough of this and other nutrients for optimal crop growth material. Beneficial effects on soil physical and chemical properties can make the use of heavy compost applications worthwhile. On the other hand, these benefits are not always apparent in the short term. Disease suppression and other beneficial biological effects of composts and compost extracts have also been widely reported, largely because decomposable organic amendments can serve as an excellent food base for beneficial soil microorganisms. In using heavy application of compost, more P is likely to be applied than the plant can take up, and therefore this element is likely to accumulate in the soil. The water quality impacts of potential P losses in runoff must therefore be taken into consideration.

4.2.3 *Anion sorption*

P deficiency is commonly associated with acid soils because of the strong sorption or fixation of inorganic P, Fe and aluminium oxide, under low-pH conditions. High P fixation is a major constraint to crop production in ca. 5 per cent of the world's agricultural soils and in 13 to 14 per cent of soils in regions with warm humid and sub-humid climates. Increasing SOM not only provides a source of P from mineralization but also can reduce the capacity of acid soils to lock up P by fixation.

4.2.4 *Metal mobility*

SOM also affects the mobility of heavy metals that may (*e.g.*, Zn) or may not (e.g., Cd) serve as plant nutrients. Addition of OM to soil can decrease or increase metal availability, solubility, and plant uptake.

4.2.5 *Soil pH buffering and amelioration*

SOM exerts a major influence on the pH buffering of surface soils, both because it contributes much of the soil CEC and because of the dissociation of carboxylic acid functional groups on the SOM molecules.

4.3 Effect on biological properties

Soil organic matter is both a raw material for and a product of biological activity in soils. Increased SOM can enhance the diversity of the microbial community as well as its biomass. The critical roles that microbial biomass performs in the soil system can be summarized as follows:

- A labile source of C, N, P, and S
- An immediate sink of C, N, P, and S
- An agent of nutrient transformation and pesticide degradation
- An agent of mineral weathering and soil formation
- Includes organisms, especially in association with plant roots, that form and stabilize soil aggregates
- Includes organisms that are antagonistic to plant pathogens and parasitic nematodes
- Produces plant growth regulators

5. SOIL QUALITY ASSESSMENT: CONCEPT

Soil quality cannot be measured directly because it is a broad, integrative, context-dependent concept. Instead, we analyse a variety of proxy measurements that together provide clues about how the soil is functioning as viewed from one or more soil-use perspectives. These measurements are called soil quality indicators. A set of low-cost readily measured indicators that accurately predict soil functions of interest is called an efficient indicator set. Indicators of soil quality may include characteristics of soil solids, soil solutions, soil atmospheres, vegetation, and other soil biota, and possibly even economic analyses of land-uses or ecosystem services.

Although the quantity and quality of data may differ, the process of soil quality evaluation follows the same basic steps regardless of the method used: identification of soil use issues followed by indicator selection and interpretation. More specifically, in order to select appropriate indicators, one must first determine the land-use objectives, and then indicators must be proposed, measured and assessed across a representative set of lands and management practices. An efficient indicator set should be used to inform land management decisions at specific sites and then be used to monitor trends in soil function after changing practices and over time (Doran, J.W., 2002)

5.1. Soil quality functions

These functions are performed on different levels and are determined by inherent soil characteristics (e.g. texture, OM content, pH, CEC, porosity etc.) and external environmental (climate, terrain, hydrological, biological) and anthropogenic (soil-use and management) factors.

Table 1. Ecological functions of soil and their indicators

Ecological functions of soil	Indicators of proper functioning
Production function	High levels of crop yields and incomes
Biotic environmental function/ living space function	High levels of species richness and functional dominance of beneficial organisms – High levels of crop yields and incomes, and high quality food and habitation
Climate-regulative function/storage function	High levels of C stocks and slow rates of greenhouse gas emissions
Hydrologic function	Adequate availability of water/reduced risks floods
Waste and pollution control function	High levels of crop yields and incomes, and high quality food and habitation

What does soil do?

- *Regulating water–* Soil helps to control where rain, snowmelt, and irrigation water goes. Water and dissolved solutes flow over the land or into and through the soil.
- *Sustaining plant and animal life–* The diversity and productivity of living things depend on soil.
- *Filtering potential pollutants–* The minerals and microbes in soil are responsible for filtering, buffering, degrading, immobilizing, and detoxifying organic and inorganic materials, including industrial and municipal by-products and atmospheric deposits.

- *Cycling nutrients*– C, N, P, and many other nutrients are stored, transformed, and cycled through soil.
- *Supporting structures*– Buildings need stable soil for support, and archaeological treasures associated with human habitation are protected in soils.

5.2. Soil degradation threats

Soil is essentially a non-renewable resource with possible high rate of degradation and extremely slow rate of regeneration processes. Degradation deteriorates soil quality by partially or entirely damaging one or more of its functions. The focus of policy actions is the reduction of risk of soil degradation. Risk of soil degradation depends on soil and terrain properties which make the soil inherently receptive of degradation. The Thematic Strategy for Soil Protection declares that for sustainable development, soils (soil functions) need to be protected from degradation. The main threats to soil functioning abilities are identified as –

1. Decline in OM
2. Soil erosion
3. Compaction
4. Salinization
5. Landslides
6. Floods
7. Contamination and
8. Sealing

Threats are area (and soil) specific in their appearance, therefore, they require additional spatial consideration during soil conservation planning. For each area-dependent threat, the following conditions have been examined in order to define:

- Identification of factors/hazards related to threat ('external' factors) – characterization of receptor ('internal' attributes),
- Performance specification, model selection (with data requirements).

In order to identify and describe areas at risk to soil threats in the "Common Criteria" document, three types of approaches:

1. *Qualitative approach*: Land use in combination with "sensitive soils", or other political boundaries using other combined criteria, *e.g.* nitrate pollution, intensive cropping areas, urban areas, etc.;
2. *Quantitative approach*: Comparative assessment, dynamic assessment, regression analysis, pedotransfer functions, scoring functions and principal component analysis.

3. *Model approach*: The model approach needs to be eventually supplemented by a quantitative approach: not only for model validation and calibration, but also in order to detect the area where the degradation actually occurs, and to observe the trend after the implementation of measures. Models can also help in approach to regionalize soil information, from the plot-level to the area/region.

6. METHODS OF SOIL QUALITY ASSESSMENT

A variety of methods or approaches are currently used to measure and assess soil quality. The methods discussed in this guide range from primarily qualitative to purely quantitative. They are as follows:

- Soil health card
- NRCS soil health card template (NRCS template)
- Soil quality test kit
- Laboratory analysis

These methods provide important information about soil quality, whether the goal is to determine changes in soil health over time or to compare management effects on soil quality in different fields or pastures (Laishram, *et al.,* 2012). Various combinations of these methodologies may be used. No single one is inherently better or more effective.

6.1 Why assess soil quality

Soil quality is evaluated to learn about the effects of management practices on soil function. Reasons for evaluating soil quality fall into three categories:

- Awareness and education
- Evaluation of practice effects and trouble-shooting
- Evaluation of alternative practices

6.1.1 *Awareness and education*

The soil quality concept emphasizes an ecological approach to land management. Management actions don't have simple, single effects in complex systems, such as soil. Management has multiple effects, both direct and indirect. For example, tillage is used to loosen surface soil, prepare the seedbed, and control weeds and pests. But tillage can also break up soil structure, speed the decomposition and loss of OM, increase the threat of erosion, destroy the habitat of helpful organisms, and cause compaction. Understanding the trade-offs that exist for the range of management options is a first step towards improved land management and public policy. Assessment as an Educational Tool includes one-on-one and field day use of in-field testing tools.

6.1.2 *Evaluation of practice effects and trouble-shooting*

Soil quality is often referred to as "Soil Health" because of objectives similar to the monitoring and maintenance of human health. Doctors monitor health indicators and watch for irregularities or declines in status. The set of health indicators measured during a check-up is familiar to all of us: temperature, pulse, blood pressure, heartbeat, urine samples, etc. Monitoring of these indicators may reveal potential problems even before painful symptoms occur; the earlier problems are observed, the easier they are to treat.

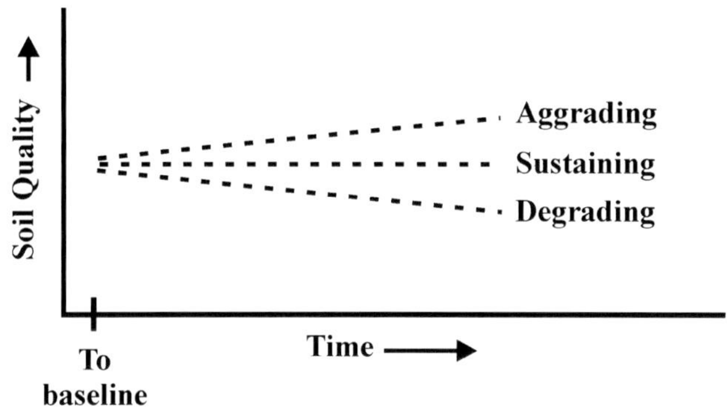

Fig. 2. Dynamic soil quality changes over time

6.1.3 *Evaluation of alternative practices*

Beyond awareness and evaluating current practices, soil quality assessment methods provide a framework for comparing management trade-offs and deciding which management options provide the greatest good, whether for one's farming operation or at a watershed or regional scale.

7. SOIL QUALITY: INDICATOR OF SUSTAINABLE LAND MANAGEMENT

Soil quality appears to be an ideal indicator of sustainable land management. Soil is the foundation for nearly all land uses. Soil quality, by definition, reflects the capacity to sustain plant and animal productivity, maintain or enhance water and air quality, and promote plant and animal health. By reflecting the basic capacity of the soil to function, it integrates across many potential uses. Nonetheless, few land managers have adopted soil quality as an indicator of sustainable land management. There are a number of constraints to adoption. Most could be overcome through a concerted effort by the research community. Specifically, we need to address the following issues: (1) demonstrate causal relationships between soil quality and ecosystem functions, including biodiversity conservation, biomass production and conservation of soil and water resources. True calibration of soil quality requires more than merely comparing values across management systems; (2) increase the power of soil quality indicators to predict response to disturbance. Although there are many indicators that reflect the current capacity of a soil to function, there are few that can predict the capacity of the soil to continue to function under a range of disturbance regimes. Both resistance and resilience need to be considered; (3) Increase accessibility of monitoring systems to land managers. Many existing systems are too complex, too expensive, or both; (4) Integrate soil quality with other biophysical and socio-economic indicators. Effective early-warning monitoring systems will require not just the inclusion of both biophysical and socio-economic indicators, but also the development of models that incorporate feedbacks between soil quality and socio-economic conditions and trends and (5) Place soil quality in a landscape context. Most ecosystem functions depend on connections through time across different parts of the landscape (Arshad and Martin, 2002). In conclusion, soil quality is a necessary but not sufficient indicator of sustainable land management. Its value will continue to increase as limitations are diminished through collaboration between scientists, land managers and policymakers. Developing sustainable land management systems are complicated by the need to consider their utility to humans, their efficiency of resource use, and their ability to maintain a balance with the environment that is favourable to both humans and most other species (Barrios *et al.*, 2006)

In particular, we are challenged to develop agricultural management systems that balance the needs for production of food and fibre with those for maintenance of the environment. Soil quality is conceptualized as the major linkage between the strategies for agricultural conservation management practices and achievement of the major goals of sustainable agriculture. In short, the assessment of soil quality or health, and direction of change with time, is the primary indicator of sustainable land management. Although soil's contribution to plant productivity is widely recognized, soil condition also impacts water and air quality. The quality of surface and sub-surface water has been jeopardized in many parts of the world by intensive land management practices and the consequent imbalance of C, N, and water cycling in soil. Soil management practices such as tillage, cropping patterns, and pesticide and fertilizer use influence water quality. In addition, these management practices can influence atmospheric quality through changes in the soil's capacity to produce or consume important atmospheric gases, such as carbon dioxide, nitrous oxide, and methane. The present threat of global climate change and ozone depletion, through elevated levels of greenhouse gases and altered hydrological cycles, necessitates a better understanding of the influence of land management on soil processes. In summary, the quality and health of soil determine agricultural sustainability, environmental quality, and as a consequence of both, plant, animal, and human health. Scientists make a significant contribution to sustainable land management by translating scientific knowledge and information on soil function into practical tools and approaches by which land managers can assess the sustainability of their management practices. Specifically, assessment of soil quality/health is needed to identify problem production areas; make realistic estimates of food production, monitor changes in sustainability and environmental quality as related to agricultural management, and to assist government agencies in formulating and evaluating sustainable agricultural and land-use policies.

8. MAJOR FACTORS AFFECTING SOIL QUALITY

- Difference between synchronize the nutrient demand and supply.
- Imbalanced fertilizer use and high harvest the nutrient from soil.
- High deficiency of the micronutrient and secondary nutrient.
- Insufficient source of OM.
- Heavy metal toxicity.
- Acidification and alkalinity.
- Rapid industrialization and urbanization.
- Soil sickness due to lack of beneficial microbial population.

9. THE SOIL HEALTH FRAMEWORK

The concept of soil health and soil quality has consistently evolved with an increase in the understanding of soils and soil quality attributes (Douglas, *et al.*, 1996). Soil quality cannot be measured directly, but soil properties that are sensitive to changes in management can be used as indicators. Soil health indicators are needed that help smallholder farmers who understand the chain of cause and effect that links farm decisions to ultimate productivity and health of plants and animals. The soil health approach is better applied when specific goals are defined for a desired outcome from a set of decisions (Kirkby *et al.*, 2004). Therefore we can think of the soil health quality as an evaluation process which consists of a series of actions –

1. Selection of soil health indicators
2. Determination of a minimum data set (MDS)
3. On-farm assessment and validation
4. Development of an interpretation scheme of indices

9.1 Selection of soil health indicators

The quality of soil is rather dynamic and can affect the sustainability and productivity of land use. It is the end product of soil degradative or conserving processes and is controlled by chemical, physical, and biological components of a soil and their interactions. Indicators, however, will vary according to the location, and the level of sophistication at which measurements are likely to be made. Therefore, it is not possible to develop a single short list which is suitable for all purposes.

Biological– Identification of biological indicators of soil quality is reported as critically important because soil quality is strongly influenced by microbiological mediated processes (nutrient cycling, nutrient capacity, and aggregate stability). Biological indicators of soil quality that are commonly measured include SOM, respiration, microbial biomass (total bacteria and fungi,) and mineralizable N. Soil organic matter plays a key role in soil function, determining soil quality, water holding capacity and susceptibility of soil to degradation.

Chemical– In order to achieve high crop yields, small holder farmers have to provide soil nutrients in large quantities. Therefore it is possible to alter the pool of available nutrients by adding inorganic fertilizers, incorporating cover crops, and using other organic materials in form of manures and composts. Results of chemical tests are soil quality indicators which provide information on the capacity of soil to supply mineral nutrients, which is dependent on the soil pH. Soil pH is an estimate of the activity of hydrogen ions in the soil solution. It is also an indicator of plant available nutrients. High activity is not desirable and the soil may require liming with base cations Ca or Mg in order to bring the solution back to neutral.

Physical– Soil physical properties are estimated from the soil's texture, bulk density (a measure of compaction), porosity, water-holding capacity. The presence or absence of hard pans usually presents barriers to rooting depth. These properties are all improved through additions of OM to soils. Therefore, the suitability of soil for sustaining plant growth and biological activity is a function of its physical properties (porosity, water holding capacity, structure, and tilth).

There are several criteria to consider when selecting soil health and soil quality indicators.

In general, appropriate indicators should be:

- Easy to asses.
- Able to measure changes in soil function both at plot and landscape scales.
- Assessed in time to make management decisions.
- Accessible to many farmers.
- Sensitive to variations in agro-ecological zone.
- Representative of physical, biological or chemical properties of soil.
- Assessed by both qualitative and/or quantitative approaches.

9.1.1 *Soil health indicator categories*

Chemical indicators can give you information about the equilibrium between soil solution (soil water and nutrients) and exchange sites (clay particles, OM); plant health; the nutritional requirements of plant and soil animal communities; and levels of soil contaminants and their availability for uptake by animals and plants. Indicators include measures of:

- Electrical conductivity
- Soil nitrate
- Soil reaction (pH)

Physical indicators provide information about soil hydrologic characteristics, such as water entry and retention that influences availability to plants. Some indicators are related to nutrient availability by their influence on rooting volume and aeration status. Other measures tell us about erosional status. Indicators include measures of:

- Aggregate stability
- Available water capacity
- Bulk density
- Infiltration
- Slaking
- Soil crusts
- Soil structure and macro-pores

Biological indicators can tell us about the organisms that form the soil food web that are responsible for decomposition of OM and nutrient cycling. Information about the numbers of organisms, both individuals and species, that perform similar jobs or niches, can indicate a soil's ability to function or bounce back after disturbance (resistance and resilience). Indicators include measures of:

- Earthworms
- Particulate organic matter
- Potentially mineralizable N
- Respiration
- Soil enzymes
- Total organic C

9.2 Determination of a minimum data set

A minimum data set was proposed to measure soil quality and its changes due to management practices through selection of key indicators, such as soil texture, OM, pH, nutrient status, bulk density, electrical conductivity and rooting depth. Collecting a minimum data set helps to identify locally relevant soil indicators and to evaluate the link between selected indicators and significant soil and plant properties. It is a minimum set of indicators required to obtain a comprehensive understanding of the soil attributes evaluated. More importantly, they serve as a useful tool for screening the condition, quality, and health of soil. For smallholder farmers these tools need to be simple measures of soil health and soil quality such as consistency, colour and workability. For extension and policy personnel, they provide basic information needed to arrive at management decisions. For researchers, there is need to conduct sufficiently detailed tests while controlling for variation in order to develop meaningful assessments of soil status, often expressed as an index of soil quality (Evah *et al.*, 2000).

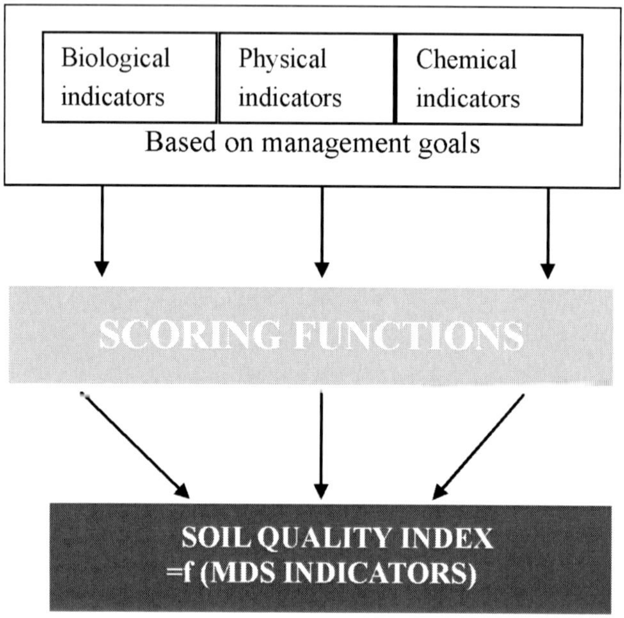

Fig. 3. Soil quality index relationship with MDS indicators (Adapted from Karlen *et al.*, 2003)

Due to these complexities, there is need to develop diagnostic measures and indicators of soil health and soil quality changes in order to derive classifications from the minimum data set (Table 2) that can better assist farmers and inform research and extension to target solutions at farm level.

Table 2. Proposed minimum data set (MDS) of physical, chemical, and biological indicators for assessing soil quality

Indicator	Rationale for assessment
Biological	
Microbial biomass C and N	Describes microbial catalytic potential and repository for C and N.
	Provides an early warning of management effects on OM.

Indicator	Rationale for assessment
Potentially mineralizable N	Describes soil productivity and N supplying potential. Provides an estimate of biomass.
Soil respiration	Defines a level of microbial activity. Provides an estimate of biomass activity.
Chemical	
SOM	As a proxy for soil fertility and nutrient availability.
Ph	Biological and chemical activity thresholds.
Electrical conductivity	Plant and microbial activity thresholds.
Extractable N, P, and K	Describes plant-available nutrients and potential for N loss. Indicates productivity and environmental quality.
Physical	
Soil texture	Indicates how well water and chemicals are retained and transported. Provides an estimate of soil erosion and variability.
Soil depth and rooting	Indicates productivity potential. Evens out landscape and geographic variability.
Infiltration and soil bulk density (ρb)	Describes the potential for leaching, productivity, and erosion. ρb is used to correct soil analyses to volumetric basis.
Water holding capacity	Describes water retention, transport, and erosion. Available water is used to calculate soil bulk density and OM.

(*Source*: Doran and Parkin, 1994)

9.3 On-farm soil health assessment

On-farm assessment of soil quality and health is recommended to assist farmers to evaluate the effects of their management decisions on soil productivity. Also, this approach permits interaction between researchers, extension and policy personnel when providing interpretation to link on-farm knowledge to soil health information (Adélia *et al.,* 2011). The main challenge is to develop soil quality and soil health standards to assess changes which are practical and useful to farmers. For instance, linking soil health measurements with farmer perceptions of soil quality can bridge the gap in interpretation of complex data sets (Table 3). Several studies show that by using local knowledge, smallholder farmers are able to accurately predict soil quality differences of productive and non-productive fields.

Table 3. Example of an interpretation framework for soil health indicators under agricultural land uses

Indicator	Ranking		
	Low	**Medium**	**High**
Total OM content (organic C % × 1.8)	Poor pore structure, hard workability (<1.7%)	Friable, but poor workability (1.7–2.6%)	Extremely friable and easy workability (> 2.6%)
Light fraction OM	Noticeable fine root fragments and weed seeds	Mixtures of root and leaf litter fragments	Dominated by large leaf litter fragments.
Organomineral fraction OM	Deep red with discoloured brown flakes of clay particles	Deep red with consistent brown colored clay particles	Near pitch-dark organic with mixtures of consistent red mineral flakes
Soil pH	High acid < 5.5	Medium acid to neutral 5.5 to 7.0	Neutral to basic 7.0 to 8.0
Soil CEC	< 10 mmolc kg^{-1}	10 to 20 mmolc kg^{-1}	> 20 mmolc kg^{-1}
Soil aggregate stability	Water stable aggregation of 50–60% indicates weak structure highly erodible	Water stable aggregation of 60–80% indicates stable structure but still susceptible to erosion	Water stable aggregation of >80% indicates highly stable structure and little susceptibility to erosion

9.4 Development of an interpretation scheme of indices

The use of scorecards for on-farm soil quality assessment is emphasized where qualitative observations of soil health are scored to obtain an overall measure of soil quality and soil health (Fig. 4). These cards may be 7 developed to evaluate soil health through farmer observations of soil physical, chemical and biological properties. These soil characteristics are classified in terms of descriptive indicators which are interpreted on a graded scale. Examples such as observations on earthworm numbers can yield a general index of biological activity in the soil.

9.4.1 *Soil quality test kits*

Assessment tools such as soil quality test kits focus on farmer-based evaluations and education regarding various soil and smallholder management practices. Further, they aim to produce an educational tool to increase public awareness of the importance of soil quality. Such kits, which are commercially available, provide for measurements of soil respiration, infiltration, bulk density, water content, electric conductivity (EC), pH, soil nitrates, aggregate stability, slake test and earthworm numbers (Andrews *et al.,* 2004).).

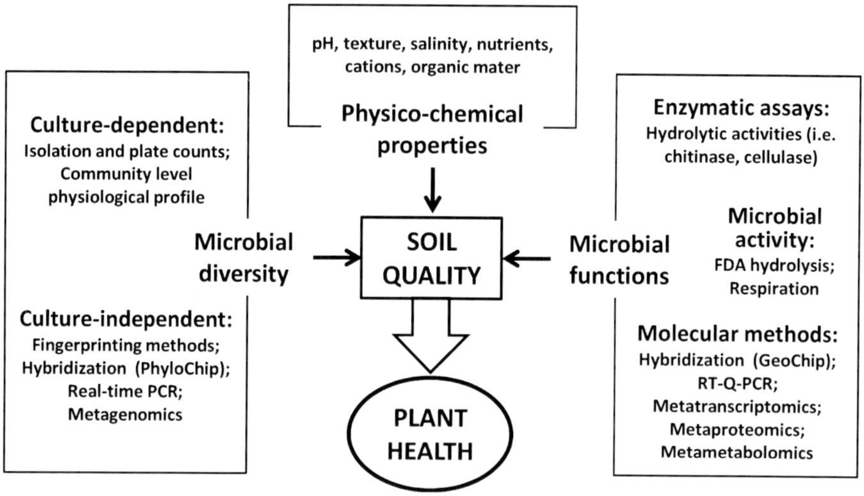

Fig 4. Relationship between the soil properties and soil quality

9.4.2 *Soil quality indices*

Finally, various soil quality indexing approaches are available and can be applied to derive a range of critical test values within which soil quality and soil health assessments can be defined.

9.4.3 *Indicator transformation (scoring)*

After determining the variables for the MDS, every observation of each MDS indicator was transformed for inclusion in the SQI methods examined. Two techniques were compared: linear scoring or non-linear scoring.

9.4.3.1 *Linear scores*

Indicators were ranked in ascending or descending order depending on whether a higher value was considered "good" or "bad" in terms of soil function. For 'more is better' indicators, each observation was divided by the highest observed value such that the highest observed value received a score of 1. For 'less is better' indicators, the lowest observed value (in the numerator) was divided by each observation (in the denominator), such that the lowest observed value receives a score of 1 (John *et al.*, 2002).

9.4.3.2 *Non-linear scores*

For this method, indicators were transformed using non-linear scoring functions constructed using Curve Expert version 1.3 shareware (http:/www.ebicom.net/~dhyams/cvxpt.htm). The shape of each decision function, typically some variation of a bell-shaped curve ('mid-point optimum'), a sigmoid curve with an upper asymptote ('more is better'), or a sigmoid curve having a lower asymptote ('less is better'), was determined according to agronomic and environmental function using literature review and consensus of the collaborating researchers.

9.4.4 *Indicator integration into indices*

Three soil quality indices was compared:

1. Additive SQI (ADD SQI);
2. Weighted additive SQI (WTD SQI)
3. Decision support system (DSS SQI)

1. *Additive index–* The additive index was a summation of the scores from MDS indicators. From these summed scores, the ADD SQI treatment means and standard deviations were calculated.

2. *Weighted additive index–* Once transformed, the MDS variables for each observation were weighted using the PCA results. Each PC explained a certain amount (%) of the variation in the total data set. This per centage, standardized to unity, provided the weight for variables chosen under a given PC. We then summed the weighted MDS variable scores for each observation and calculated the treatment means and standard deviations.

3. *Decision support system SQI–* This technique applied the additive value function method to solve hierarchical multi-attribute problems. To create the importance order hierarchy for the DSS SQI, the results of an informal survey completed by Central Valley farmer collaborators (unpublished data) were used. Like the other SQIs, the DSS used scored indicator values from either the PCA or EO selected MDSs. The DSS used indicator scores for the treatment means and reported a median and range of outcomes that are not statistically comparable. Instead, dominance among alternatives is established.

10. SOIL QUALITY MANAGEMENT: KEY STRATEGIES FOR AGRICULTURAL LAND

How can you manage land in a way that improves soil productivity, water quality, and other soil benefits? Start with these six components of soil quality management. If you want to assess the effect of your existing land management practices, begin by asking which of your practices fall into each of these categories. Choosing specific practices within each category depends on your situation because different kinds of soil respond differently to the same practice.

10.1 Enhance organic matter

Organic matter management: Whether your soil is naturally high or low in organic matter, adding new organic matter every year is perhaps the most important way to improve and maintain soil quality. Regular additions of organic matter improve soil structure, enhance water and nutrient holding capacity, protect soil from erosion and compaction, and support a healthy community of soil organisms. Practices that increase organic matter include: leaving crop residues in the field, choosing crop rotations that include high residue plants, using optimal nutrient and water management practices to grow healthy plants with large amounts of roots and residue, growing cover crops, applying manure or compost, using low or no tillage systems, using sod-based rotations, growing perennial forage crops, and mulching.

10.2 Avoid excessive tillage

Tillage management: Reducing tillage minimizes the loss of organic matter and protects the soil surface with plant residue. Tillage is used to loosen surface soil, prepare the seedbed, and control weeds and pests. But tillage can also break up soil structure, speed the decomposition and loss of organic matter, increase the threat of erosion, destroy the habitat of helpful organisms, and cause compaction. New equipment allows crop production with minimal disturbance of the soil.

10.3 Manage pests and nutrients efficiently

Chemical management: An important function of soil is to buffer and detoxify chemicals, but soil's capacity for detoxification is limited. Pesticides and chemical fertilizers have valuable benefits, but they also can harm non-target organisms and pollute water and air if they are mismanaged. Nutrients from organic sources also can pollute when misapplied or over-applied. Efficient pest and nutrient management means testing and monitoring soil and pests; applying only the necessary chemicals, at the right time and place to get the job done; and taking advantage of non-chemical approaches to pest and nutrient management, such as crop rotations, cover crops, and manure management.

10.4 Prevent soil compaction

Compaction management: Compaction reduces the amount of air, water, and space available to roots and soil organisms. Compaction is caused by repeated traffic, heavy traffic, or traveling on wet soil. Deep compaction by heavy equipment is difficult or impossible to remedy, so prevention is essential. Subsoil tillage is only effective on soils with a clearly defined root-restricting plow pan. In the absence of a plow pan, subsoil tillage to eliminate compaction can reduce yield. Prevention, not tillage, is the way to manage compaction.

10.5 Keep the ground covered

Residue management: Bare soil is susceptible to wind and water erosion, and to drying and crusting. Ground cover protects soil, provides habitats for larger soil organisms, such as insects and earthworms, and can improve water availability. Ground can be covered by leaving crop residue on the surface or by planting cover crops. In addition to ground cover, living cover crops provide additional organic matter, and continuous cover and food for soil organisms. Ground cover must be managed to prevent problems with delayed soil warming in spring, diseases, and excessive build-up of phosphorus at the surface.

10.6 Diversify cropping systems

Diversity management: Diversity is beneficial for several reasons. Each plant contributes a unique root structure and type of residue to the soil. A diversity of soil organisms can help to control pest populations, and a diversity of cultural practices can reduce weed and disease pressures. Diversity across the landscape can be increased by using buffer strips, small fields, or contour strip cropping. Diversity over time can be increased by using long crop rotations. Changing vegetation across the landscape or over time not only increases plant diversity, but also the types of insects, microorganisms, and wildlife that live on your farm.

11. FUTURE DIRECTIONS

Rapidly increasing population in countries that can least afford it have made them food insecure. With inadequate inputs in agriculture, developing countries are degrading their lands rapidly and destroying ecosystems. Currently, soil resource assessment and monitoring is entering a new era, in terms of quality of information produced by new information technologies through the innovative use of Geographic Information Systems and remote sensing and will significantly improving the acceptance and use of soil health information. The use of electronic technology will significantly increase the demand for and ability to process more data. Further innovations will result in model approaches in soil genetic studies that will demonstrate the integral role of soils in ecosystems. Particularly the use of ICT including on-line delivery content through wikis, blogs and social book marking is set to push sharing of soil health information to a new level. At the more fundamental level, basic research will be needed in order to select and develop proper indicators, applicable at different farmer scales. Innovations will be required in setting up effective study programs, which would guarantee the accumulation of the necessary baseline soil data in order to develop appropriate minimum datasets.

REFERENCES

Adélia NN, António C and Celeste OAC (2011) Impacts of land use and cover type on runoff and soil erosion in a marginal area of Portugal. *Applied Geography* **31:** 687–699.

Andrews SS, Douglas LK and Cynthia AC (2004) The soil management assessment framework: a quantitative soil quality evaluation method. *Soil Sci. Soc. Am. J.* **68:** 1945–1962.

Arshad MA and Martin S (2002) Identifying critical limits for soil quality indicators in agro-ecosystems. *Agriculture, Ecosystems and Environment* **88:** 153–160.

Barrios E, Delve RJ, Bekunda M, Mowo J, Agunda J, Ramisch J, Trejo MT and Thomas RJ (2006) Indicators of soil quality: A South-South development of a methodological guide for linking local and technical knowledge. *Geoderma* **135:** 248–259.

Doran JW and Parkin TB (1994) Definery and assessory soil quality. In: Doran, JW, Coleman DC, Berdick, DF and Stewort BA, Eds. Defining soil quality for a sustainable environment, *Soil Science Society of americ Journal,* Madison 3–31.

Doran JW (2002) Soil health and global sustainability: translating science into practice. *Agriculture, Ecosystem & Environment* **88:** 119–127.

Douglas ER, Jason MG and Robin FH (1996) Farmer-based assessment of soil quality: A soil health scorecard, University of Wisconsin. Madison.

Evah WM, Nancy KK, Paul CS and Paul LW (2000) Diagnostic indicators of soil quality in productive and non-productive smallholders' fields of Kenya's Central Highlands. *Agriculture, Ecosystems and Environment* **79:** 1–8.

John WD Stamatis (2002) Soil health as an indicator of sustainable management. *Agriculture, Ecosystems and Environment* **88**(2): 107–110.

Karlen DL, Mausbach MJ, Doran JW, Cline RG, Harris RF and Schuman GE (1997) Soil quality: a concept, definition and framework for evaluation. *Soil Science Society of America Journal* **61:** 4-10.

Kirkby MJ, Jones RJA, Irvine B, Gobin A, Govers G, Cerdan O and Huting J (2004) Pan-European Soil Erosion Risk Assessment, *ISPRA*. 73.

Laishram J, Saxena KG, Maikhuri RK and Rao KS (2012) Soil quality and soil health - a review. *International Journal of Ecology and Environmental Sciences* **38(1):** 19-37.

■■■■

17

Organic Farming: An option for Food Quality and Environmental Benefits

Gangadhar Nanda[1], D.K. Singh[2], B.L. Meena[3] and Uadal Singh[4]

[1]Ph. D. Scholar, Department of Agronomy, G. B. Pant University of Agriculture and Technology, Pantnagar, Uttarakhand – 263 145
[2]Professor, Department of Agronomy, G. B. Pant University of Agriculture and Technology, Pantnagar, Uttarakhand – 263 145
[3]ICAR-Central soil Salinity Research Institute, Karnal – 132001
[4]Associate Professor (Horticulture), College of Agriculture, Lalsot, Distt: Dausa (Raj.) – 303 511
E-mail: gangadhar.nanda4@gmail.com

ABSTRACT

Adoption of Green revolution technologies in our country led to self sufficiency in food grain production. However, this intensive cultivation has not only resulted in pollution of soil, surface and underground water but also affected the human health. So, protection of environment and human health along with production sustainability is the concern now-a-days. Organic farming is a form of sustainable agriculture which can be an alternative to this intensive cultivation. Organic farming methods rely on organic inputs and recycling of farm waste for nutrient supply, emphasize cropping rotation and soil biological processes for pest management and ban applications of synthetic fertilizers and pesticides, and hence may reduce negative effects attributed to conventional farming. During initial years, the yield levels are low, but after the conversion period, better soil biological activity leads to augmented yield levels. Reduced availability of nitrogen and no use of synthetic pesticides create stress on plants that enhance the synthesis of active phyto-chemicals, such as phenols, flavonoids and anti-oxidants which enhance the quality of food. Improvement in soil health, soil carbon sequestration, bio-diversity protection, reduced energy requirement per unit area, reduced nitrate leaching and emission of nitrous oxide are also achieved due to organic farming. Therefore, adoption of organic farming would promote agricultural diversification and sustainable production of healthy food, augment family income as well as can solve environmental issues.

Keywords: Organic farming, Productivity, Food quality, Environmental benefits

1. INTRODUCTION

Green revolution has increased the food grain production of our country to a state of sufficiency due to use of high yielding varieties, fertilizers, pesticides and other inorganic inputs. But this intensive cultivation has not only led to pollution of soil, water bodies, underground water and the environment, but also affected the human beings. Further with increased demographic pressure and shrinking resource base due to enhanced industrialization and urbanization, it seems difficult to feed the future population in a sustainable manner with environmental protection. So, there is a need to go for a truly sustainable farming system on a wide scale to meet this challenge and ensure food and ecosystem security. A system is called sustainable only if it gives sufficient amount of quality food, increases the natural resource base, protects the environment, is economically viable and contributes to the wellbeing of farmers and their community (Reaganold and Watchter, 2016). In this regard, organic farming seems to be the better alternative to conventional farming. Organic farming is a production system that promotes environmentally, socially and economically sound production of food and fibres, and excludes the use of synthetically compounded fertilizers, pesticides, growth regulators, livestock feed additives and genetically modified organisms. The overall goal of organic farming is to provide food of high quality with minimum environmental impacts in a sustainable manner (Leifeld and Fuhrer, 2010). It is often called as organic agriculture, ecological or biological agriculture. Many researchers believe that organic farming is ideologically driven and inefficient, and in comparison to conventional farming, it requires more area to produce same amount of food as conventional agriculture. They also hypothesise that mass adoption of organic farming will lead to destruction of forests, wetlands and grasslands of the world (Trewavas, 2001 and Emsley, 2001). Some still consider it as an inefficient approach towards food security and question its relevance in the coming future (Connor and Mínguez, 2012 and Pickett, 2013). But the area, amount of funding related to organic farming and the market size of organic foods have steadily increased (Willer and Lernoud, 2015). Organic farming systems covered an area of 50.9 million hectares of land in 179 countries worldwide at the end of 2015 with organic food market value of 81.6 billion US Dollars. India is the country with largest organic producers (5.85 lakhs) in the world where it is cultivated in an area of 1,80,000 hectares of land (The World of Organic Agriculture 2017, FiBL). Recently, with enhanced awareness of global food safety and environment consciousness, the demand for organic food has increased many folds. Also, high quality, nutritious and safe organic food has generated greater interest among consumers towards organic farming and organic foods (Suja *et al*., 2012).

2. FEATURES OF ORGANIC CROP PRODUCTION

Organic farming methods rely on organic inputs and recycling of farm waste for nutrient supply, emphasize cropping system design and soil biological processes for pest management, and ban applications of synthetic fertilizers and pesticides

(Rigby and Cáceres, 2001). They may thus reduce negative effects attributed to conventional farming (Reganold *et al.*, 1987; Mäder *et al.*, 2002; Oehl *et al.*, 2004). Organic farmers also use improved varieties of crops and modern equipments and adopt soil and water conservation practices (Reaganold and Wachter, 2016). Furthermore, due to utilisation of locally available resources, utilizing the family member for labour intensive operations and selling the farm product at premium prices will give economic advantage to them, while adopting organic farming (Giovannucci, 2006). Organic farming differs significantly from conventional farming in terms of input cost for plant protection and labour cost for manuring and weeding operations resulting in significant increase in total labour cost and cost of cultivation of organic farming than conventional counterpart though both were comparable in terms of gross cost and net returns per hectare (Tashi and Wangchuk, 2015).

Effect on crop productivity

In general, during the initial years of transition from conventional to organic farming, a decline in yield is seen which can be attributed to requirement of time for the adaptation of soil biological activity to a new situation (Sacco *et al.*, 2015). The major factor for reduced productivity of organic production systems is attributed to reduced nitrogen availability in soil (Berry *et al.*, 2002; Clark *et al.*, 1998, 1999). On the other hand, high levels of reactive nitrogen (NH_4^+, NO_3^-) in soils may contribute to the emission of nitrous oxides, which is a major source of agricultural emissions. Appropriate selection of organic manure, fertiliser and amendments will help to supply the crop with nitrogen is depended upon the pattern and timing of N mineralisation and its synchronisation with the N demand of the crop with minimal environmental pollution (Poudel *et al.*, 2002 and Sacco *et al.*, 2015). Grain yield of rice and the recovery, and agronomic efficiency of nitrogen reduce significantly whereas, its physiological and partial factor productivity increases under organic farming than conventional one (Huang *et al.*, 2016). From a 6-year study, Venugopalan *et al.* (2010) found that cotton, soybean and chickpea yields were either similar or higher in organic farming systems than conventional systems in Central India. Elephant foot yam corm yield increased by 20 per cent and net return by 28 per cent under organic farming over conventional one (Suja *et al.*, 2012). According to a meta-analysis consisting of 115 studies in variety of crops showed reduction in organic yields by 19.2 per cent lower under organic farming than conventional farming. However, the diversification practice of multiple cropping and crop rotation substantially reduce the yield gaps when these methods are adopted in organic systems (Poinsio *et al.*, 2015).

Food quality

Food quality is a complex function of the nutrients in a food, the concentration of those nutrients, the form in which they exist, free from mycotoxins and chemical contaminants and organoleptic quality (taste, flavour, aroma, appearance, mouth feel). Besides traditional nutritional components (fat, protein, carbohydrate, calories, vitamin and mineral levels), a wide range of polyphenols, flavonoids and anti-oxidants are also of great importance in promoting healthy growth and in preventing diseases and ageing (Benbrook, 2009). Several past investigations suggest that production system can affect crop quality. Reduced availability of nitrogen and no use of pesticides create stress on plants that enhance the synthesis of active phyto-chemicals, such as phenols, flavonoids and anti-oxidants which enhance the quality of food. With respect to food quality, Oliveira *et al.* (2013) noticed an increment of 55 per cent vitamin C and 139 per cent total phenolics in tomato fruits from organic farming than from conventional farming. Higher hydrolysable polyphenols in peel and pulp of organically grown oranges and papaya and higher antioxidant in banana, papaya and mango peel over conventional farming have also been noticed (Faller and Fialho, 2010). Milled rice from organic mode took less cooking time with less gruel solid loss and the values for length:breadth ratio, chewiness and gumminess were higher than conventional ones (Kaur *et al.*, 2015). Singh *et al.* (2012) found higher amylose and carbohydrate content, and lower protein content in organic milled rice than those from conventional and integrated mode of production. The organic foods are safe and of high quality has a great demand in the international markets.

3. ENVIRONMENTAL BENEFITS

The potential environmental benefits of organic farming are in the areas of better soil health, carbon sequestration, maintenance of bio-diversity, reduced energy requirement per unit area, reduced nitrate leaching and emission of nitrous oxide.

Effect on soil properties

Soils under intensive cultivation are prone to degradation and require higher attention with the use of fossil fuel in terms of fertilizers, pesticides and irrigation to offset the process of degradation (Kendall and Pimentel, 1994). But these manipulations result in pollution of environment, health problems, destruction of natural habitats, higher energy consumption and ultimately lead to unsustainable agricultural production system (Pimentel *et al.*, 1995).

Physico-chemical properties

Addition of organic manures, incorporation of legume component in space and time lead in organic farming system to build up of organic carbon which helps in reducing bulk density, increasing water holding capacity and formation of water stable aggregates. This further increases infiltration rate so, there is less soil erosion as compared conventional farming system. Organic farming also lowered bulk density by 2.3 per cent, improved water holding capacity and porosity of soil by 28.4 and 16.5 per cent, respectively over conventional farming (Suja *et al.*, 2012). From a 6-year study, Venugopalan *et al.* (2010) reported that organic farming improved the organic carbon and available Zn content in soil whereas, conventional system led to higher pH and exchangeable sodium in Central India.

Biological properties

Adoption of organic farming leads to building of organic matter which is a source of energy for micro-organisms and hence greater diversity and number of micro-organisms is found in organic farming system which is reflected by their biomass and greater enzymatic activities and respiration in soil. Some of the enzymes are also helpful in solubilising the nutrient and making them available to plants. Benbi *et al.* (2016) noted significantly higher total carbon, soil organic/inorganic carbon, microbial biomass carbon, soil organic carbon stock and microbial activities under organic farming than its conventional counterpart at different depths (0–7.5 and 7.5–15 cm). Thus, better soil fertility and greater microbial bio-diversity in organic farming may reduce the dependence on external inputs (Mader *et al.*, 2002).

Soil carbon sequestration

Soil carbon sequestration refers to the capture of atmospheric CO_2 in soil in passive pools of organic matter making it resistant to rapid oxidation. In organic farms, nutrients to crops are supplied through green manure, farmyard manure/compost utilizing the on-farm residues of crops and livestock, but in conventional systems, only mineral fertilizers are applied to supply nutrient to the crops. Higher sequestration of soil organic carbon in organic system is probably due to addition of higher amounts of organic manure and adoption of crop rotation including pastures and legume crops. So, this practice may counterbalance emissions of green house gases like nitrous oxide and methane from agriculture sector (UNFCCC, 2008).

4. ENERGY USE

Compared to conventional farming, in organic farming, the energy use (input) per unit area is reduced due to avoidance of chemical sources of nutrients and pesticides which require energy for their synthesis and due to reduced tillage requirement as a result of increased organic matter which saves the fossil fuel. But in terms of output when calculated per product unit, organic farming produces lesser energy as the yields are often lower than conventional ones. From a 21-year study, Mader *et al.* (2002) reported that adoption of organic farming led to reduction of fertilizer, pesticide and energy by 34 per cent, 97 per cent and 53 per cent, respectively with yield reduction of 20 per cent than conventional one. In winter wheat-maize-soybean rotation system, Dal Ferro *et al.* (2017) observed reduced net energy requirement by 33.41 per cent per unit area under organic farming. Greater soil organic matter and lesser energy use in organic systems make them better in addressing the climate change which is the prime concern nowadays.

5. BIO-DIVERSITY

Biodiversity is the key player for agro-ecosystem stability and it is also true for getting stable supply of foods. Bio-diversity in organic farms are maintained or enhanced by several ways including adoption of diversified crop rotations, production of crop and livestock in a single farm. The practice of intercropping and crop rotation help to build up natural enemies whose population increase under organic farming due to avoidance of chemical pesticides. Organic soil management results in diverse species and greater population of bacteria, fungi, actinomycetes and earthworms that help in building fertile soil with stable soil aggregate which is less prone to erosion by soil.

6. NITRATE LEACHING AND EMISSION OF NITROUS OXIDE

Nitrate form nitrogen is prone to leaching loss and when it reaches ground water causes ground water pollution and when used for drinking purpose creates human health problems. Nitrous oxide (N_2O) is potential green house gas with global warming potential of about 300 times than carbon dioxide (CO_2). Both nitrate leaching and nitrous oxide emission are less in organic farming due to limited nitrogen availability. There was 32 per cent reduction in nitrous oxide emission under organic farming than conventional farming (Benoit *et al.*, 2015).

7. LINE OF RESEARCH FOR COMPARISON OF CONVENTIONAL VS. ORGANIC FARMING SYSTEM

- Study of dynamics of nutrient uptake by plant and nutrient release by soil.
- Studies on nutrient losses (nitrate and ammonium leaching, ammonium volatilization, denitrification) under two systems.
- Studies on greenhouse gas emission (nitrous oxide, carbon dioxide, methane *etc.*)
- Soil carbon stocks/soil carbon sequestration
- Species diversity of soil microbes and earthworms
- Energy budgeting of organic vs. conventional system and calculation of energy indices.
- Studies on quality of produce under the two systems in terms of appearance, sensory score, nutritive quality, anti-oxidant and secondary metabolite content, post-harvest shelf life, toxic pesticide residues and nitrate content in leafy vegetables
- Comparison of soil erosion in organic vs. conventional farming system
- Comparison of organic vs. inorganic farming system in terms of different indicators of sustainability (productivity, profitability, effect on environment and social wellbeing)

8. RESEARCH NEEDS IN ORGANIC FARMING FOR SUSTAINABLE CROP PRODUCTION

- Crop variety selection for organic farming or breeding for varieties under organic farming system.
- Studies on management aspects (weed management, disease and insect management).
- Development of organic fertilizer and their registration and standardization of dose and timing for different crops under different agro-ecosystems.
- Studies on nitrogen conservation after legume cropping.
- Increasing nutrient use efficiency in organic systems with high yield.
- Isolation of elite strains of microbes for P, K, Zn and Si solubilisation and multi-nutrient solubilising plant growth promoting rhizo-microbes.

9. CONCLUSION

Adoption of organic farming would promote agricultural diversification and can potentially contribute towards increasing family income as well as can solve environmental issues. Normally, organic yields are lower than conventional yields and the adoption of diversification practice of multiple cropping and crop rotation must be adopted in organic farm to reduce the yield gap. From the above studies and considering 25 per cent premium price for produce, organic farming may be economical and practicable. Considering the multiple benefits of organic farming and ensure sustainable, and quality food production without harming the environment. Its practice should be adopted in large scale with enforcement of proper guidance, financial support, market development involving all key partners from farmers to input dealers, government agencies bankers, NGOs and agencies engaged in different levels of agri-marketing.

REFERENCES

Benbi DK, Sharma S, Toor AS, Brar K, Sodhi GPS and Garg AK 2016. Differences in soil organic carbon pools and biological activity between organic and conevtionally managed rice-wheat fields. *Organic Agriculture*. https://doi.org/10.1007/s13165-016-0168-0

Benbrook C 2009. The Impacts of Yield on Nutritional Quality: Lessons from Organic Farming. *Hort Science* **44(1):** 12–14.

Benoit M, Garnier J, Billen G, Tournebize J, Gréhan E and Mary B 2015. Nitrous oxide emissions and nitrate leaching in an organic and a conventional cropping system (Seine basin, France). *Agriculture, Ecosystems and Environment* **213:** 131–141.

Berry PM, Sylvester-Bradley R, Philipps L, Hatch DH, Cuttle SP, Rayns FW and Gosling P 2002. Is the productivity of organic farms restricted by the supply of available nitrogen. *Soil Use Manage.* **18:** 248–255.

Clark MS, Horwath WR, Shennan C and Scow KM 1998. Changes in soil chemical properties resulting from organic and low-input farming practices. *Agron. J.* **90:** 662–671.

Clark MS, Horwath WR, Shennan C, Scow KM, Lantni WT and Ferris H 1999. Nitrogen, weeds and water as yield-limiting factors in conventional, low-input and organic tomato systems. *Agric. Ecosyst. Environ.* **73:** 257–270.

Connor DJ and Mínguez MI 2012. Evolution not revolution of farming systems will best feed and green the world. *Global Food Security* **1:** 106–113.

Dal Ferro N, Zanin G and Borin M 2017. Crop yield and energy use in organic and conventional farming: A case study in north-east Italy. *European Journal of Agronomy* **86:** 37–47.

Emsley J 2001. Going one better than nature? *Nature* **410:** 633–634.

Faller ALK. and Fialho E 2010. Polyphenol content and antioxidant capacity in organic and conventional plant foods. *Journal of Food Composition and Analysis* **23(6):** 561–568.

Giovannucci D 2006. Evaluation of organic agriculture and poverty reduction in Asia. IFAD.

Huang L, Jun Y, Yang J, Zhang R, Bai Y, Sun C and Zhuang H 2016. Relationships between yield, quality and nitrogen uptake and utilization of organically grown rice verities. *Pedosphere* **26(1):** 85–97.

Kaur M, Kaur N, Kaur M and Sandhu KS 2015. Some properties of rice grains, flour and starches: A comparison of organic and conventional modes of farming. *LWT - Food Science and Technology* **61(1):** 152–157.

Kendall HW and Pimentel D 1994. Constraints on the expansion of the global food supply. *Ambio* **23:** 198–205.

Leifeld J. and Fuhrer, J. 2010. Organic farming and soil carbon sequestration: what do we really know about the benefits? *Ambio* 39: 585-599.

Mäder P, Fließbach A, Dubois D, Gunst L, Fried P and Niggli U 2002. Soil fertility and biodiversity in organic farming. *Science* **296:** 1694–1697.

Oehl F, Sieverding E, Mäder P, Dubois D, Ineichen K, Boller T and Wiemken A 2004. Impact of long-term conventional and organic farming on the diversity of arbuscular mycorrhizal fungi. *Oecologia* **138:** 574–583.

Oliveira AB, Moura CFH, Gomes-Filho E, Marco CA, Urban L and Miranda MRA 2013. The Impact of Organic Farming on Quality of Tomatoes Is Associated to Increased Oxidative Stress during Fruit Development. *PLoS ONE* **8(2):** e56354. https://doi.org/10.1371/journal.pone.0056354

Pickett JA 2013. Food security: intensification of agriculture is essential, for which current tools must be defended and new sustainable technologies invented. *Food and Energy Security* **2:** 167–173.

Pimentel D, Harvey C, Resosudarmo P, Sinclair K, Kurz D, McNair M, Crist S, Shpritz L, Fitton L, Saffouri R and Blair R, 1995. Environmental and economic costs of soil erosion and conservation benefits. *Science* **267:** 1117–1123.

Ponisio LC, M 'gonigle LK, Mace KC, Palomino J, De Valpine P and Kremen C 2015. Diversification practices reduce organic to conventional yield gap. *Proceedings of the Royal Society B: Biological Sciences* **282(20141396)**: 1–7.

Poudel PD, Horwath WR, Lanini WT, Temple SP and van Bruggen AHC 2002. Comparison of soil nitrogen availability and leaching potential crop yields and weeds in organic, low-input and conventional farming systems in northern Carolina. *Agric. Ecosyst. Environ.* **90:** 125–137.

Reganold JP and Wachter JM 2016. Organic agriculture in the twenty-first century. *Nature Plants* **2:** 1-8.

Reganold JP, Elliott LF and Unger YL 1987. Long-term effects of organic and conventional farming on soil erosion. *Nature* **330:** 370–372.

Rigby D and Cáceres D 2001. Organic farming and the sustainability of agricultural systems. *Agr. Syst.* **68:** 21–40.

Sacco D, Moretti B, Monaco S and Grignani C 2015. Six-year transition from conventional to organic farming: effects on crop production and soil quality. *European Journal of Agronomy* **69:** 10–20.

Singh DK, Singh G, Gupta S, Arora M and Verma S 2012. Yield sustainability and quality of basmati rice as influenced by conventional, organic and integrated modes of cultivation. *Oryza* **49(2):** 102–107.

Suja G, Sundaresan S, John KS, Sreekumar J and Misra RS 2012. Higher yield, profit and soil quality from organic farming of elephant foot yam. *Agronomy for Sustainable Development* **32(3):** 755–764.

Tashi S and Wangchuk K 2016. Organic vs. conventional rice production: comparative assessment under farmers' condition in Bhutan. *Organic Agriculture* **6(4):** 255–265. https://doi.org/10.1007/s13165-015-0132-4.

The World of Organic Agriculture 2017 available at http://www.fibl.org/fileadmin/documents/en/news/2017/mr-world-organic-agriculture-2017-english.pdf.

Trewavas A 2001. Urban myths of organic farming. *Nature* **410:** 409–410.

UNFCCC 2008. Challenges and opportunities for mitigation in the agricultural sector. Technical paper FCCC/TP/2008/8, Bonn, 21 Nov 2008.

Venugopalan MV, Rajendran TP, Chandran P, Gowswami SN, Challa O and Damre PR 2010. Comparative evaluation of organic and non-organic cotton (*Gossypium hirsutum*) production systems. *Indian Journal of Agricultural Sciences* **80(4):** 287-292.

Willer H and Lernoud J (eds). The World of Organic Agriculture: Statistics and Emerging Trends 2015 (FiBL-IFOAM, 2015).

■■■■